Ice-Age Science Recapitulation

Science Exploration by Rolf A. F. Witzsche

© Text Copyright Rolf A. F. Witzsche 2018
all rights reserved

This book contains the transcript with images of the exploration video with the above title:
see: http://www.ice-age-ahead-iaa.ca/

A comprehensive Ice-Age-Science exploration
Lead in:

The video explores 4 time-frames, from 500 million years ago to 30 years into the future.
It explores cosmic climate-change dynamics, freshwater, energy, agriculture, and culture.

The Strategic Defense of Humanity?
We are, potentially, 5 to 15 years away from Canada, Europe, and Russia loosing its agriculture - the foundation for their physical existence as nations, and are possibly as close as the 2050s, together with all nations outside the tropics, loosing their territory to the resuming cold, snow, and ice of the near Ice Age.
While we have the resources and technologies on hand to build us a New World that the recurring Ice Age cannot seriously affect, nothing is being built to secure our food supply and future existence. The great challenge before us is simply ignored, almost universally. To say that we are in a critical situation, because of the ongoing denial, is a gross understatement.

What do we know, because we have physically measured it?
In order to better understand the climate in our time, and to enable us to respond to its imperatives, let's review what we actually know about the climate on Earth in terms of what we have physically measured or photographed, because what we know is amazingly big.

Leadership is required in the Campaign to Secure Our Future with the Science Paradigm of Truth.

Science opinions, that are often intentional Imperial science traps, are fast becoming displaced by the 'voice' of extensive physical measurements that establish a science paradigm based on the 'truth' of physical measurements in numerous diverse disciplines. It is unique that the measurements place the world's climate scene squarely and directly into the court of the Sun, including that of Global Warming, which had peaked in the 1990s and is now collapsing back again in the shadow of diminishing solar activity that may take us all the way to the start of the next Ice Age in the 2050s.

This video, a wide-ranging three-months project, is my most-comprehensive production on the Ice Age subject to date. The video presents an overview of the astrophysical dynamics, most of which have been carefully hidden for political objectives, and explores the severity of the Ice Age consequences for humanity. It also opens a window to the grand opportunities for humanity, by it coming out of its small-minded shell, scientifically, and technologically, in the face of the magnitude of the challenge that the near Ice Age imposes. Grasping the opportunity that it opens up for humanity, inspired by scientific imperatives versus impotent dreaming, has the potential to sweep all the lesser problems off the table, including nuclear war, terror, empire, economic collapse, and poisoned political environments.

This is so, because the Ice Age Dynamics present greater imperatives for humanity to rouse itself to its inherent potential, and to build itself a New World, than all the lesser issues have combined that have festered for decades with few breakthroughs made, and continued to do so all the way to the present.

Table of Contents

What we know? How do we respond?..19

Part 1 - Climate-Change History - 500 million years 20

Known climate history of the Earth 21

We have measured the changing historic climate 22

Oxygen atoms 23

Two extra neutrons attached..24

The concentration of heavy water molecules 25

Rainout adds to the concentration of heavy molecules 26

Ratio preserved for us in the calcite shells of micro organisms 27

Calcite a combination of water with carbon dioxide 28

By analyzing deep sea sediments 29

We see two long climate cycles evident...30

The mainstream theory about the Sun is incorrect..31

The Sun is an extremely variable factor 32

One false theory sees the solar system oscillating...33

Another false theory sees the Sun in an orbit 34

Electro-dynamic Galaxy and Sun 35

NASA made a critical discovery 36

NASA discovered evidence on the galactic scale 37

NASA's x-ray and gamma-ray explorations..38

The NASA photographed geometry of the plasma phenomenon.......................................39

Our Sun onto a totally different conceptional platform 40

Seeing the true Sun 41

NASA's discovery defines our Sun as a plasma star 42

Our entire galaxy formed at a node point 43

The two plasma streams act together in a dynamic system 44

Potential plasma connections 45

The proof lies in the observed motions of stars 46

Dogma that the stars orbit gravitationally around the galactic center 47

The measured speed of the stellar motion disprove 48

According to the laws of orbital mechanics 49

Motions of stars accords with electro-magnetic principles 50

The entire galaxy rotates as one unit 51

The principle of the plasma galaxy 52

The origin of the cosmic plasma streams 53

The implicate and explicate order 54

David Bohm sees cosmic space as a sea of latent energy 55

Explicate like ripples on a surface of water 56

A very distant origin 57

In the deep intergalactic realm 58

The electrodynamic nature of the galaxies 59

Cosmic dynamics and the solar-system dynamics 60

We cannot afford to ignore the dynamic nature of the Universe 61

On any other basis than plasma electrodynamics 62

Climate background the coldest in 440 million years 63

Plasma density in the galaxy is present at its weakest state 64

At the bottom end of a long diminishing slope 65

Antarctica froze up, thawed out again, and froze up once more 66

The Plasma-Sun Principle 67

The Plasma Sun is a large sphere of plasma 68

Swarming electrons tend to migrate away from the center of gravity 69

The migration leaves the protons less shielded 70

A very large sphere of plasma with little density at its core 71

Our Sun is a thousand times too light 72

The low solar mass density proves 73

When we look through the umbra of sunspots 74

The giant star UY Scuti 75

A plasma-star doesn't produce its own energy 76

All known atomic elements are synthesized 77

The surface of the Sun is a sea of reactions cells 78

The resulting rich spectrum in sunlight 79

The Sun's light spectrum would be rather meagre 80

Since the plasma fusion process is externally powered 81

That's how Antarctica froze up when the Sun became 'colder' 82

Dynamics of the Pleistocene Epoch 83

The Pleistocene created a new phase shift on Earth 84

We see the last 500,000 years expanded 85

We see the same large-spike climate oscillations 86

The Pleistocene climate is of a type that affects us hugely 87

We call the current warm period, the Holocene epoch 88

The Pleistocene Ice Age epoch will continue 89

Part 2 - Climate-Change History in the last 5 million years 90

  timeframe in which human development started 91

  Celebrating the near Ice Age 92

  Ice Ages periods of high volumes of cosmic-ray flux 93

  When cosmic-ray particles pass through us 94

  The evidence is found in cultural development 95

  Cosmic-ray flux larger during the last Ice Age 96

- Around a million years ago 97
- The amazing proliferation of the human species 98
- The Pleistocene is our cradle 99
- That's something to celebrate and to move forward with 100
- China, which had developed one of the earliest civilizations 101
- Echoed in China's Belt and Road initiative 102
- Chinese people, aided by their cultural background 103
- How the Sun generates Ice Ages 104
- Expanded view of the last half-million years 105
- For 85% of this period the climate was cold 106
- The remaining 15% spaced between the glacial times 107
- The same interglacial spikes that we have measured in the ice layers 108
- But what is it that we see ? 109
- The same type of dynamics that NASA has photographed 110
- The evidence that the Sun is located at a node point 111
- Between two long interstellar plasma streams 112
- The principle is the same 113
- NASA measured the evidence 114
- Ulysses measured a void in the solar-wind pattern 115
- The sharp climate pulses 116
- Pulses represent the period in which the Sun is in a high-powered mode 117
- If the galactic plasma density wasn't at the deep low level 118
- We are the product of the ice ages 119
- False and true Ice Age theories 120
- The Milankovitch Cycles Theory 121
- Exposure to the Sun at 65 degrees North 122
- Ice Age glaciation phenomenon into the court of the Sun 123

- Solar hibernation measured 124
- Dramatically increased solar cosmic-ray flux 125
- The bottom line is that Ice Ages are solar caused events 126
- Still enough density for our Sun to be fully powered 127
- This doesn't mean that the Sun is actually dying 128
- During its low-powered hibernation mode 129
- Glaciation is caused by the reduced energy 130
- The glaciation will last until the next resonance effect 131
- The Asymmetric Interglacial 132
- We are 4,400 years past the point of symmetry 133
- The forming of the Primer Fields 134
- Primer fields focus interstellar plasma unto a Sun 135
- David LaPoint explored what happens 136
- The concentrated plasma surrounds the Sun like a mantle 137
- Whatever portion of the in-flowing plasma is not consumed 138
- The principle of the complimentary primer fields 139
- Our Sun being a rather mediocre star among stars 140
- The Sun is classified as a G class star 141
- High-powered stars are larger in size 142
- Hanging by a fine thread 143
- We are 4,400 years past the symmetry point 144
- The primer fields will then collapse 145
- During the last 4,400 years 146
- On the steepening slope of the last 1,000 years 147
- Recognize why the cyclical events are diminishing 148
- Without this recognition it is tempting to assume 149
- Warming pulses started with intervals of 1,300 years 150

- Minimum cycles started with 260 years between them 151
- Nested Plasma Structures - and they are shrinking 152
- The long-ago theorized Oort Cloud system 153
- The nested plasma structures have both been shrinking 154
- Interval between the warming pulses has shrunk 155
- Intervals between the minima been shrinking 156
- Shorter intervals less-inflated, plasma structures 157
- We are on a path that cannot reverse 158
- We face the full Ice Age at the end of the line 159
- When the fine thread breaks 160
- At the present time, no Plan-B exists 161

Part 3 The 50-years boundary zone 162

- The boundary time-zone to the next Ice Age 163
- The solar wind is diminishing 164
- Steam being boiled off from a heated kettle 165
- The solar dynamics are similar 166
- When the solar system weakens 167
- The Sun is our climate master 168
- The cosmic-rays affect the Earth's climate enormously 169
- Cosmic-ray flux flows from the Sun 170
- Cosmophysical Factors 171
- The researcher Simon Shnoll 172
- Each time the results were different 173
- In one case the reactivity spiked tremendously 174
- It is generally assumed that all cosmic-ray flux is galactic 175
- Simon Shnoll's experiment disproves this assumption 176
- Voyager-1 spacecraft penetrated the heliosphere 177

In comparison with the attenuation that the heliosphere affords	178
The CERN lab's CLOUD experiment	179
Demonstrated in laboratory experiments	180
When artificial cosmic-rays were injected	181
Cloudiness affects everything	182
Phase-2 of the solar collapse	183
After the solar wind stops	184
So does the Sun	185
All the way through Phase-2, the Sun will get colder	186
Phase-3 of the solar collapse	187
Phase-3 begins when the primer fields can no longer form	188
Measurements of Phase-2 to Phase-3 transition	189
Phase-3 is the phase of the hibernating Sun	190
We cannot respond in a reactive mode any longer	191
Phase-1 of the solar collapse is the most critical phase	192
Plan-B while we still have a chance	193
But what about Manmade Global Warming - can't it save us?	194
The last meaningful historic climate recovery	195
Small as the last warming pulse was	196
Compare the absorption coefficients	197
Water vapor is effective across a vastly wider band	198
The CO2 effect is no bigger than a cat	199
Wonderful if manmade global warming was possible	200
To create a Manmade New World	201
The Strategic Defence of Humanity	202
Objective is to block the recognition of the near Ice Age	203
Renamed to Manmade Climate Change	204

Similar in effect to Illuminati 205

The Manmade Global Warming effect without a possible solution 206

An existential challenge without an identified face 207

Against the imposed, reduced-mentality background 208

The impositions are forced 209

Killing upwards to 100 million people a year with starvation 210

The effect is - and this is intended 211

The Greek playwright Aeschylus 212

To raise his audience to a higher-level self-perception 213

Manmade Climate Change is physically impossible 214

The Strategic Defence of Humanity requires 215

Under the freedom of the truth 216

However, the truth poses a big challenge for scientific recognition 217

When the greenhouse-effect becomes reduced 218

Consequently nothing gets build 219

More than 50,000 protest 'signatures' had been uttered 220

Just imagine the boundless freedom that this simple science recognition 221

Earth's greenhouse effect is diminished by cosmic-ray increase 222

As the result of the reduced greenhouse-moderation 223

In Australia wheat harvests fell 40% 224

The removal of a dozen floors from the towers 225

The result of the shrinking greenhouse effect 226

Now consider precipitation diminishes by 80% 227

That's like taking 80 stories off the buildings 228

The diminishing greenhouse effect may come to our aid 229

None of the climate transformations are manmade 230

The entire climate scene is wrapped up in dreams 231

The last Global Warming pulse that occurred in the 1700s 232

The climate warming pulses are of short duration 233

The fast progression puts the onus on us for fast-action responses 234

Most people consider the wonderful warm climate to be normal 235

A fortunate anomaly had rescued us from the Little Ice Age 236

Plan-B is critical for our 7-billion world population 237

What had enabled the phenomenal world population growth 238

The product of science and technologie 239

With the climate collapsing like a falling stone 240

Our power to create climate independent agriculture 241

The total eradication of nuclear weapons will happen 242

When a country like China looses a portion of its food crops 243

Canada, Europe, and Russia will likely take the lead 244

When critical physical imperatives begin to force the issues 245

Part 4 - Climate-Change, 5 to 15 years into the future 246

Huge winter blizzards in late spring 247

Massive snowfall when spring planting should be in progress 248

At the same time droughts ravished the American Southwest 249

Snow in North Africa in January 2017, in the Sahara 250

Wheat crop losses in Australia in the same season 251

When climate collapse narrows the growing season 252

Measuring the solar dynamics in real-time 253

High-flux periods between the solar cycles 254

If we look at the full cycle 24, to-date 255

The very heartbeat of the solar system itself, is slowing down 256

Radio-flux measurements have collapsed by half for cycle 24 257

Now, the Sun has lost its 'top hat' 258

The Sun's polar magnetic fields have diminished 259

A dynamic feature of the Sun has suddenly failed 260

How close to the End are we? 261

The start-up of Ice Age glaciation caused by the Sun changing states 262

The Younger Dryas re-glaciation 263

Measured Beryllium ratios obsolete numerous theories 264

The swarm of the increasing fringe effects that we encounter 265

Even the Earth's rotation may be affected 266

Gravitational drag by the moon on the Earth 267

The weakening electrodynamics of the primer fields and their effects 268

Dramatic magnetic pole drift has been measured since the 1800s 269

Evident on the Sun itself 270

Now, 11 years later, when the entire face of the Sun is peppered with holes 271

Coronal holes are warning signs that we should heed 272

The giant flash flood event in 2015 273

Distracting climate effects 274

Concerns are raised over the thinning of Arctic sea ice 275

Arctic-warming the result of global cooling 276

Cold air is heavier than warm air 277

The colder the temperature is, the heavier is the air 278

The circulation is termed 'The Polar Mobile Anti-Cyclone' 279

The stronger the Polar Mobile Anti-Cyclone operates 280

In Russia and Europe the cold flow, named the 'Beast from the East,' 281

The southward component of the cold flow 282

The anticyclone flow sweeps across Europe and predominantly Norway 283

Just the beginning, with the climate collapse now accelerating 284

The Earth is getting colder year after year, until the Sun falls back into glacial hibernation 285

The world is already beginning to experience 286

To build a new world that the collapsing climate cannot touch 287

This issue is so real and so big 288

Bigger than all the silly games that politicians, bankers, and media play 289

Only the effects can be prevented - not the cause 290

Fear over Arctic warming should ring as warning bells to rouse society 291

Future winter blizzards in July or August, instead of just in April? 292

Plan-B anyone? 293

Nobody in living experience has ever faced Ice Age conditions 294

When the phase shift happens that takes us back to normal 295

Ice core records speak to us of a 70% less-radiant Sun 296

Ice core records also speak of world with 80% less rain 297

Part 5 To rouse the living, wake the dead 298

How does one rouse society to the truth ? 299

Great science projects have measured the Sun and its effects 300

The measurements also tell us that the final phase shift is near 301

I have presented the scientific discoveries that have been made over the years 302

To alert society of the great danger that Canada, Europe, Russia, and the USA are facing 303

The videos hosted on my website are categorized 304

The videos are also hosted on YouTube 305

Transcripts of the videos, complete with images, are also available 306

One form of the transcripts is interactive 307

All text and the images on a single page also as a PDF file 309

Transcripts also available in printed form 310

The transcript books are indexed by their cover image 311

The book detail pages include descriptions 312

Also available on the transcripts index page 313

The index page lists all my videos and available transcripts 314

A major focus on physical science is evidently needed 315

Society has become lost in a dream where nothing is real 316

Consider the Bering Strait tunnel project 317

The other flaw is that the tunnel isn't designed to connect with anything 318

Now compare the tunnel link with my World Bridge infrastructure 319

My part in the video productions is focused on drawing the field together 321

Part 6 - Freshwater and Energy 322

Both, freshwater and energy are needed in abundance 323

Freshwater can be drawn from the outflow of rivers 324

Distributed worldwide in large-volume arteries 325

Freshwater can also be drawn directly from the oceans 326

For a boundless energy resource we need to look no further than the ionosphere 327

Freshwater Unlimited 328

Freshwater from the Amazon River, the Orinoco River, the Parana River, the Congo River 329

Worldwide water distribution critically important during Ice Age conditions 330

The climate will then be 40 times colder 331

The colder Sun will produce 80% less precipitation 332

The tropics receive the strongest rainfall 333

The loss may be far less in the tropics than the 80% reduction 334

Tropical rivers will become the world's main source for freshwater 335

Large-scale ocean-water desalination offers a supplemental solution 336

The modern desalination process is inefficient 337

The produced volume is minuscule in comparison with major rivers 338

When the desalination filters are located in deep oceans 339

The first 270 feet would be required to overcome osmosis pressure 340

If the filters were placed 9,000 meters deep 341

The potential that the inhibiting osmosis pressure can be completely neutralized, electrically 342

Water has a slight dipole electric property 343

By the electric effect, water is drawn into a salt solution, towards the salt 344

The electric attraction lifts the salt column until equalization is achieved 345

Since the electric neutralization has not yet been achieved 346

The theoretically possible potential to break the osmotic connection is promising 347

The use of carbon nanotubes as filter elements is being explored 348

Another promising new type of filter for desalination would utilize graphene sheets 349

We could then have rivers of freshwater flowing out of the oceans 350

Desalination shouldn't be needed at the present stage 351

Energy Forever 352

One of the ghosts of pure illusion is nuclear-fusion power 353

It took scientists decades of dreaming, and billions of dollars 354

So, where do we go from here, after the dream has failed? 355

Scientists had an idea that this tiny pellet of fuel could unlock an energy-rich future 356

We have depleted our energy resources so intensively that we have only 60 to 200 years left 357

The answer is a type of source that is self-renewing 358

The source is cosmic, and the interface is the ionosphere 359

NASA has been able to photograph two electric plasma bands 360

The natural systems appear to use this resource liberally 361

The barrier is the false belief that our Sun is self-powered 362

Another very-large energy resource is available through the liquid-fluoride nuclear power reactor 363

All conventional nuclear power systems utilize only a half a percent of the nuclear fuel 364

The molten salt reactor, in contrast, utilizes nearly all of its fuel 365

Part 7 - Culture and Science 366

The series: 'The Lodging for the Rose' by Rolf A. F. Witzsche 367

A project to bridge the near universal isolation in modern society, from one another 368

I was gradually drawn to pay attention to the little that was known at the time 369

An exploration series of novels about the beauty and sublimity of our humanity 370

The Ice Age Challenge is so imperative that all the lesser challenges, like nuclear war, will be laid aside 371

I named my 12-volume epic series, 'The Lodging for the Rose' 372

And, oh, how much greater than a rose are we? 373

The title 'Winning without Victory' into a mini series of the principle of active peace 374

The Kaleidoscope project series 375

Another series I named the 'Sex and Sacrament' series 376

In addition, I have produced a large series of special focus videos 377

Another large group is available in the form of audio books 378

I also attached singe-story audio books to the front pages of my videos 379

The earliest special presentation are presented as PDF books 380

Society needs to be the leader for the politicians 381

A video production that is designed to explore the power of cultural effects 382

Society's writers, poets, musicians, scientist, and artists, own the job to uplift the humanity 383

The uplifting also needs to reach deep into the domain of physical science 384

As human beings, we have the power to reach as high as we need to 385

We haven't seen anything yet 386

With this song we will have the commitment within reach to move forward 387

More Illustrated Science Books by Rolf A. F. Witzsche 388

What we know? How do we respond?

The Strategic Defence of Humanity?

Part 1 - Climate-Change History in 'Mega' Time' (500 million years)

Part 2 - Climate-Change History in 'Macro' Time' (last 5 million years)

Part 3 - Climate-Change History in 'Micro' Time (last 50 years - boundary zone)

Part 4 - Climate-Change History in 'Nano' Time (5 to 15 years into the future)

Part 5 - "To rouse the living, wake the dead" (My science contribution)

Part 6 - Freshwater and Energy

Part 7 - Culture and Science

In order to better understand the climate in our time, and to enable us to respond to its imperatives, let's review what we actually know about the climate on Earth in terms of what we have physically measured or photographed, because what we know is amazingly big. The result is a 7-part exploration.

Part 1 - Climate-Change History - 500 million years

Part 1 - Climate-Change History in 'Mega'-Time - The past 500 million years

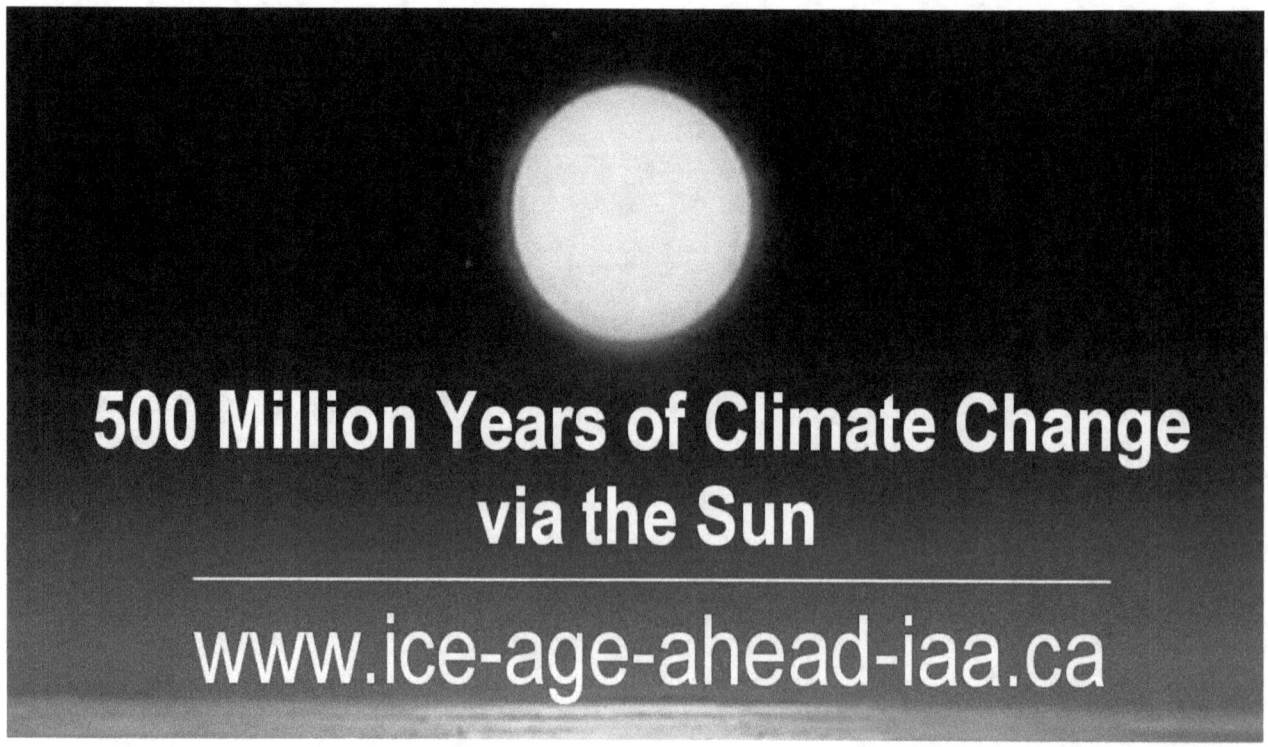

Known climate history of the Earth

The known climate history of the Earth takes us back in time a whopping 500 million years, to near the very beginning of life on our planet. This is possible, because life itself has provided the measurable proxy for the ever-changing climate on Earth.

Part 1 - 500 Million Years of Climate Change via the Sun

We have measured the changing historic climate

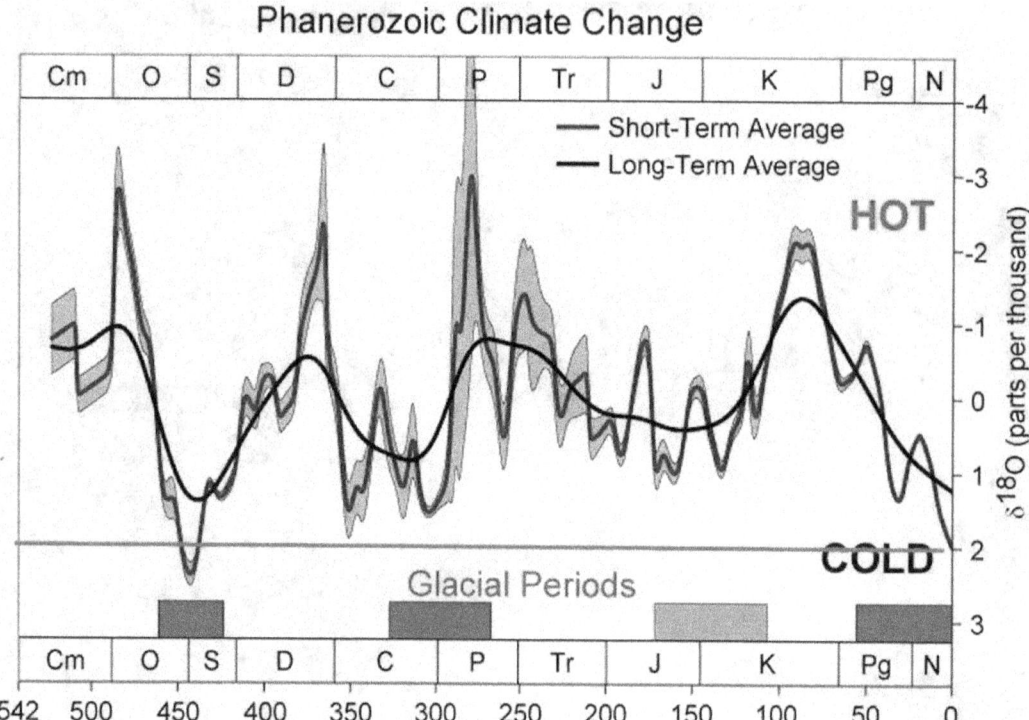

We know that the climate on Earth has been on a roller-coaster ride between extreme cold and warm climates. We know this, because we have measured the changing historic climate in proxy by measuring the ratio of a rare heavy-oxygen isotope that is found in water. The ratio of it in sea water is climate sensitive.

Oxygen atoms

We have discovered that all water, including in the oceans, contains a minute amounts of these heavy oxygen atoms in its molecular structure.

Two extra neutrons attached

An oxygen atom can have two extra neutrons attached. This makes it heavy. The heavy oxygen atom makes the water molecules that contain them, likewise heavy.

The concentration of heavy water molecules

We have further discovered that the concentration of those heavy water molecules varies on the surface of the oceans, with climate temperatures. This is so, because light water molecules evaporate more readily at cold surface temperatures than the heavy molecules, which increases the density of the heavy molecules on the surface.

Rainout adds to the concentration of heavy molecules

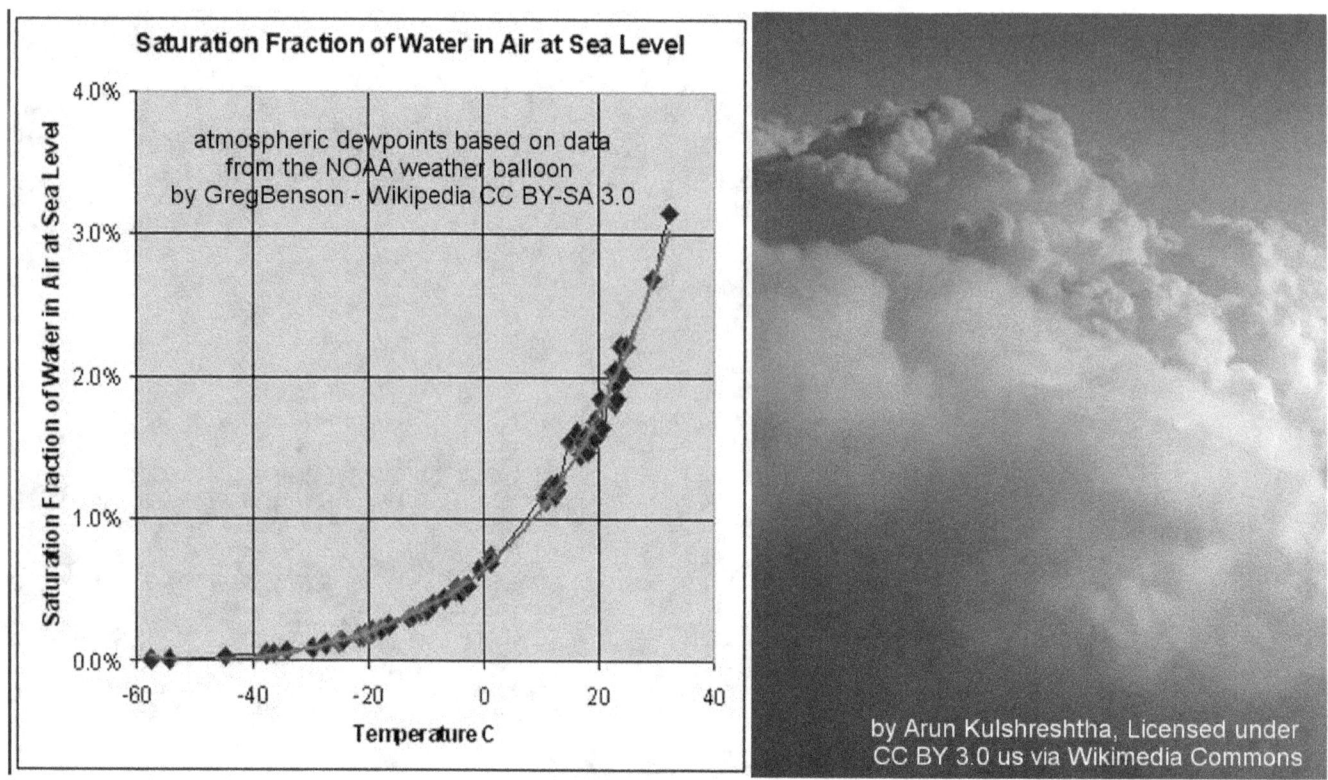

Also, the heavy molecules condense back into water more readily, in the clouds, which occurs more strongly in colder climates. The rainout thereby adds to the concentration of heavy molecules on the ocean surface in cold climates .

Ratio preserved for us in the calcite shells of micro organisms

It has been discovered that the historic ratio of the heavy-water concentration has been preserved for us in the calcite shells of micro organisms that had lived at the surface of the sea at the time.

Calcite a combination of water with carbon dioxide

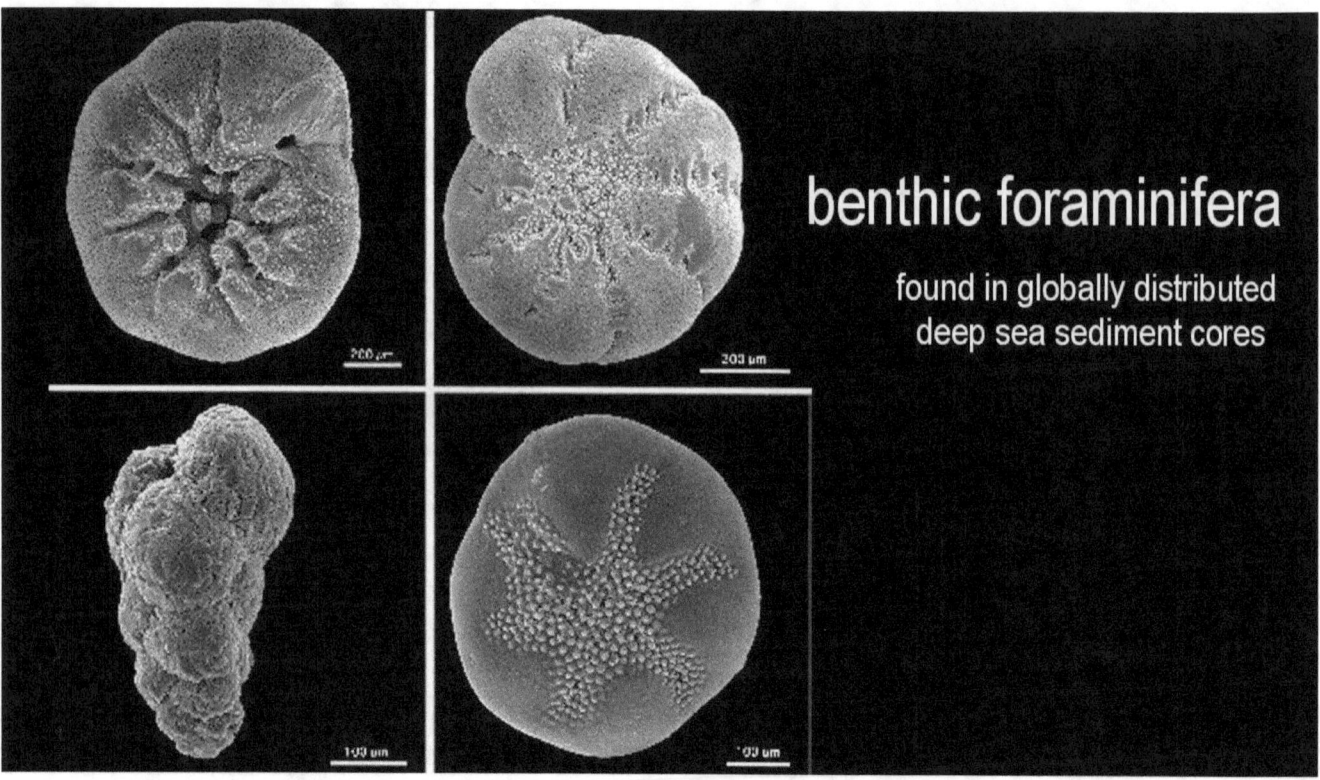

The organism create the calcite (calcium carbonate ($CaCO_3$)) of their shells with a combination of water with carbon dioxide. The ratio of the heavy oxygen atoms in the water is thereby preserved for us in the calcite shells that accumulate on the ocean floor after the organisms die.

By analyzing deep sea sediments

By analyzing deep sea sediments for their contained heavy oxygen ratio, it has become possible to physically measure the climate fluctuations that have occurred over the last 500 million years, going back in time almost to the very beginning of life on the Earth.

We see two long climate cycles evident

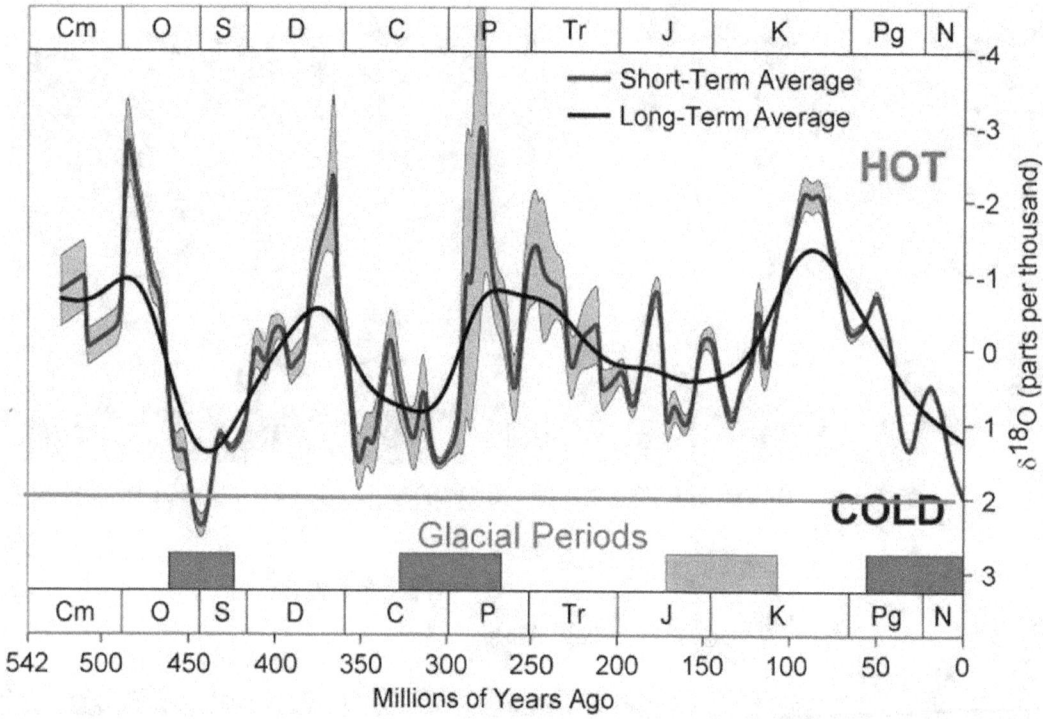

The result of the measured ratios is plotted here.

What we see is rather amazing. We see surprisingly large fluctuations in the historic record, but we don't see them as random fluctuations, as one might expect. We see a principle reflected in the fluctuations.

We see two long climate cycles evident, super-imposed on each other. One has a short cycle time of roughly 30 million years per cycle, and the other a longer cycle time of between 140 to 150 million years. But what does this mean?

The mainstream theory about the Sun is incorrect

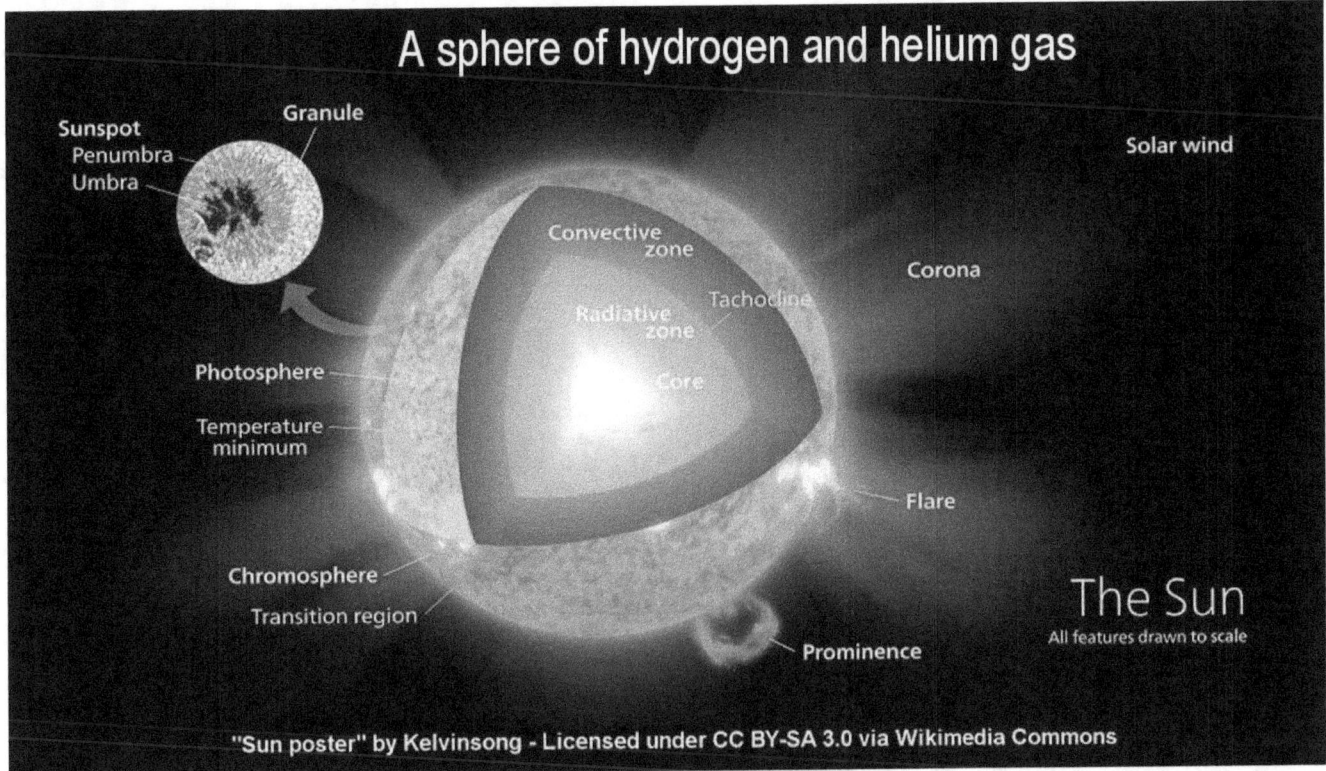

It means that the mainstream theory about the Sun is incorrect. In mainstream theory the Sun is regarded as an internally heated sphere of hydrogen gas that is powered by a process of nuclear fusion at its core, which combines hydrogen into helium. This is a process that cannot vary by the principle involved. On this basis the Sun. is regarded as in invariable constant for all climate considerations, so that climate fluctuations are deemed to be unrelated to the Sun. But this is not physically possible.

The Sun is an extremely variable factor

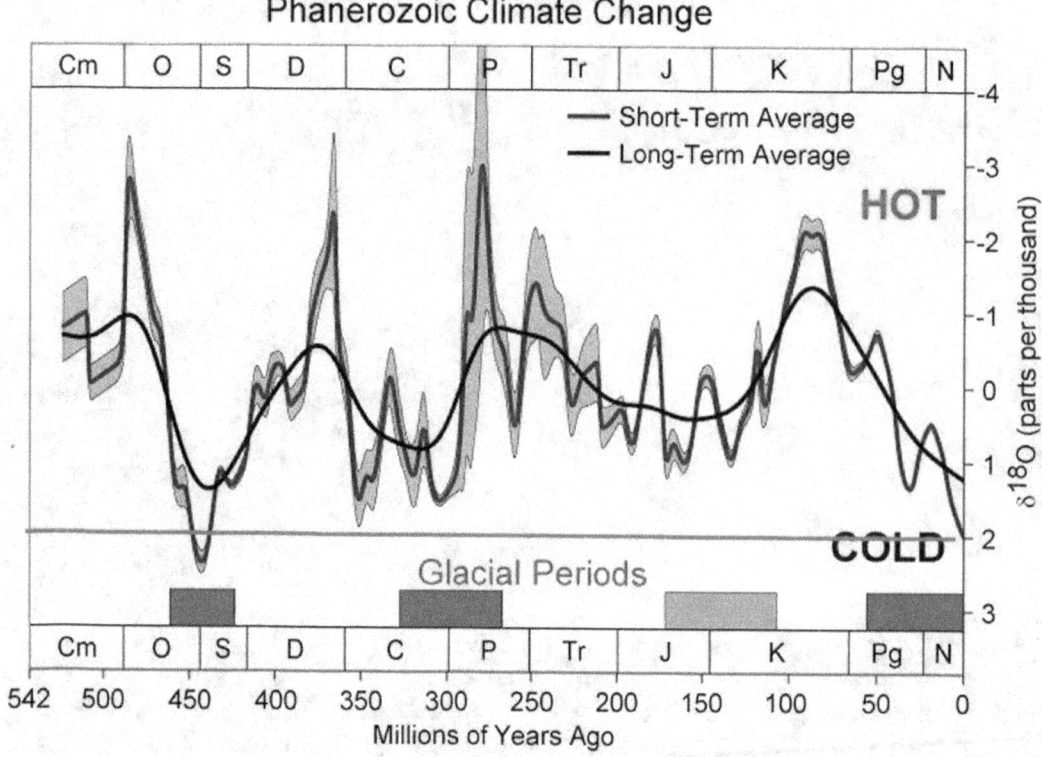

The measured climate fluctuations tell us that the Sun is an extremely variable factor, as nothing else can cause these enormous climate extremes.

Nevertheless, many theories exist that envision processes that are not physically possible, which theorize climate changes by factors other than the Sun.

One false theory sees the solar system oscillating

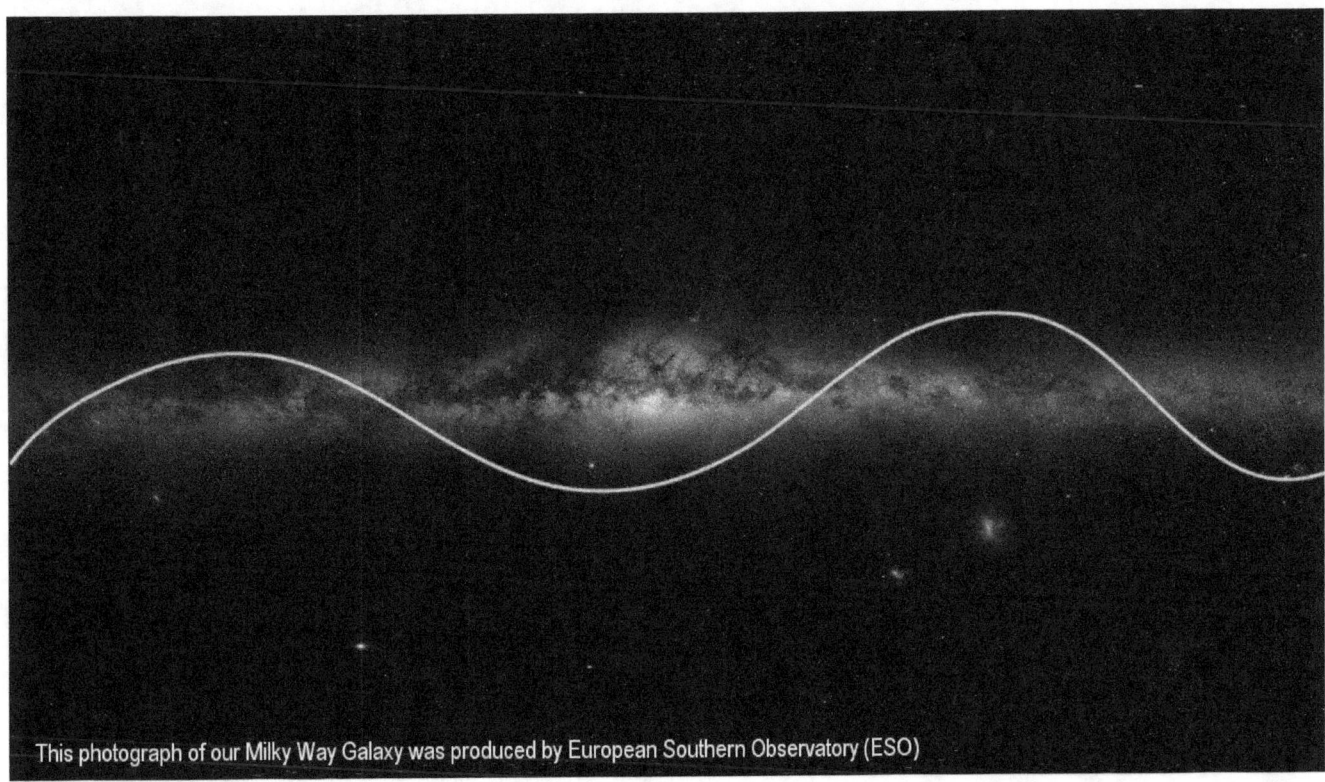

This photograph of our Milky Way Galaxy was produced by European Southern Observatory (ESO)

One false theory sees the solar system, and with it the Earth, to be oscillating above and below our galaxy's thin disc of stars, whereby the Earth is deemed to be periodically exposed to stronger cosmic radiation that is deemed to have affected its climate.

Another false theory sees the Sun in an orbit

Another false theory sees the Sun in an orbit around the galactic center on a path that is crossing the various spiral arms of the galaxy. The theory is that that the solar system is thereby exposed to changing stellar density that is also deemed to vary cosmic radiation.

The problem with this reasoning is that the physical principles don't exist that would make these types of concepts possible.

Electro-dynamic Galaxy and Sun

Electro-dynamic Galaxy and Sun

Electro-dynamic Galaxy and Sun

NASA made a critical discovery

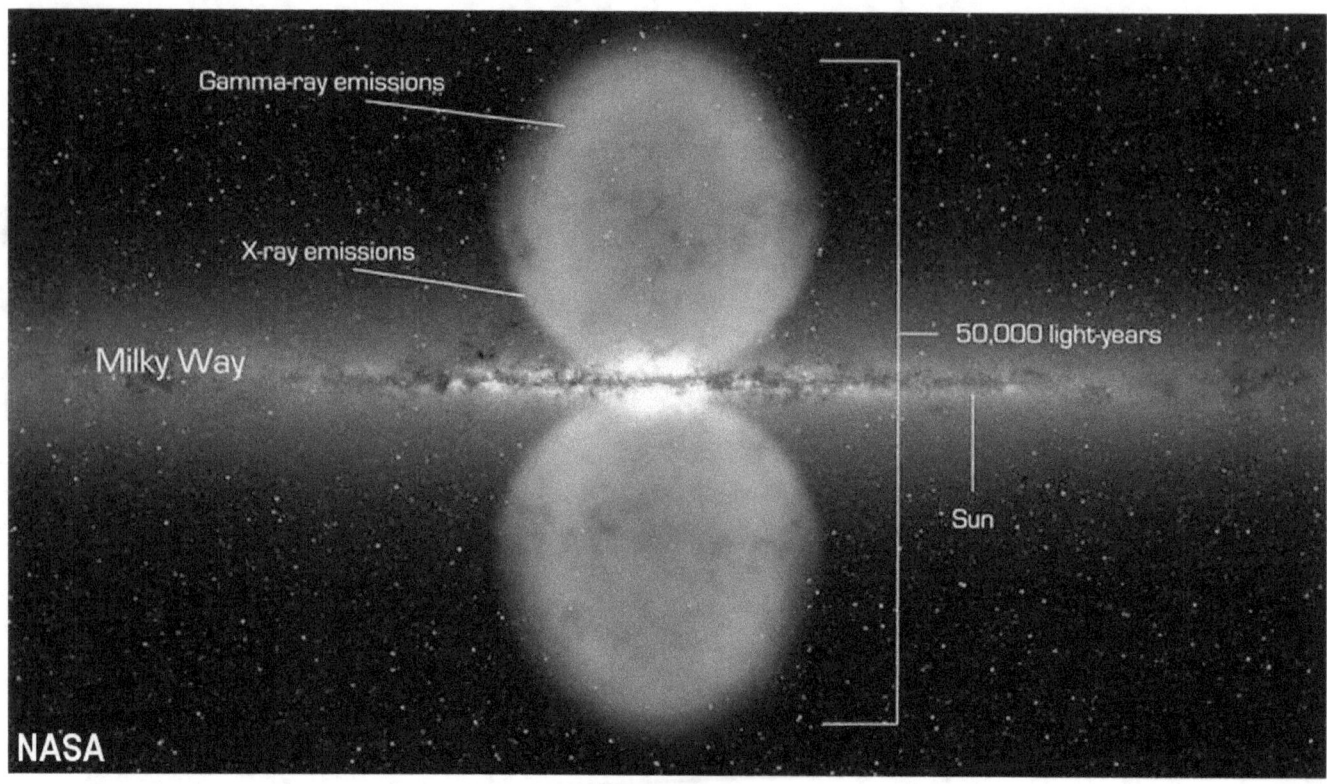

NASA solved the problem when it made a critical discovery in recent years that points to an actually possible cause for the two long climate cycles in the historic record.

NASA discovered evidence on the galactic scale

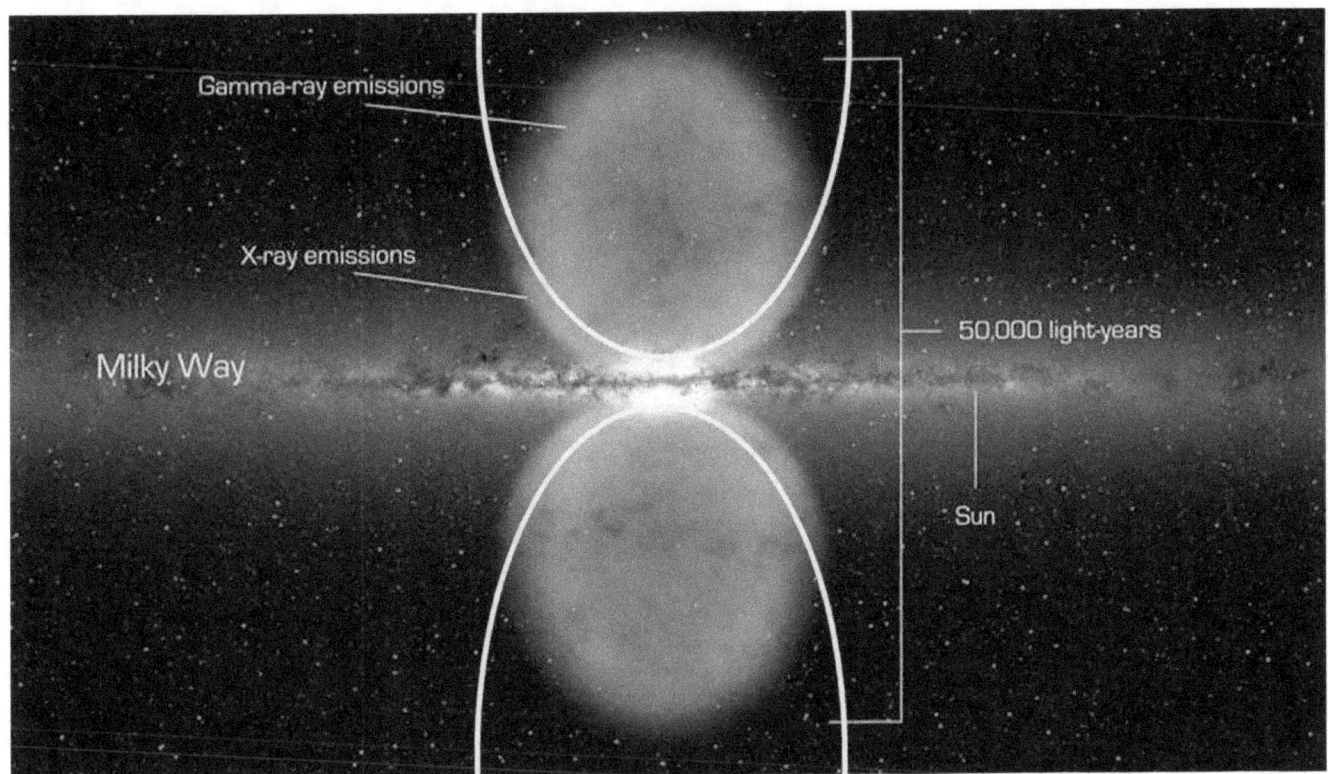

NASA discovered evidence on the galactic scale that reflects known physical principles discovered in laboratory experiments. NASA has thereby discovered the galactic scale cause that is reflected in the climate on Earth in the very long galactic scale timeframe.

NASA's x-ray and gamma-ray explorations

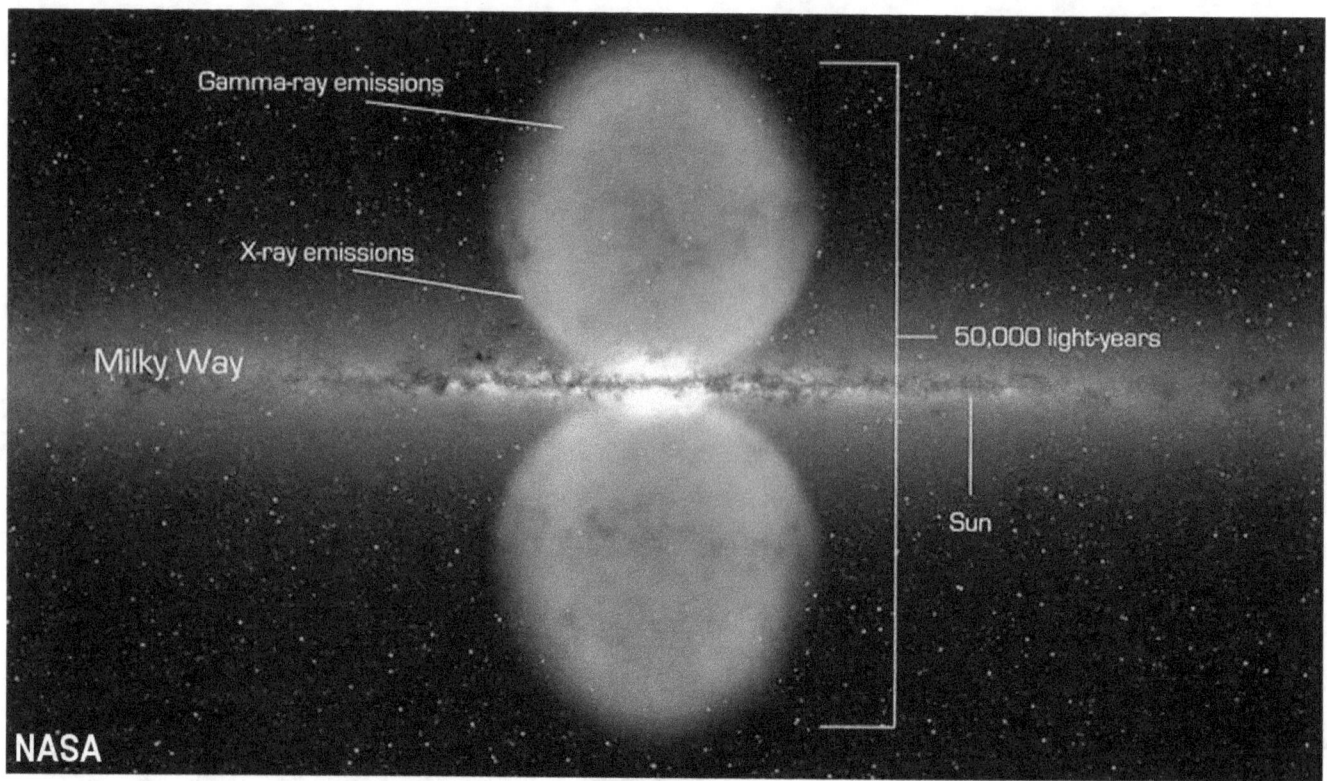

NASA's x-ray and gamma-ray explorations discovered the existence of two gigantic plasma structures extending from the center of our galaxy, perpendicular to the galactic disc for 25,000 light years into opposite directions.

The NASA photographed geometry of the plasma phenomenon

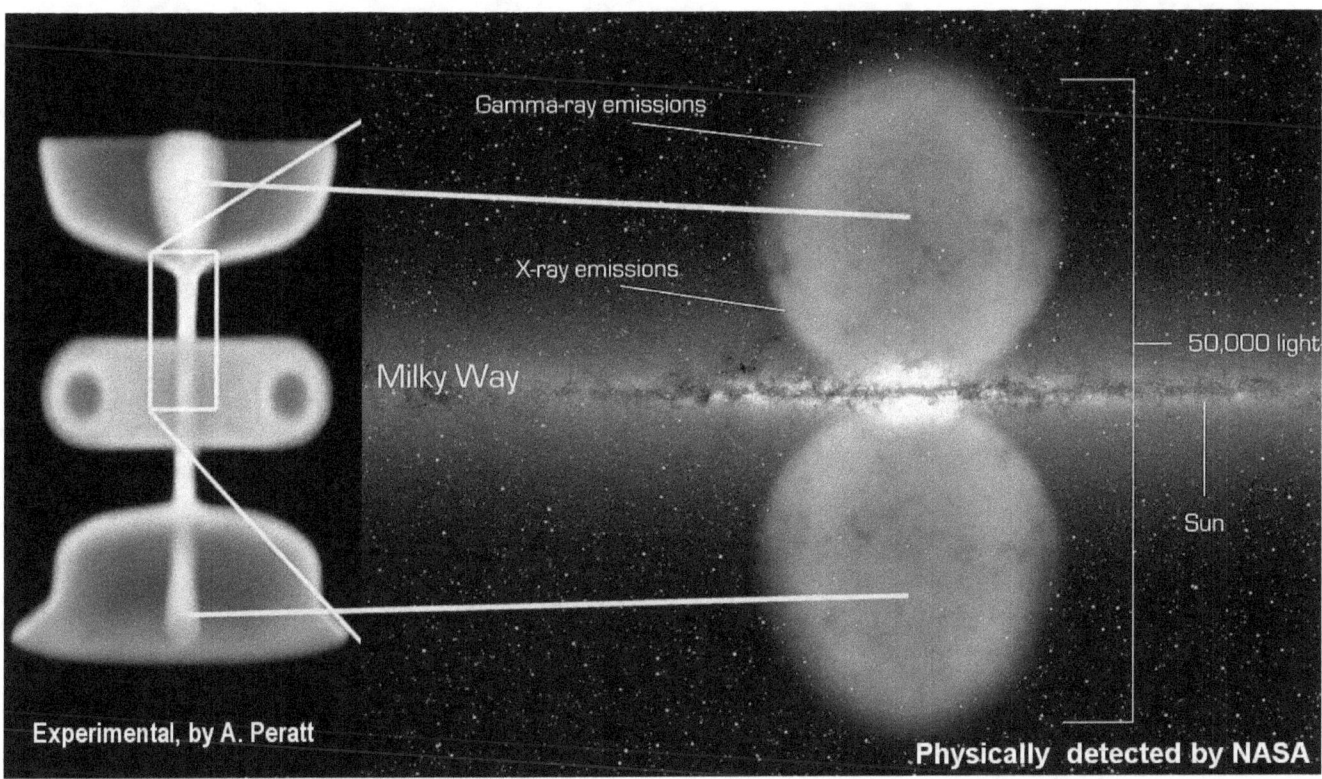

The NASA photographed geometry of the plasma phenomenon on the very large galactic scale, matches remarkably well the physical feature discovered in high-energy plasma discharge experiments conducted at the Los Alamos National Laboratory.

Our Sun onto a totally different conceptional platform

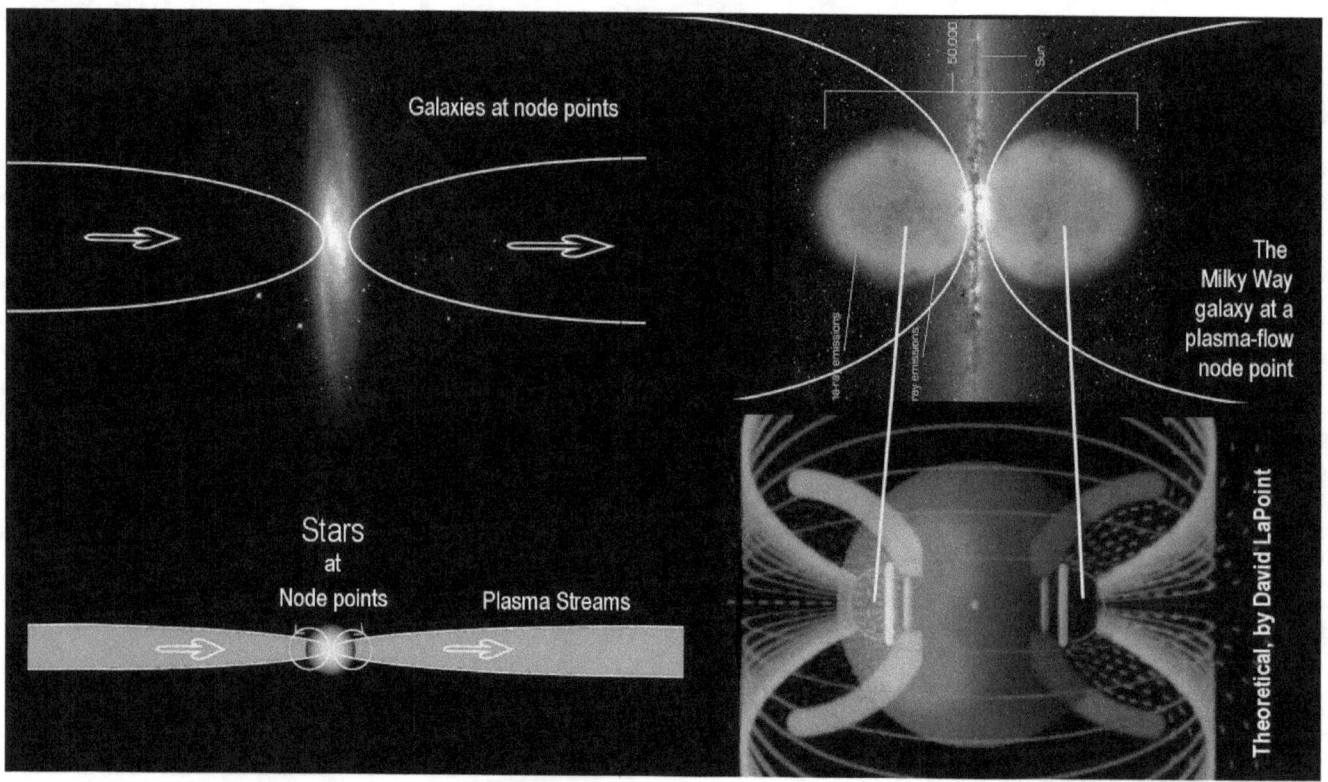

The 'photographed' discovery also places our Sun onto a totally different conceptional platform than that of the gas-fusion model of mainstream cosmology, for which no exclusive evidence actually exists.

Seeing the true Sun

Seeing the true Sun - the Sun is a Plasma Star

NASA's discovery defines our Sun as a plasma star

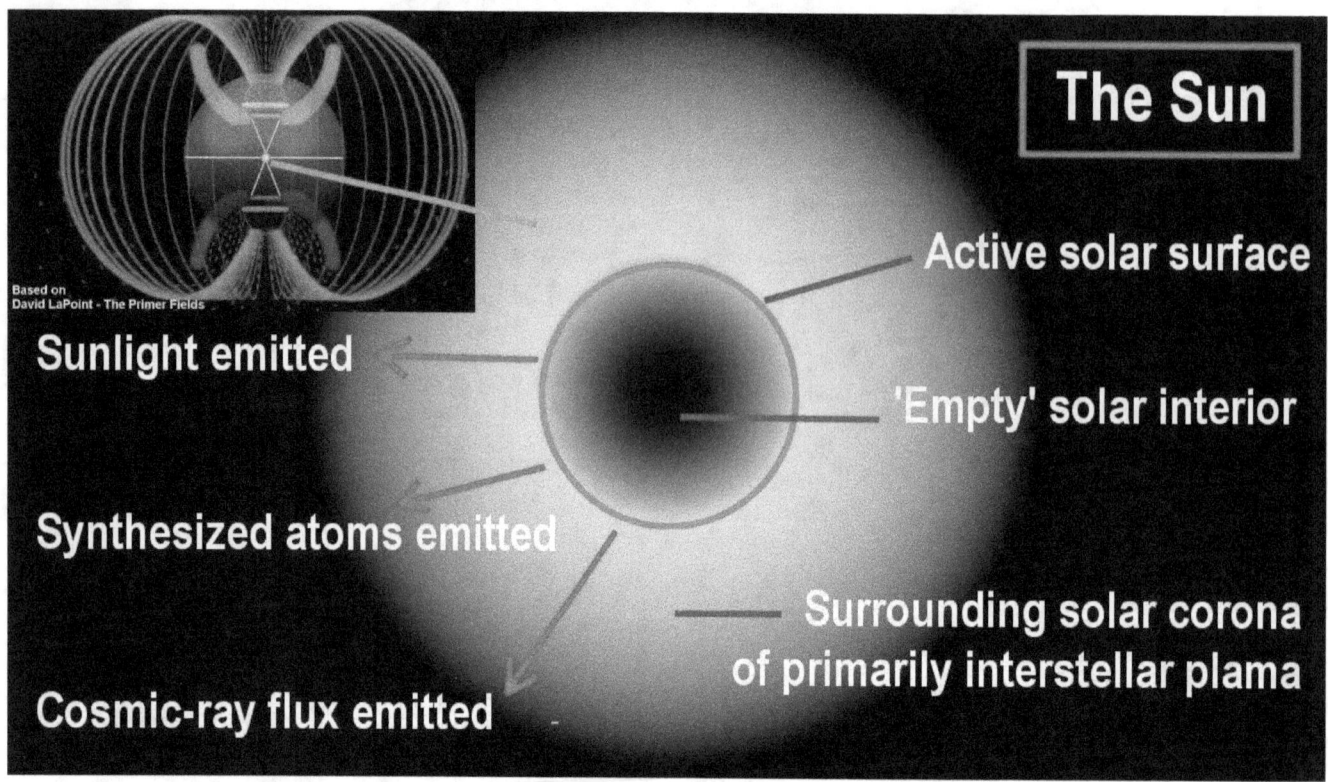

NASA's discovery defines our Sun as a plasma star that is externally powered by its interacting with interstellar plasma streams, which themselves are a part of the galaxy. This platform for the Sun, for which a lot of measured evidence exists, renders our Sun as highly variable, and as responding to changing plasma density conditions in the galaxy.

Our entire galaxy formed at a node point

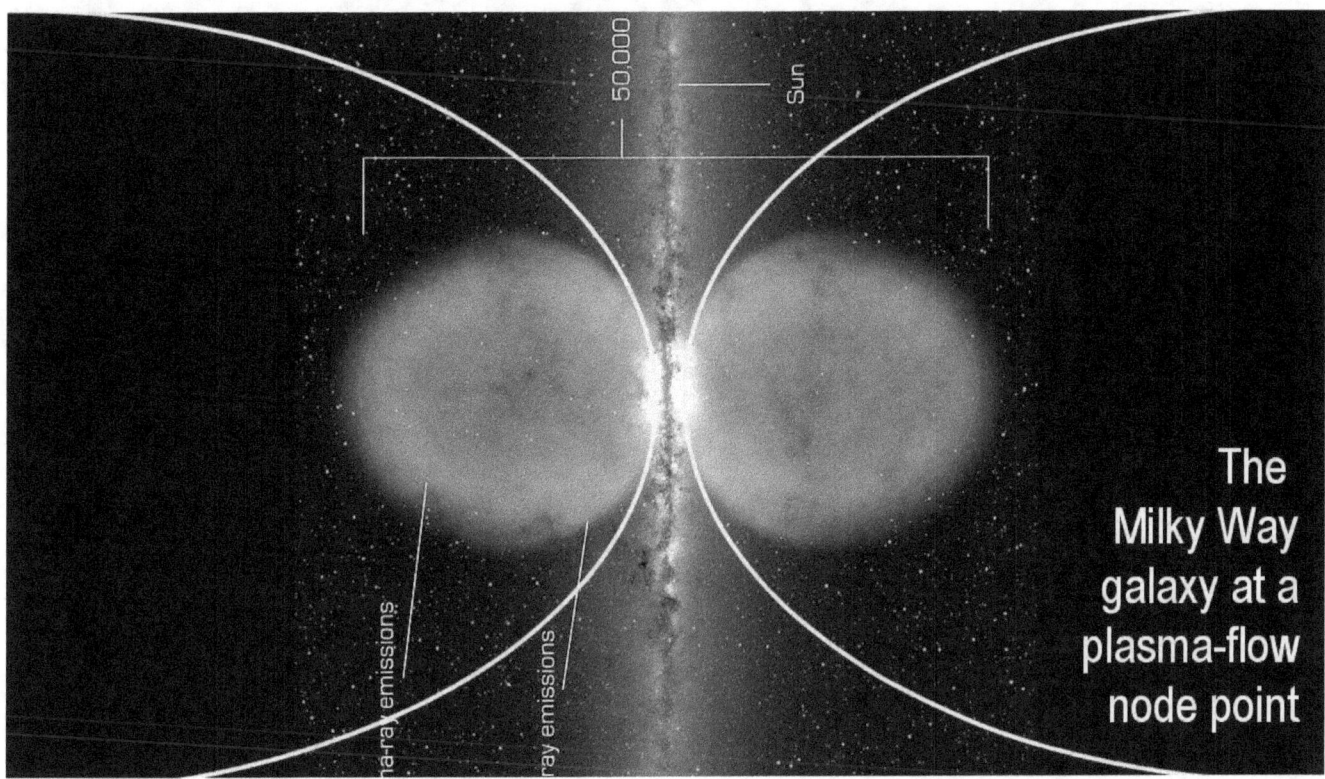

The NASA discovered feature indicates that our entire galaxy has been originally formed at a node point between extremely long intergalactic plasma streams.

The two plasma streams act together in a dynamic system

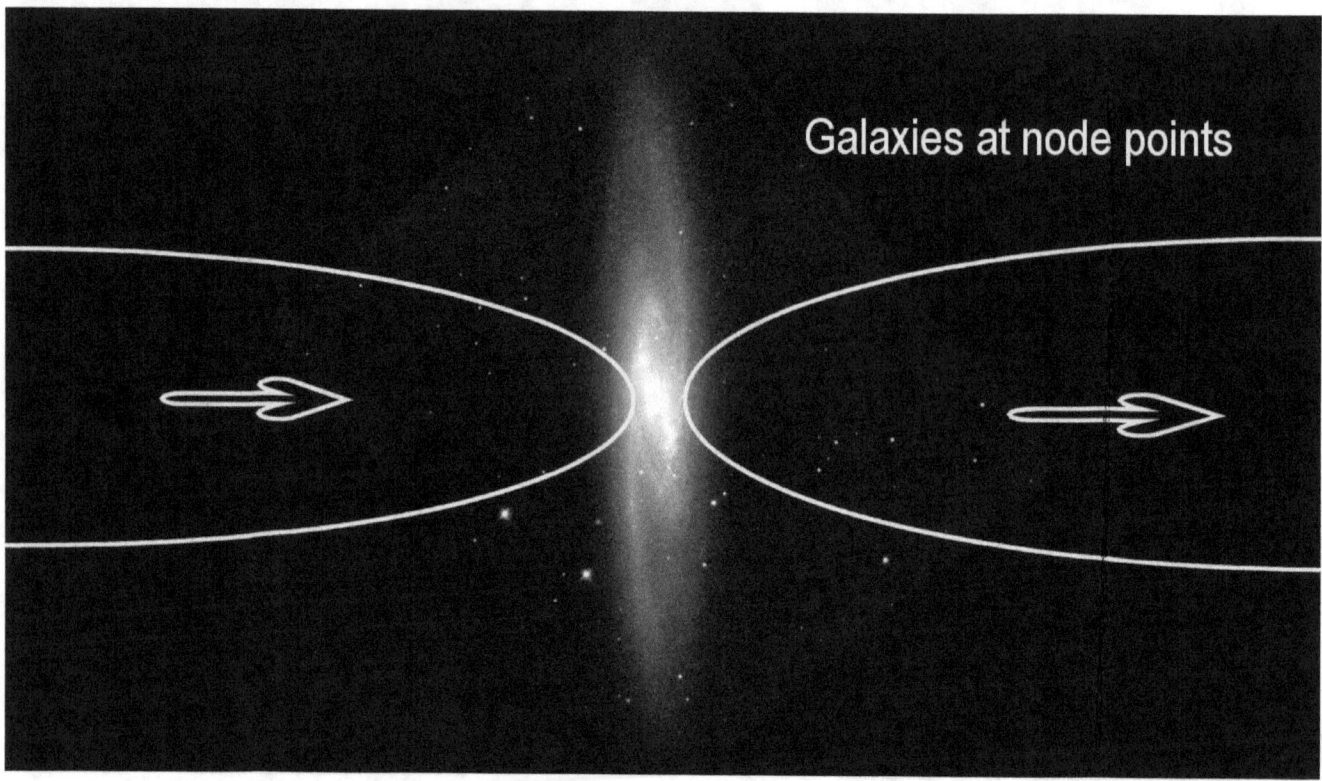

Since each of the two plasma streams has its own resonance characteristic, which act together in a dynamic system that powers the galaxy, their combined resonances are reflected throughout the galaxy. The fluctuating plasma density, is of course, also reflected on Earth, as the galactic plasma density directly affects the operating intensity of our Sun.

Potential plasma connections

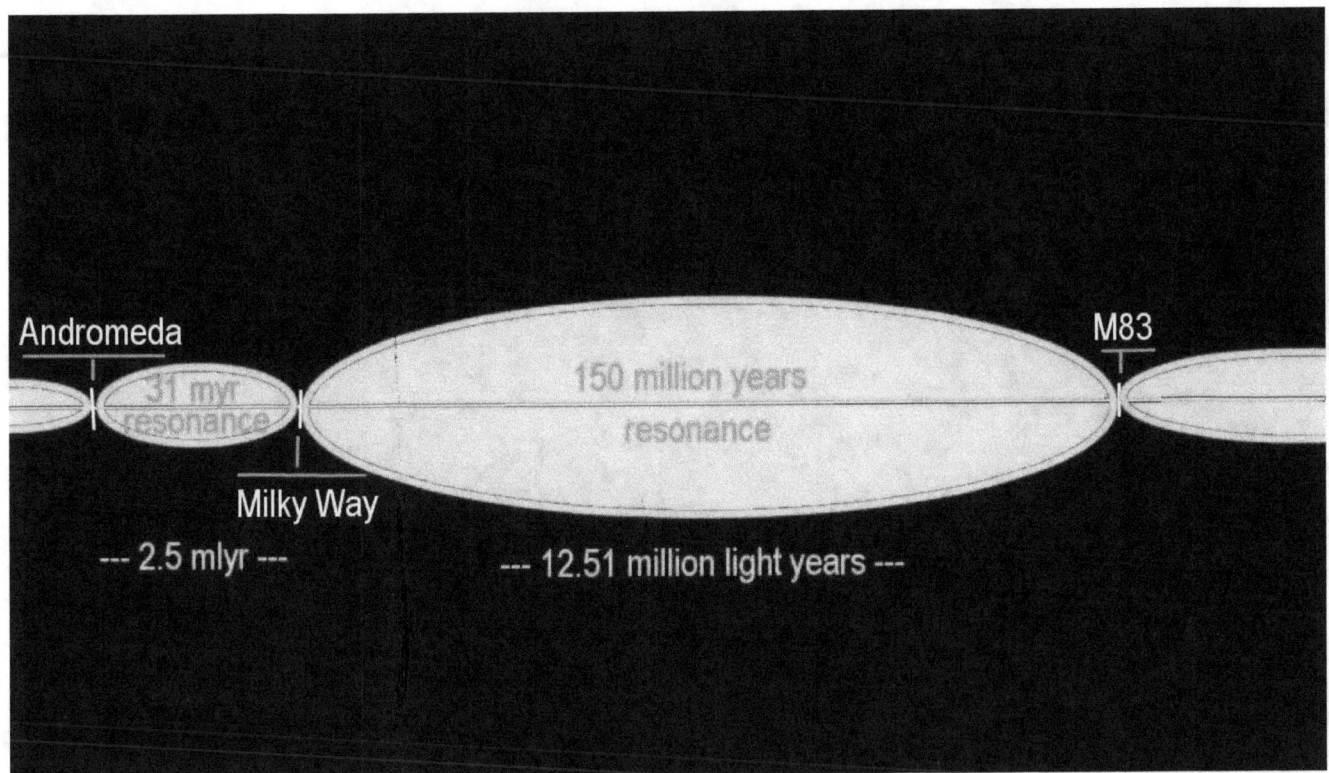

One of the potential plasma connections is with the nearby Andromeda galaxy. It gives us the 30 million years cycle. The plasma stream to the more distant galaxy, the M83 galaxy, gives us the longer, almost 150 million years, cycle.

The proof lies in the observed motions of stars

What actual proof do we have in direct physical measurements, that our galaxy is electrically powered, and therefore is electrically organized?

We have some well-known proof for this. The proof lies in the observed motions of stars.

Dogma that the stars orbit gravitationally around the galactic center

It is a widely believed dogma that the stars in a galaxy orbit gravitationally around the galactic center - which is actually, physically impossible.

The measured speed of the stellar motion disprove

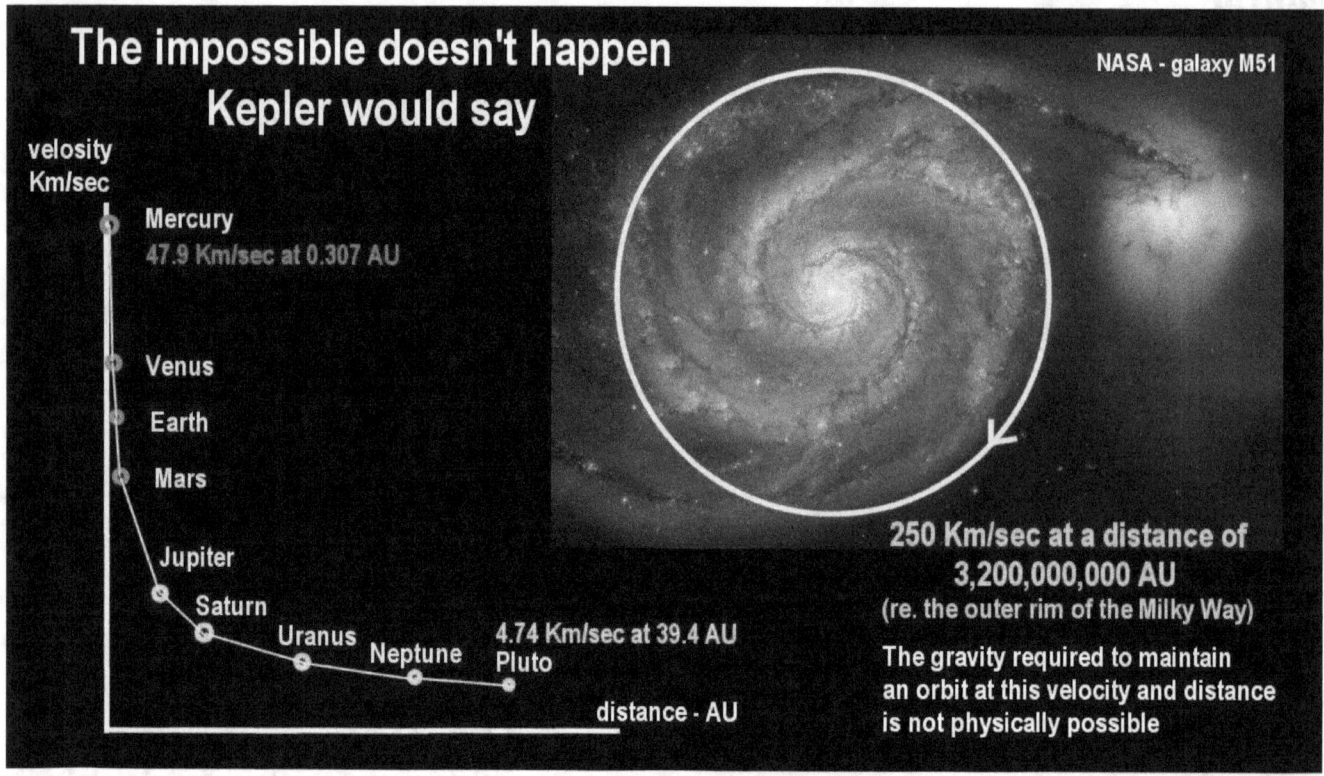

The measured speed of the stellar motion disprove this theorized possibility.

According to the laws of orbital mechanics

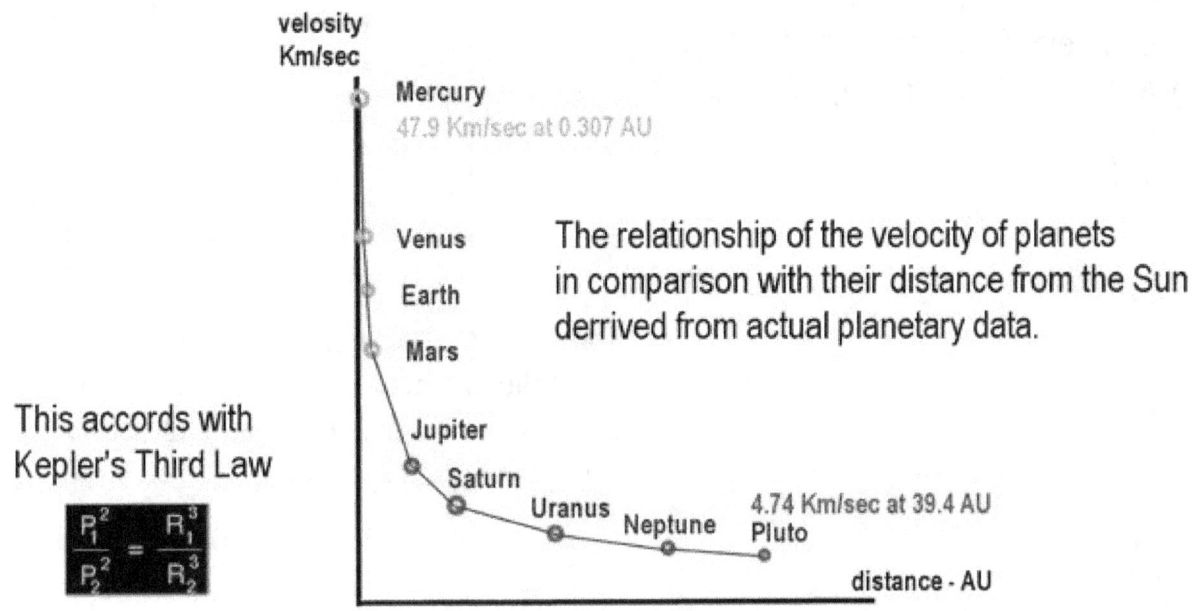

According to the laws of orbital mechanics, discovered by Johannes Kepler in the 1600s, the distant planets orbit at a dramatically lower velocity that matches the weak gravity, as gravity diminishes with the square of the distance.

Motions of stars accords with electro-magnetic principles

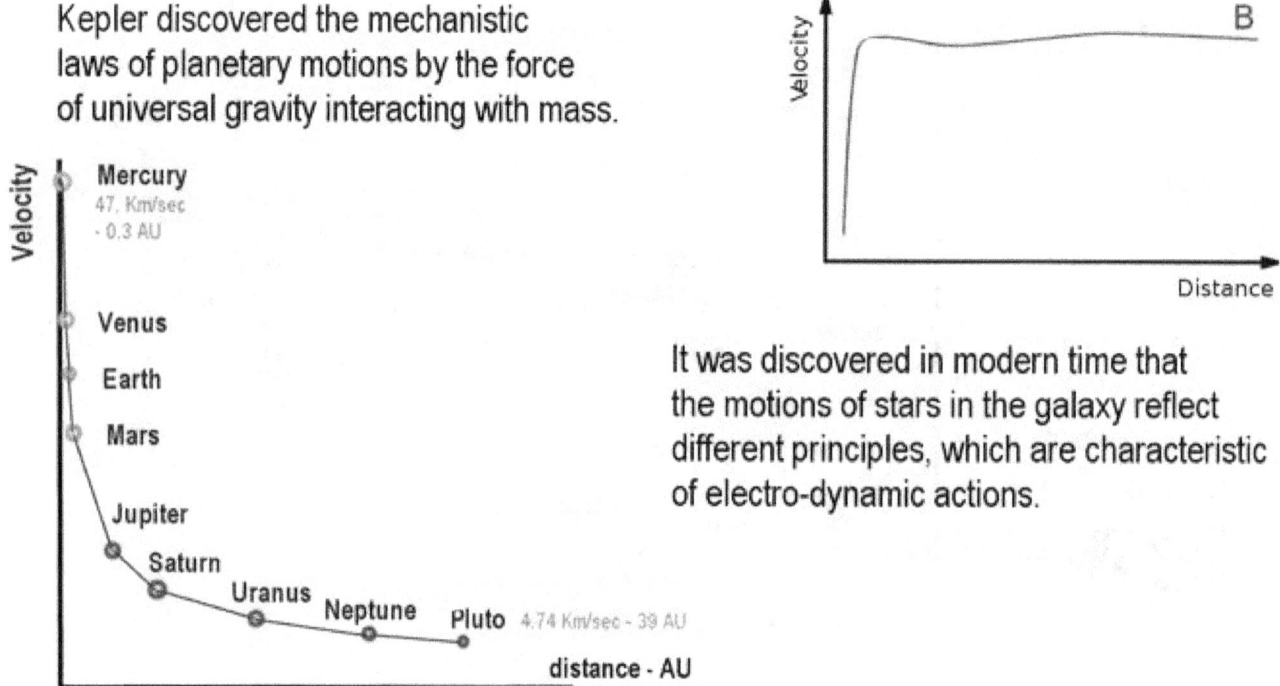

Kepler discovered the mechanistic laws of planetary motions by the force of universal gravity interacting with mass.

It was discovered in modern time that the motions of stars in the galaxy reflect different principles, which are characteristic of electro-dynamic actions.

The motion of stars, however, do not follow the physically necessary distance, gravity, and velocity relationship that enable orbital systems to function. Instead the motions of stars accords with electro-magnetic principles.

The entire galaxy rotates as one unit

By the electromagnetic principles, the entire galaxy rotates as one unit, in a similar manner as a homopolar electric motor rotates. The measured motions of the stars matches the typical effects well, of electrically motivated motion that are widely applied in technology.

Since the measured stellar motions matches the physics of electromagnetic principles, the physical measurements stand as proof that the galactic system and everything within it is electrodynamically organized and powered by streams of electric plasma particles.

The principle of the plasma galaxy

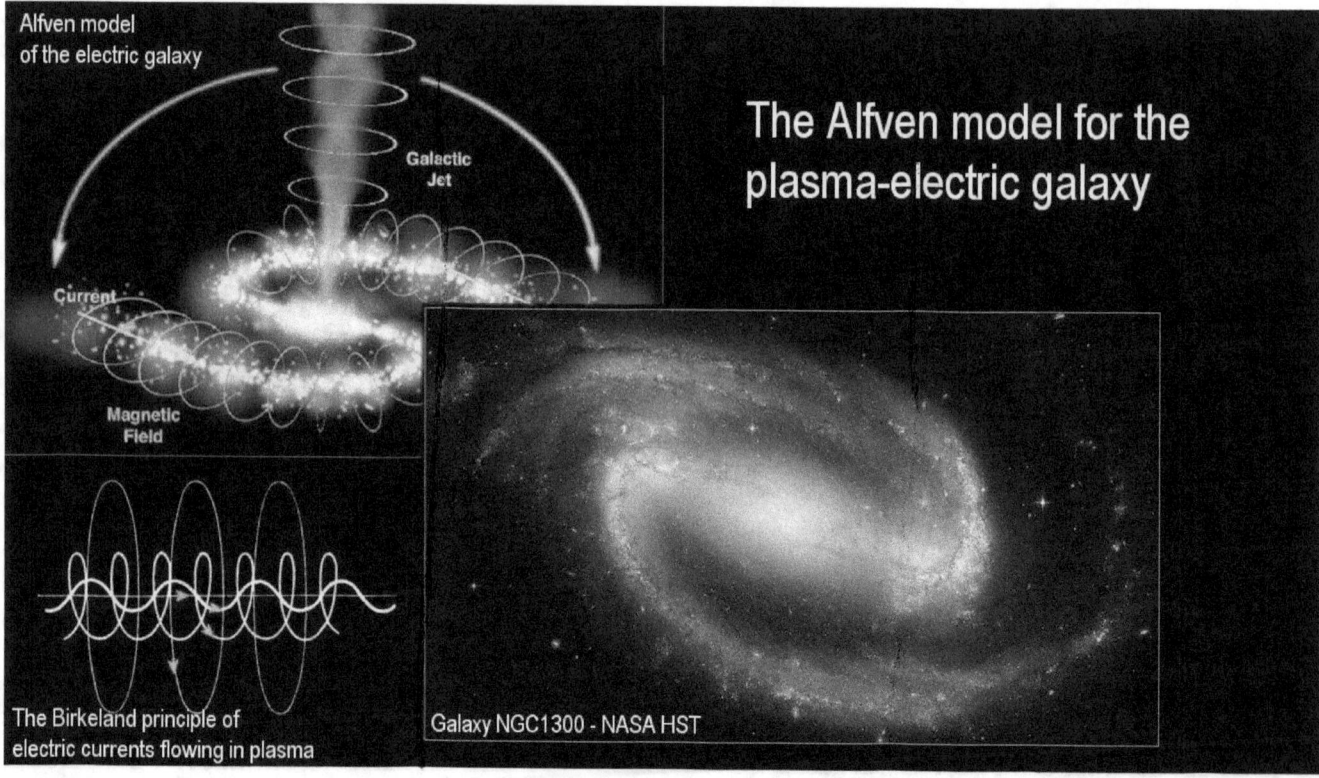

The principle of the plasma galaxy was already recognized decades ago by Hannes Alfven, the 1970 Nobel laureate for his related pioneering work in magnetohydrodynamics.

The origin of the cosmic plasma streams

While we have no measurements are available as to the origin of the cosmic plasma streams that power the the galaxies and everything within them, which researchers believe span all space, their existence may be traced to a specific quality of the universe itself,

The implicate and explicate order

which the American theoretical physicist David Bohm termed "the implicate and explicate order."

David Bohm sees cosmic space as a sea of latent energy

David Bohm, whom Albert Einstein is reported to referred to as his successor, sees cosmic space not as an empty void, but as a sea of latent energy

Explicate like ripples on a surface of water

that becomes explicate to some degree like ripples on a surface of water.

The plasma particles which pervade space, which are 100,000 times smaller than the smallest atom, may be a continuously-ongoing explicate phenomenon of the cosmic universe. No Big Bang origin theory is needed here.

Part 1 - 500 Million Years of Climate Change via the Sun

A very distant origin

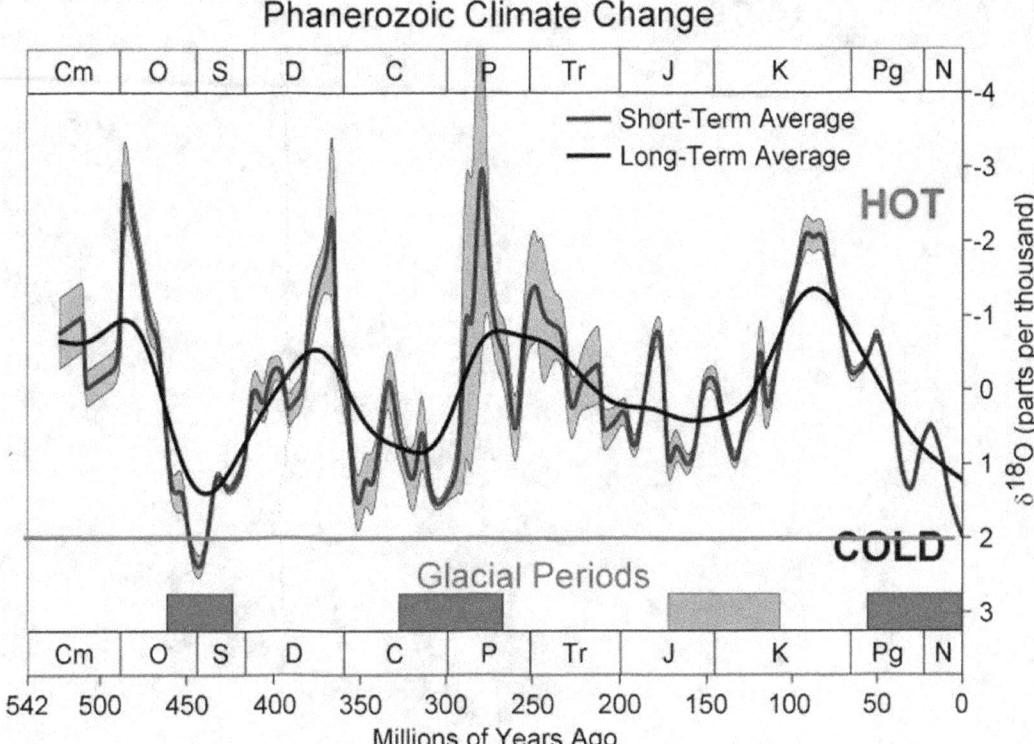

This means that what we see reflected in the Earth's long-term climate record, which affects us extremely deeply, has a very distant origin

In the deep intergalactic realm

in the deep intergalactic realm and in the resonance effects of the plasma streams that form there and are maintained there.

The electrodynamic nature of the galaxies

The recognition of the electrodynamic nature of the galaxies, the solar systems within them, all the way to the planetary systems, is critical as a basis for understanding the Earth's climate dynamics.

Cosmic dynamics and the solar-system dynamics

This is so, because the cosmic dynamics and the solar-system dynamics, extend all the way to the Earth climate dynamics, which are functionally linked by the basic nature of the universe and its principles that are expressed everywhere and on every scale.

We cannot afford to ignore the dynamic nature of the Universe

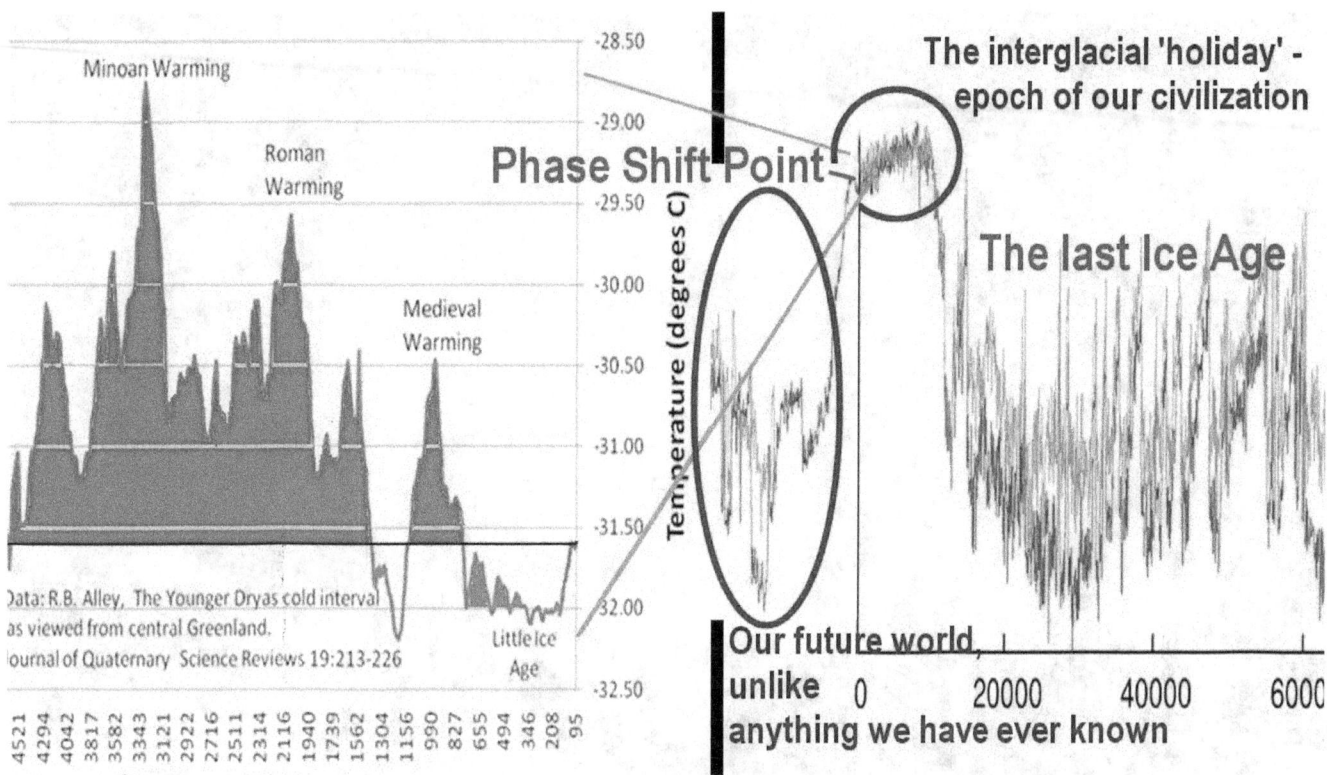

We cannot afford to ignore the dynamic nature of the Universe that we see reflected in all the physical measurements that were made, because our very existence ultimately depends on us responding intelligently to the dynamics of the universe as we progress towards a new round of the Ice Age conditions again, which have been the norm for the last few million years, but which no one has historically experienced or can even imagine.

On any other basis than plasma electrodynamics

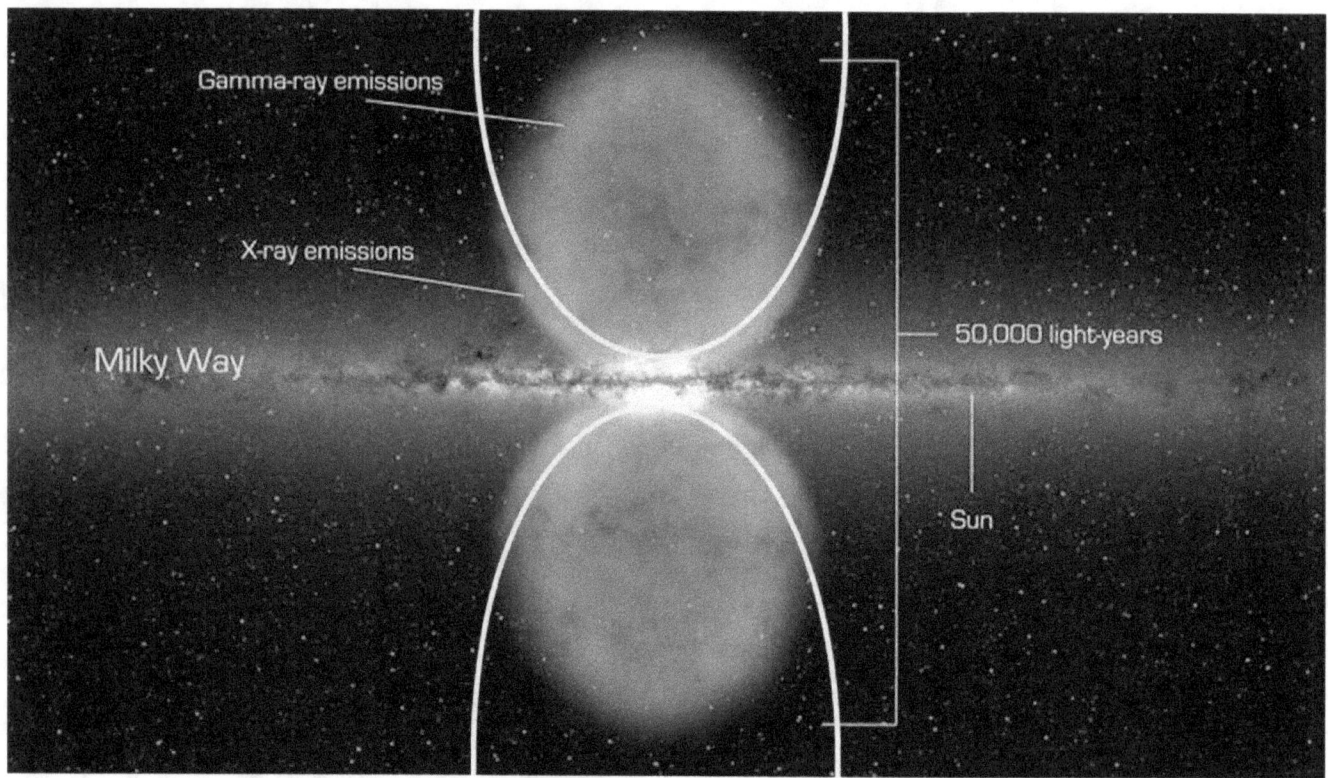

It is not possible to understand the dynamics of the ice ages, especially the gigantic historic climate fluctuations that got us into the present ice age epoch, on any other basis than the context of plasma electrodynamics that the universe is built on.

Climate background the coldest in 440 million years

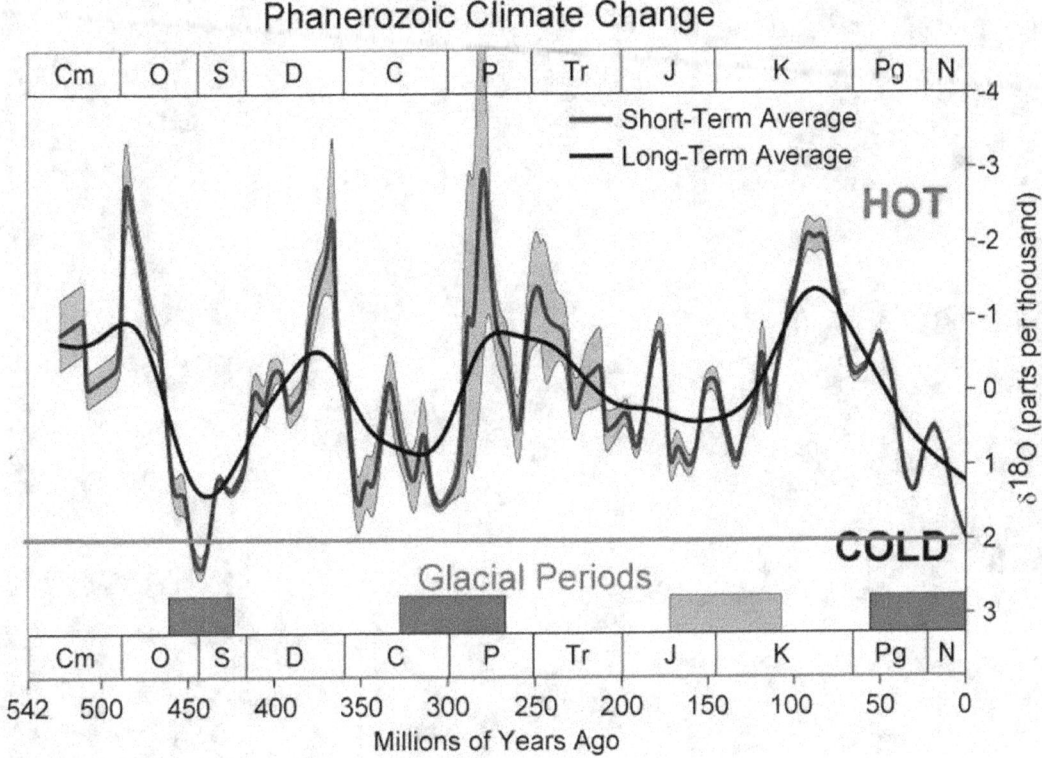

We are currently in the midst of a deep glaciation epoch, because the two very long historic climate cycles that have been measured, combine at their low point. The result is, that the climate background for the Earth is now at the coldest level it has been in 440 million years.

Plasma density in the galaxy is present at its weakest state

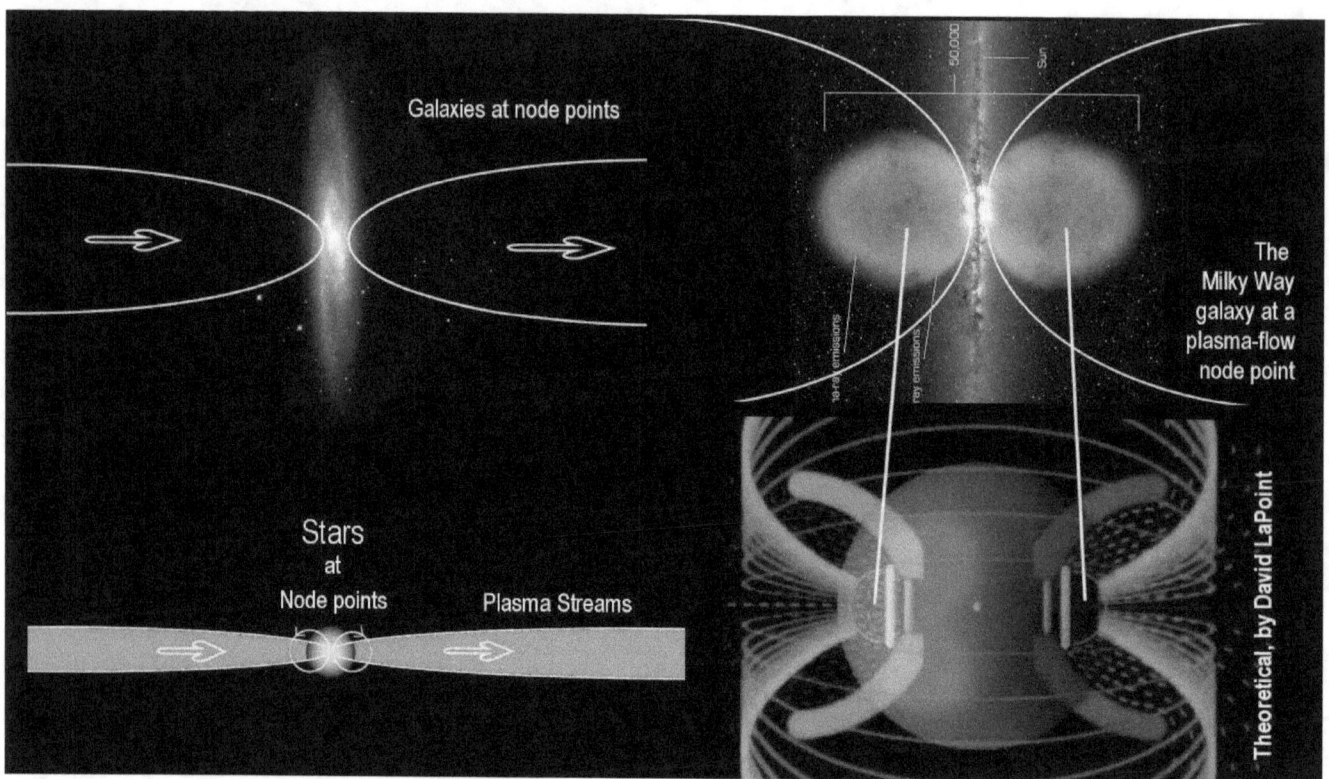

This is so, because the plasma density in the galaxy is present at its weakest state, which has dramatic effects on the operation of our Sun.

At the bottom end of a long diminishing slope

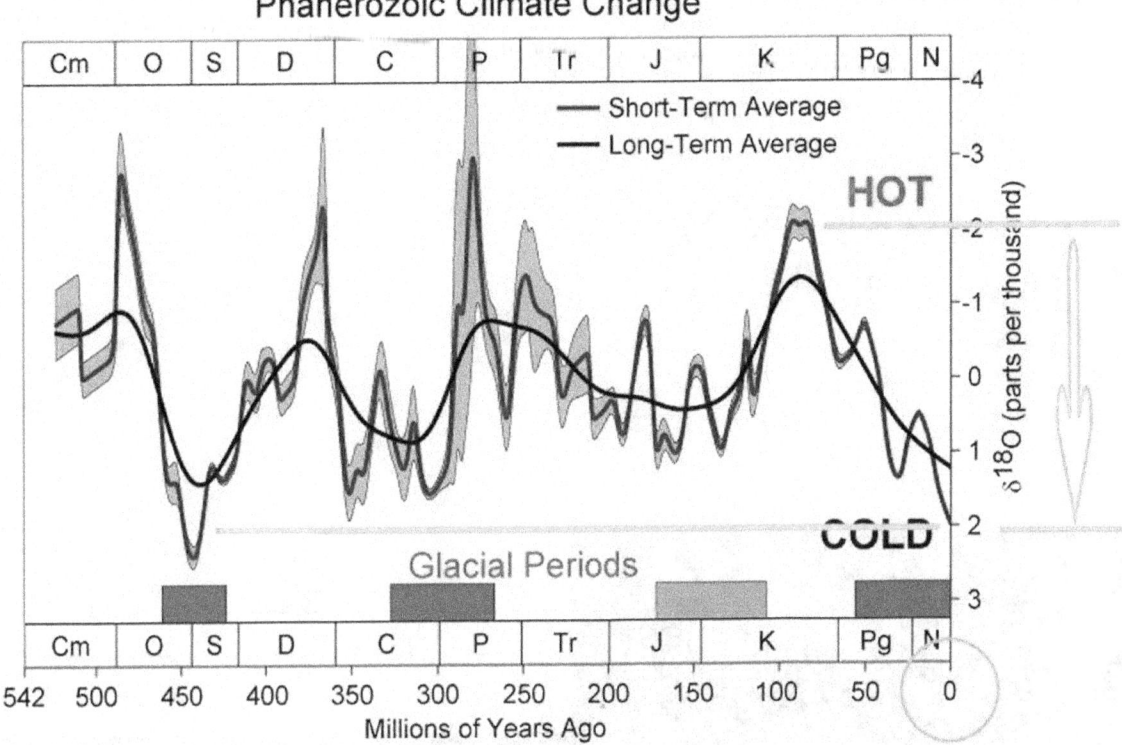

In historic terms, we are presently at the bottom end of a long diminishing slope of cosmic weakening that began 100 million years ago. It would be surprising if the present extreme galactic weakness was not reflected in the climate on Earth.

Part 1 - 500 Million Years of Climate Change via the Sun

Antarctica froze up, thawed out again, and froze up once more

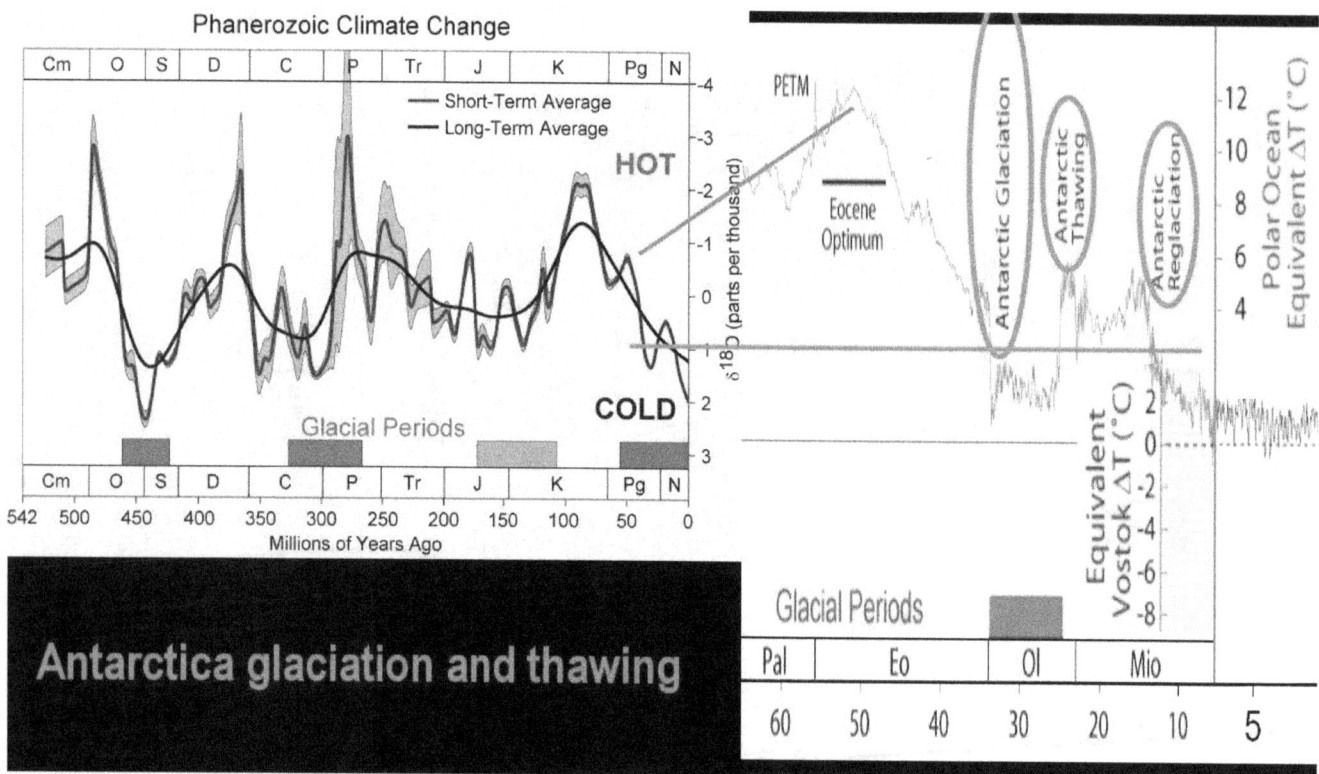

In order to get a feeling of how big the climate collapse was that occurred over those 100 million years of the big down-slope, lets focus only on the last 65 million years, in a side by side comparison, because it was in this later timeframe that Antarctica froze up, then thawed out again, and froze up once more.

If one considers that these momentous effects occurred within a narrow region of the 100-million-year down slope, it becomes evident that we are looking at gigantic climate transitions on the long historic timescale, which in their magnitude far exceed the effects that changing cosmic-ray flux could possibly have on the Earth's climate. This means that only a plasma Sun that responds to changing electrodynamic conditions in the galaxy, is able to produce the near miraculous climate fluctuations that have been measured.

The measurements put the Plasma Sun squarely onto the table, as no other cause is able to generate the measured effects.

The Plasma-Sun Principle

The Plasma-Sun Principle

The Plasma-Sun Principle

The Plasma Sun is a large sphere of plasma

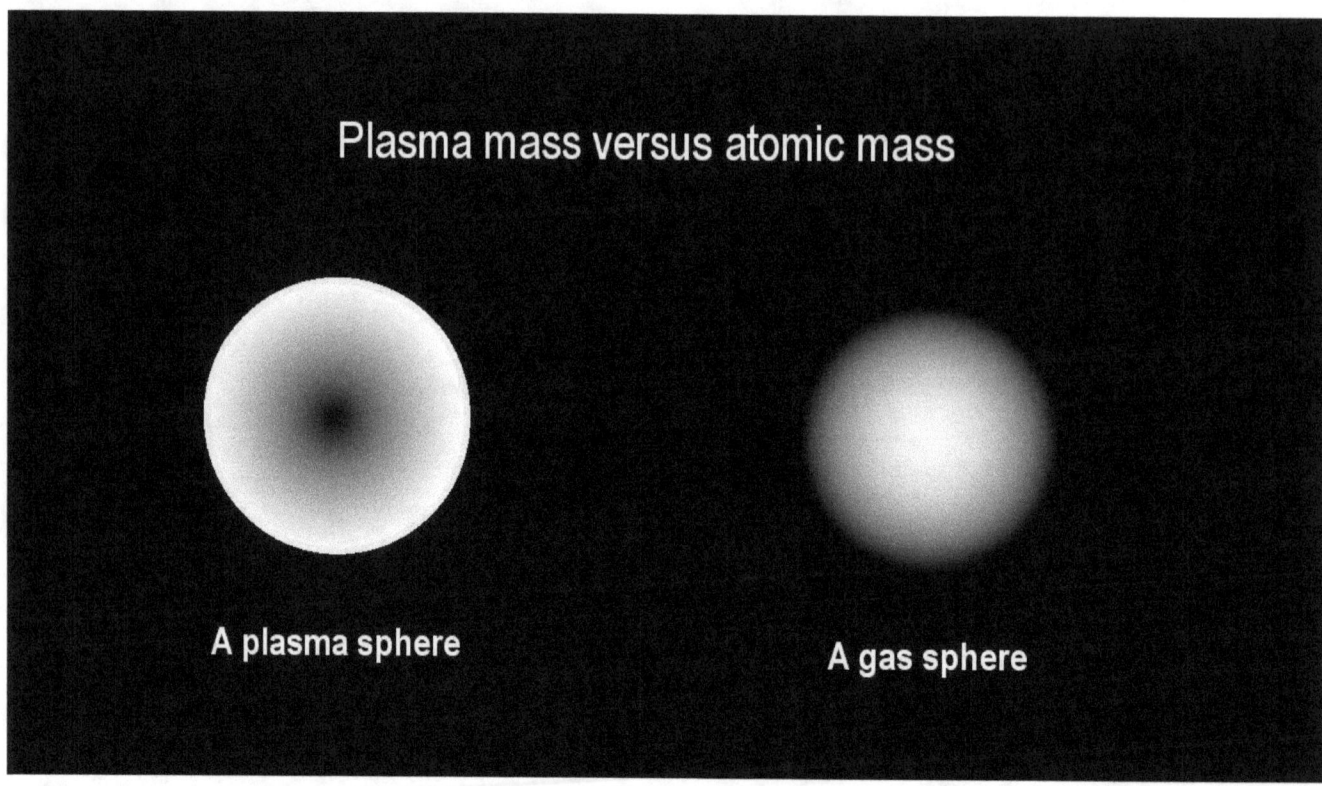

The Plasma Sun is a large sphere of plasma that is electrically active and responds in a totally different manner to the force of gravity.

Swarming electrons tend to migrate away from the center of gravity

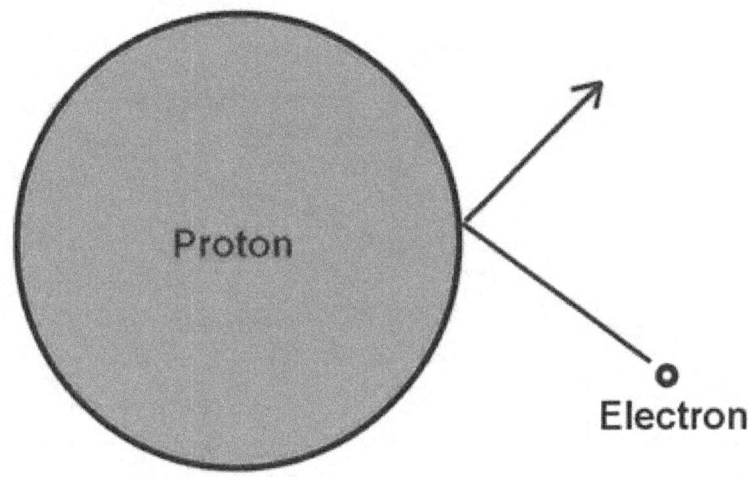

Plasma is the name for electrically charged particles, named electrons and protons, existing in unbound form. They interact by the electric force. Electrons are attracted to the protons by this force, but before they can latch onto each other the electron is repelled at close distance, only to be attracted anew in an endless dance. The swarming of the electrons also keeps the protons isolated from each other, which normally repel one another by the same electric force.

However, when a plasma sphere is large, gravity comes into play. It has the effect that the swarming electrons, which are a thousand times lighter than the protons, tend to migrate away from the center of gravity to the surface where they form a dense, electron-rich layer.

The migration leaves the protons less shielded

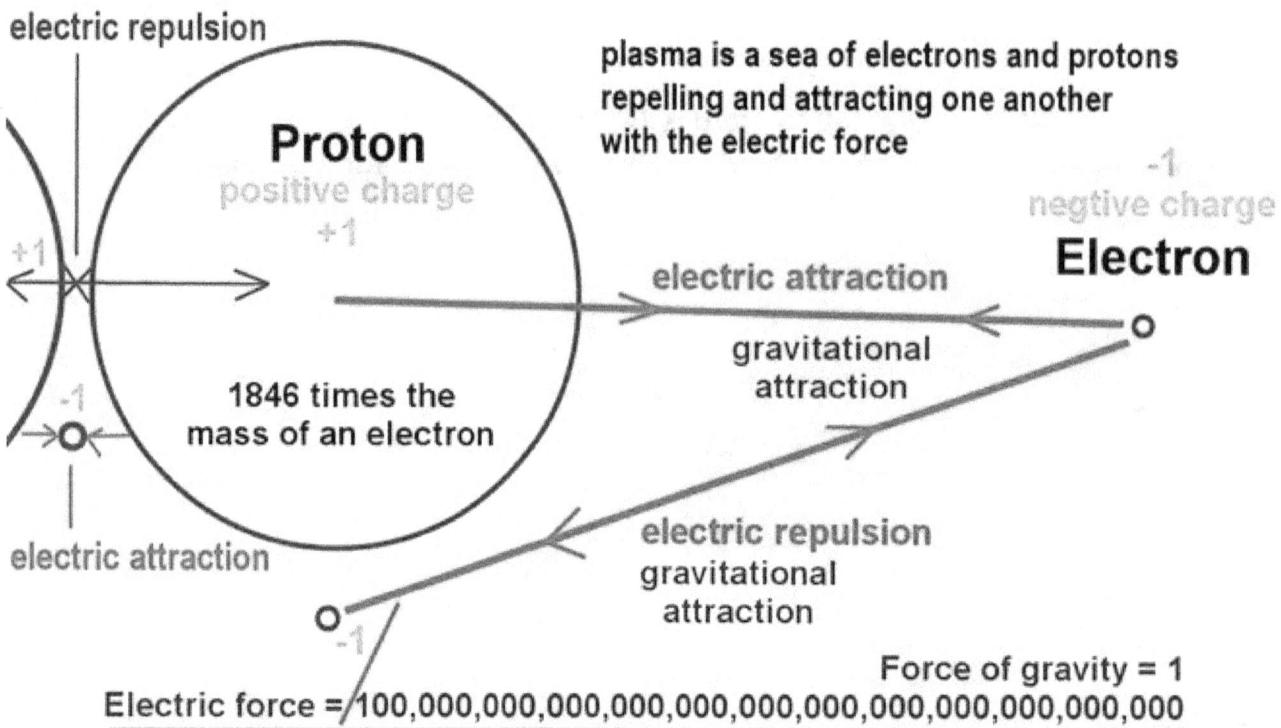

The migration leaves the protons at the center less shielded from each other, which enables them to repel each other more strongly.

A very large sphere of plasma with little density at its core

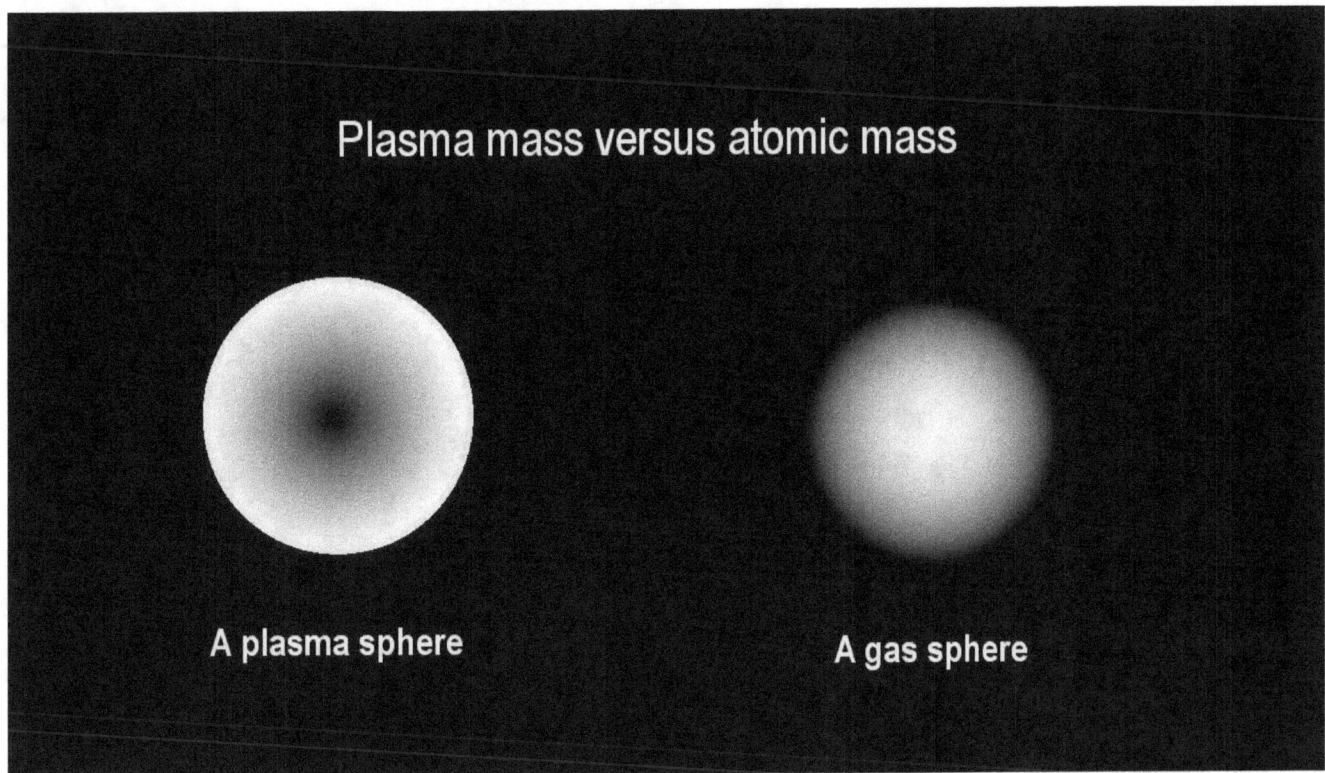

The result becomes a very large sphere of plasma with little density at its core, and low overall mass density, but with an extremely dense, and electron-rich surface. That's what interstellar plasma streams interact with. These electrodynamics principles create a type of geometry that enables large stars to exist, with a large surface area. It creates stars that could not exist as gas spheres. Gas spheres are limited in size by the maximum tolerable gravitational compression before such a gas sphere explodes in nuclear fission. This built-in limit is one of the reasons why our Sun cannot possibly be a gas sphere.

Our Sun is a thousand times too light

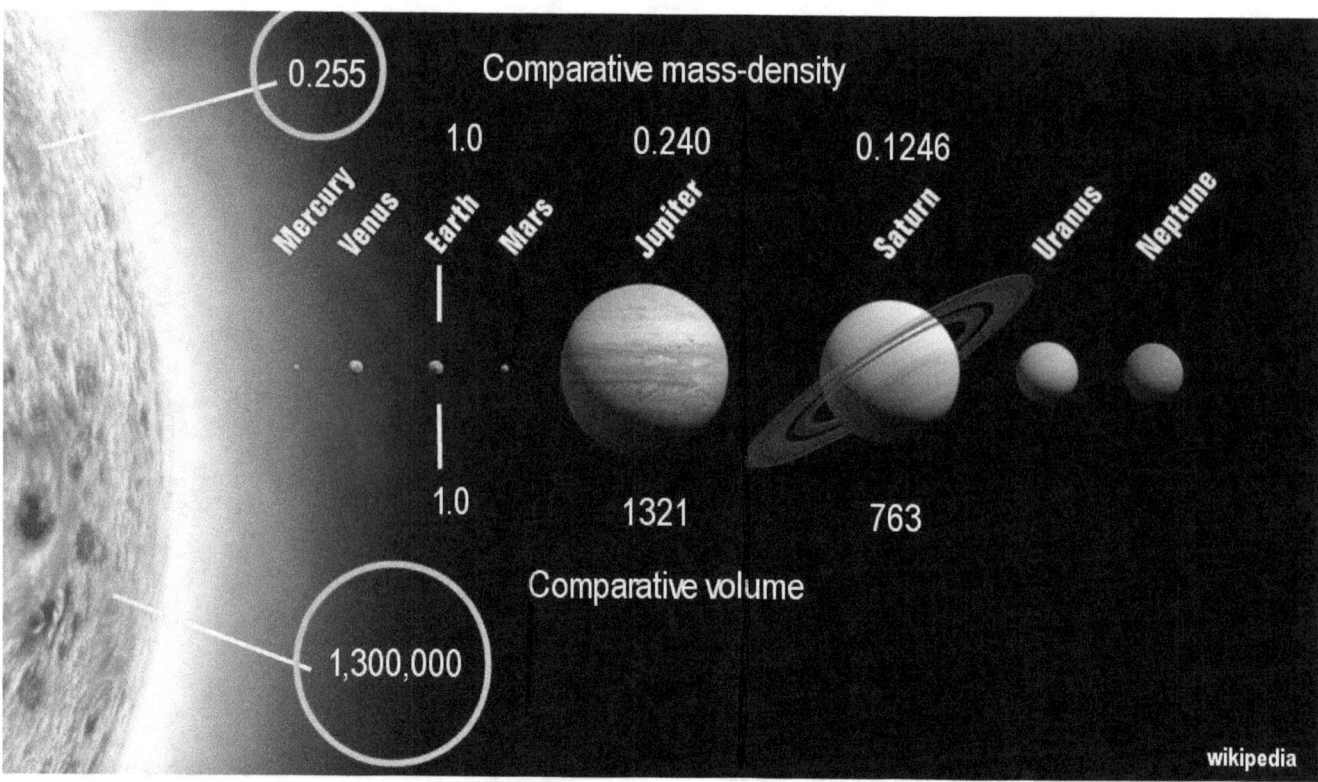

Also our Sun is a thousand times too light to be a gas sphere of its size. The comparison of its mass density with that of the big gas planets, makes this rather plain. In the comparison, Jupiter, which has twice the volume of Saturn, has twice the mass density, because of gravitational compression, while the Sun, that has a thousand times greater volume, has roughly the same mass density as Jupiter. The Sun's low mass density is only possible by the Sun being a plasma star that is a largely empty shell.

The low solar mass density proves

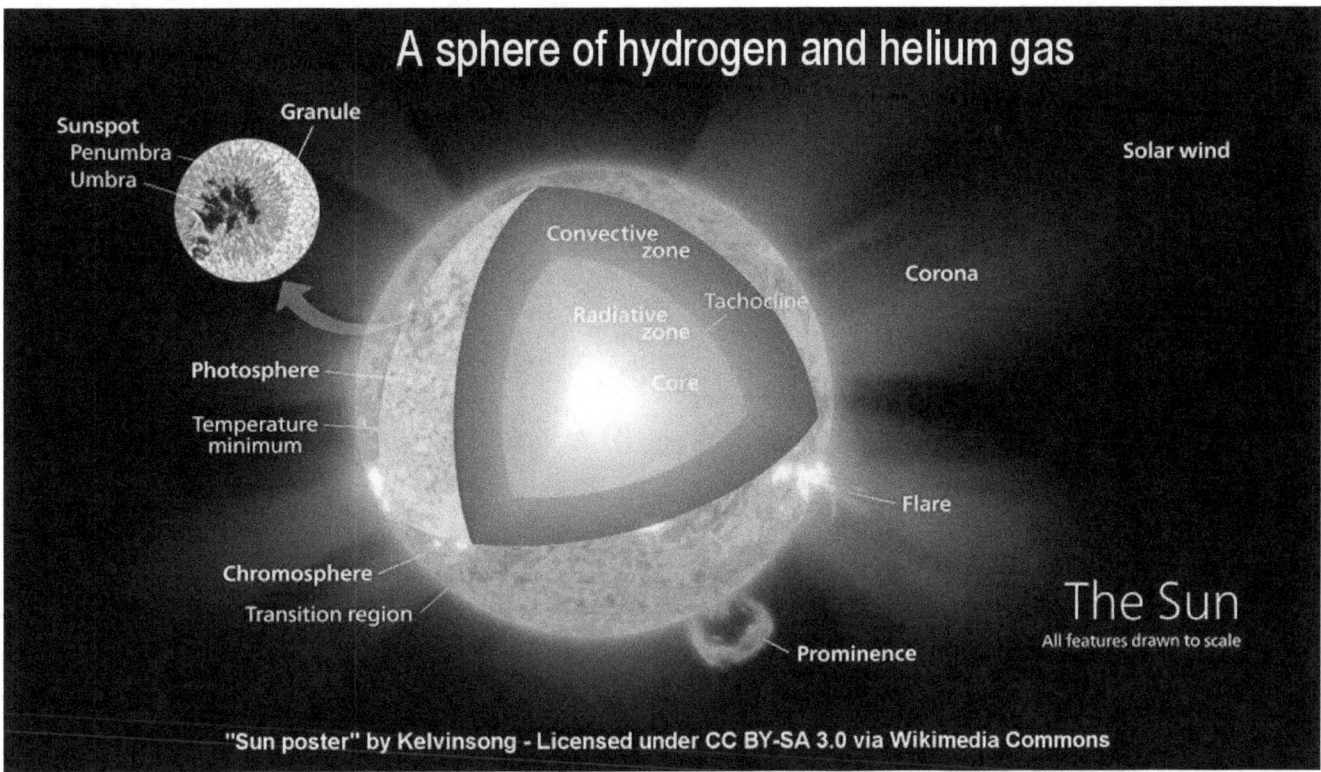

The low solar mass density proves the gas-compression hydrogen-fusion model for our Sun to be a myth, as do the sunspots.

When we look through the umbra of sunspots

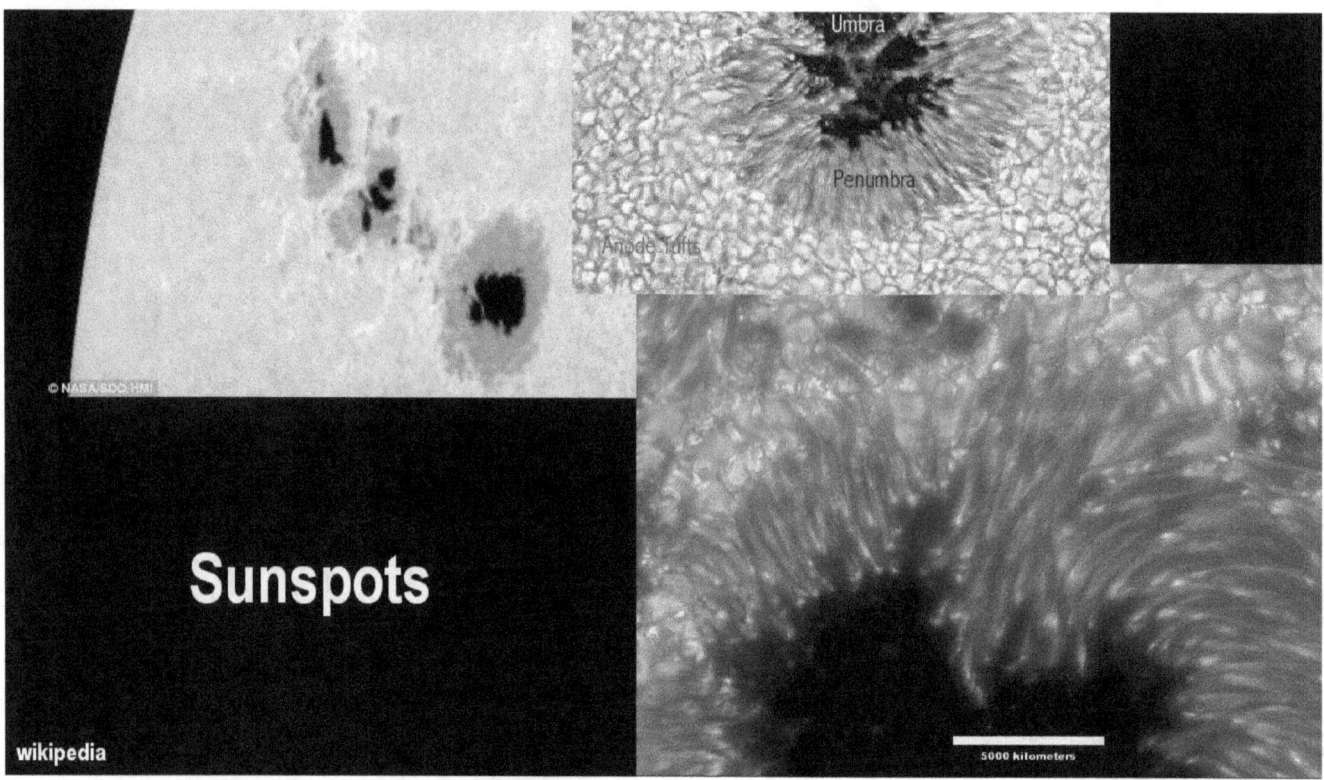

When we look through the umbra of sunspots, we see the Sun being dark inside. If the Sun was internally energized, the sunspots would be bright, not dark. They are dark, not because mythical dark energy blocks the light from within. They are dark, because there is nothing behind the sunspots, but a largely empty shell of plasma.

The giant star UY Scuti

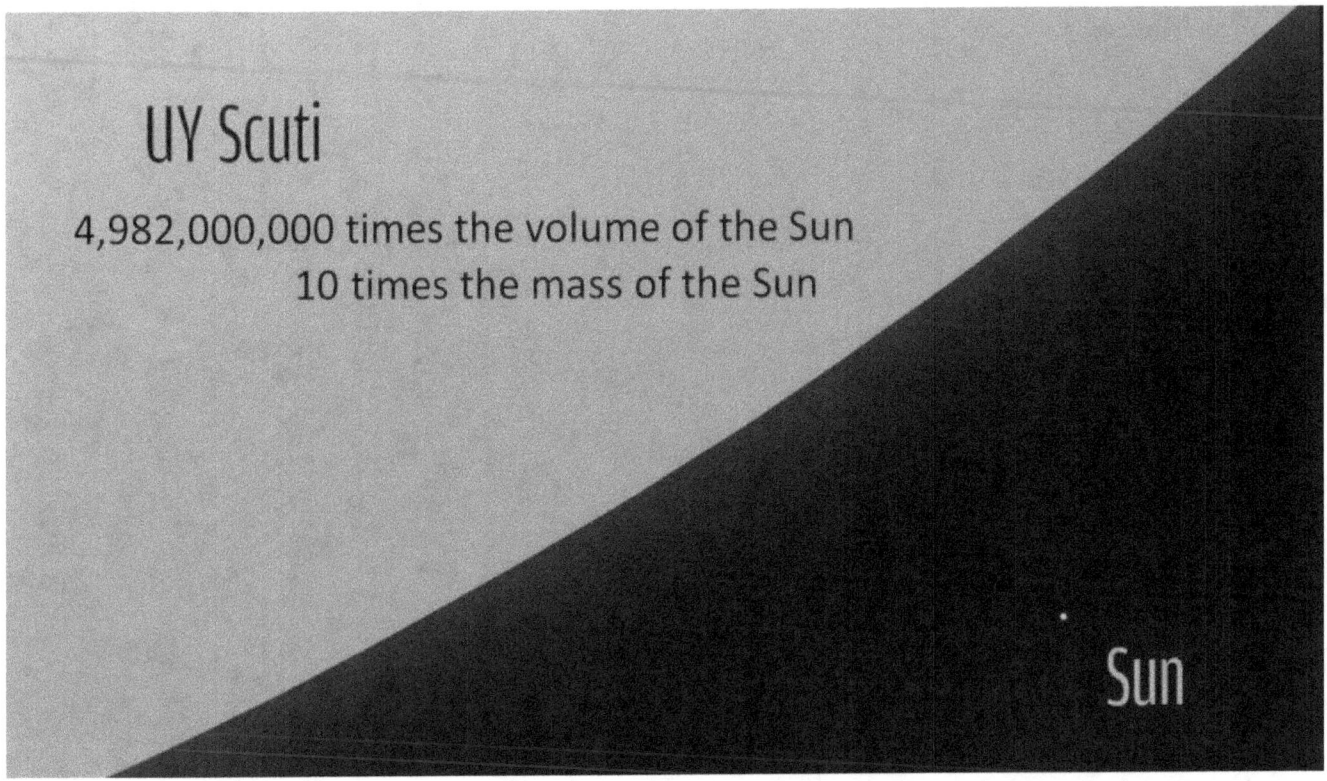

The gas-compression nuclear-fusion model is powerfully disproved by some of the big stars that can't possibly exist under this model.

The giant star UY Scuti, for example, is measured to be 1700 times larger than our Sun and to contain ten times its mass. This is not possible under the gas compression fusion model. This star has its 10 solar masses spread out across the star's 5-billion-times larger volume. As one researcher puts it, this star is so thinly spread that it is practically a vacuum.

A plasma-star doesn't produce its own energy

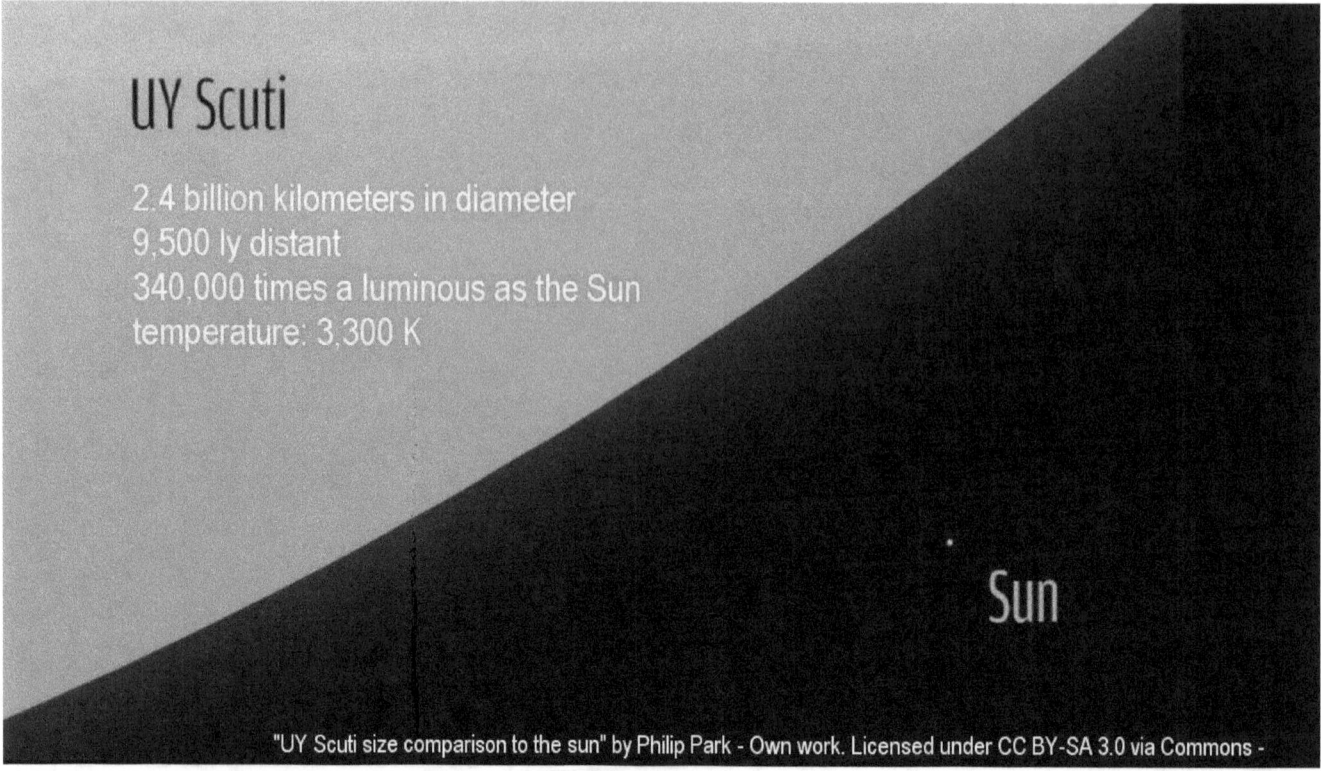

The giant star that is practically a vacuum, which should not exist under the gas-compression-fusion model, does indeed exist. It outshines our Sun 340,000-fold with is mere 10 solar masses.

As a plasma-star, which doesn't produce its own energy, but merely acts as a type of catalytic converter of the plasma flowing into it, the impossible luminance of the UY Scuti star is totally possible. The great luminance is possible, because a plasma star, like our Sun, doesn't create the energy it radiates. It merely converts it into a different form. It reflects back what flows into it.

All known atomic elements are synthesized

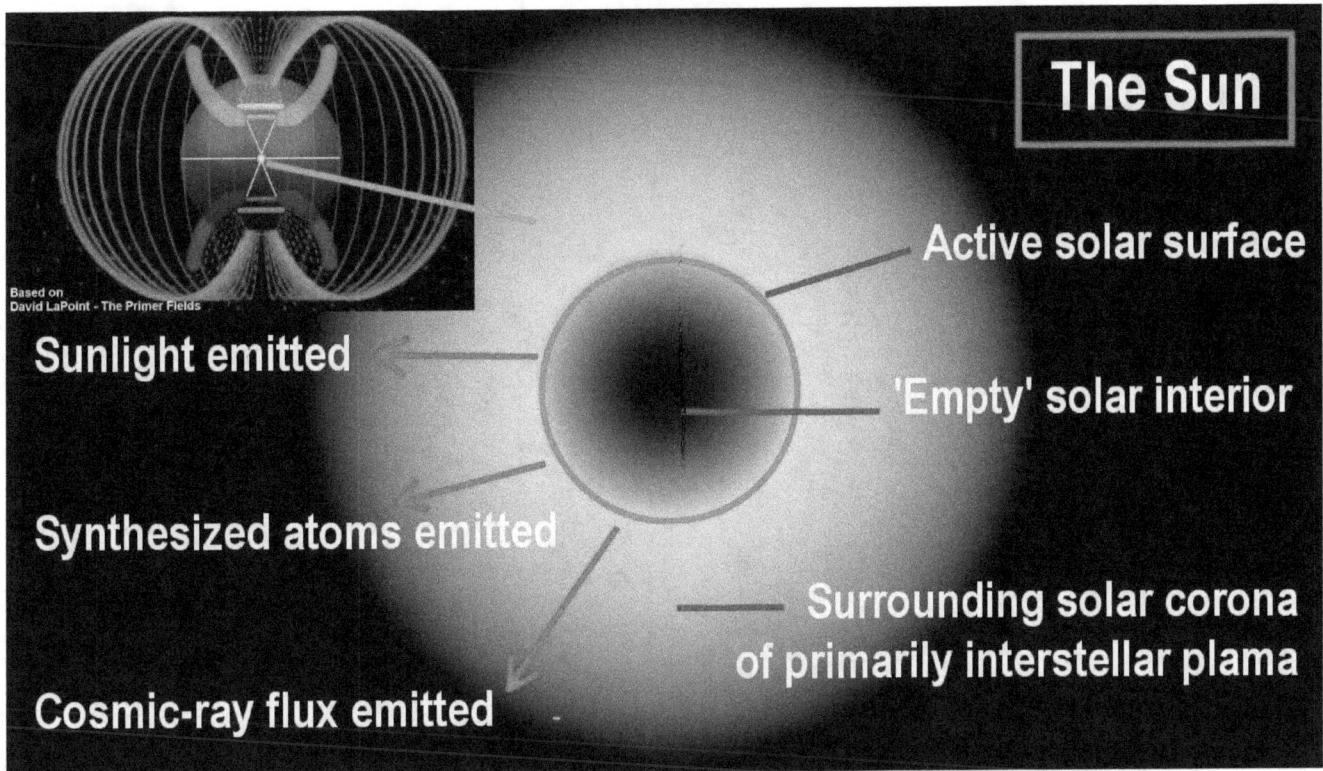

At the surface of a plasma star, by it interaction with interstellar plasma that surrounds it, all known atomic elements are synthesized. These created elements, acting together, emit the rich spectrum of sunlight that we see.

The surface of the Sun is a sea of reactions cells

The surface of the Sun is a sea of reactions cells where all known atomic elements are synthesized, each of which has its own light emission spectrum. The atomic synthesis can be theorized to be caused by the magnetic compression of flowing plasma into dense concentrations where they naturally combine into atomic structures, which, by this process become electrically neutral and flow away with the solar wind. The atomic synthesis in the fusion cells creates the sink effect that keeps the plasma streams flowing into a sun and not pile up there.

The resulting rich spectrum in sunlight

The resulting rich spectrum in sunlight, emitted by a wide range of atomic structures, makes it rather plain that our Sun cannot be anything other than a plasma star. An internally heated sphere of hydrogen gas would not be able to generate the sunlight that we see.

The Sun's light spectrum would be rather meagre

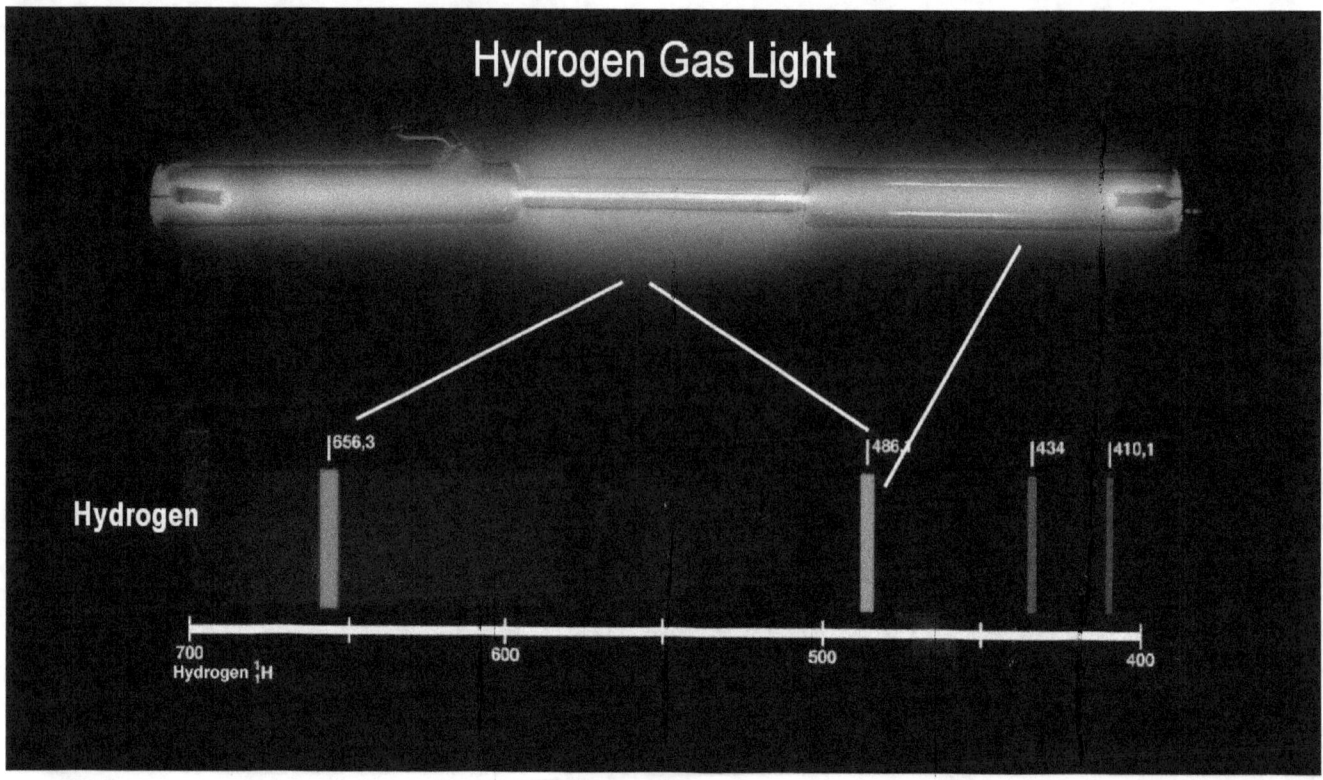

The Sun's light spectrum would be rather meagre if the Sun was a gas star, as meagre as the emission spectrum of hydrogen gas.

Since the plasma fusion process is externally powered

The richly colorful world that we live in under the Sun, is a part of the proof that our Sun is a plasma star. And since the plasma fusion process is necessarily externally powered, the sunlight also reminds us scientifically that our Sun is deeply affected by the changing plasma density in interstellar plasma streams, which are affected by the prevailing plasma density in the galaxy, which in turn is affected by plasma-resonance in intergalactic plasma streams.

That's how Antarctica froze up when the Sun became 'colder'

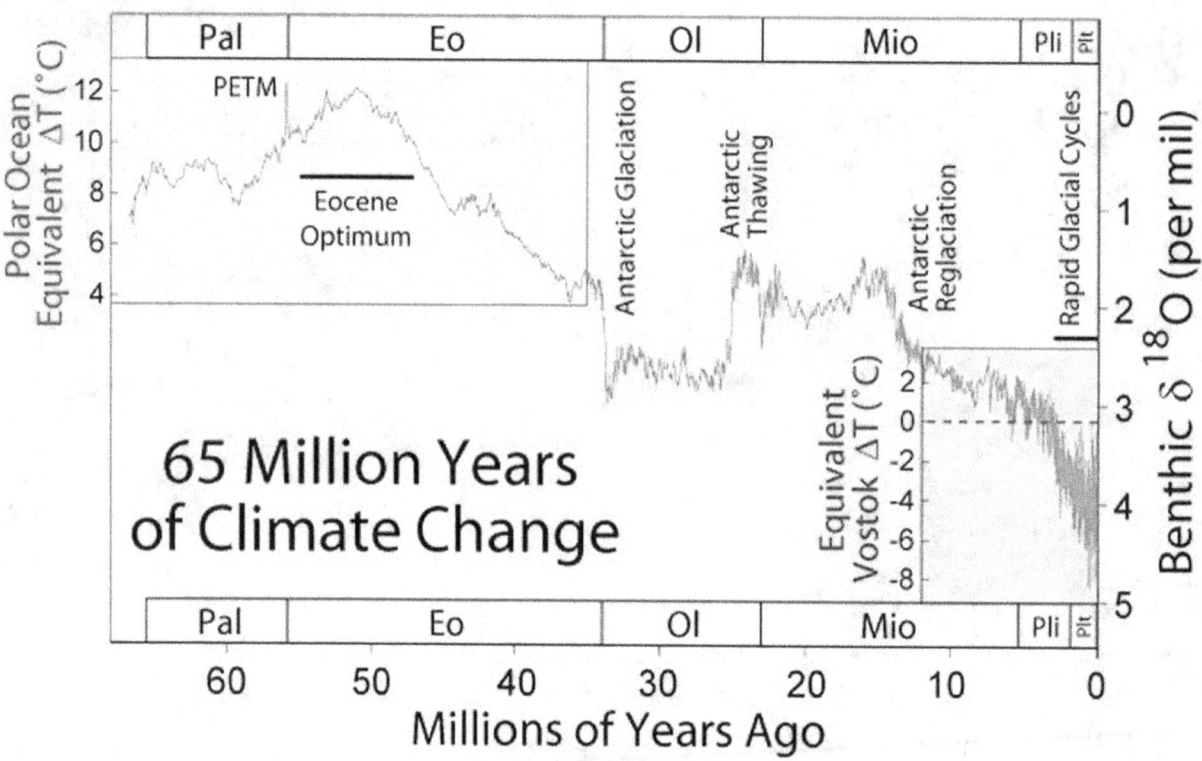

That's how Antarctica froze up, roughly 30 million years ago, when the Sun became 'colder', and then thawed out again when the Sun became 'hotter', after which Antarctica froze up once more 12 million years ago as the Sun 'cooled down' again as it had to when the galactic plasma density diminished.

Our Sun, being a plasma star, reflects as a matter of principle, the changing pattern of the galactic plasma density.

Antarctica has remained frozen as the solar activity continued to diminish on the weakening slope of the gigantic galactic climate change that is expressed in galactic plasma density, and correspondingly in the climate on Earth.

Dynamics of the Pleistocene Epoch

Dynamics of the Pleistocene Epoch

Epoch of the modern Ice Ages, ongoing

Dynamics of the Pleistocene Epoch The epoch of the modern Ice Ages, ongoing.

The Pleistocene created a new phase shift on Earth

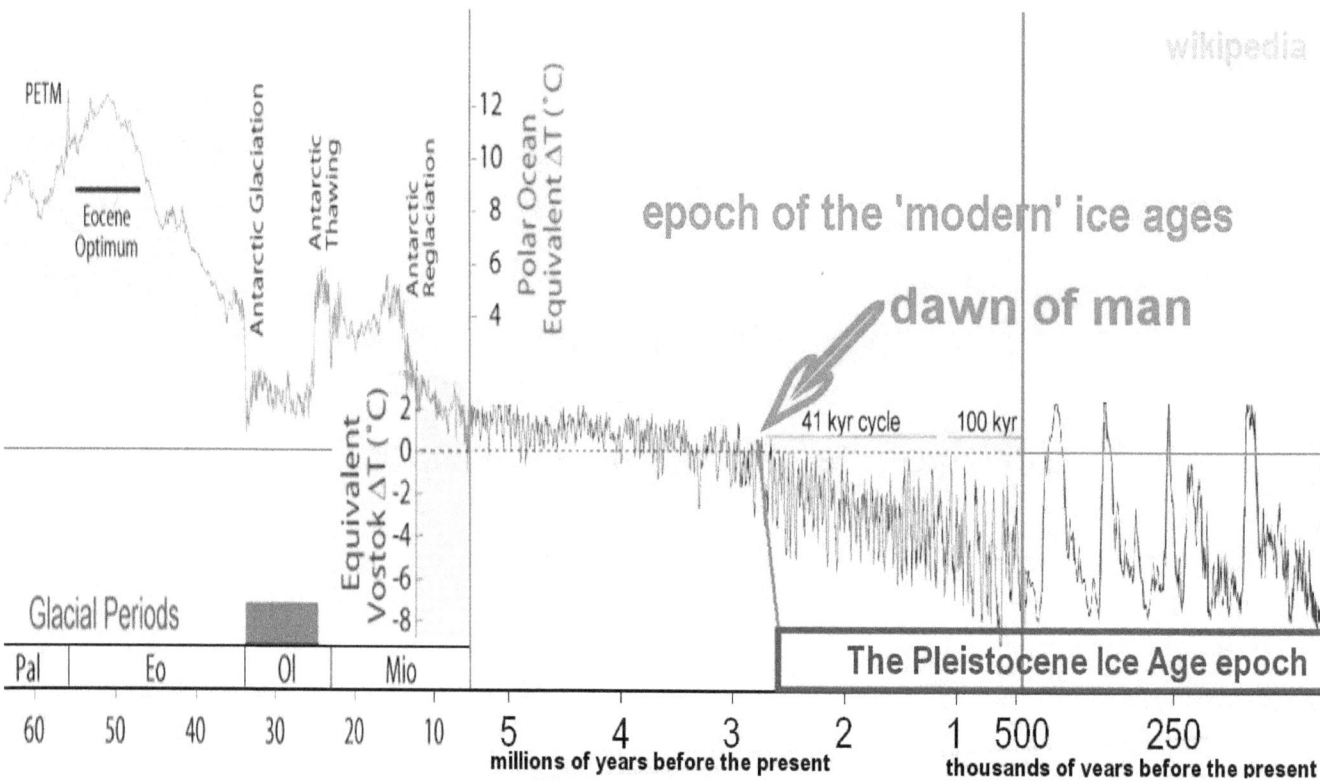

If we expand the last 5 million years still further, as it is shown in the middle of this graph, we see the climate beginning to fluctuate as it was getting still colder, in intervals of 41,000 years. With the start of the fluctuations, 2.5 million years ago, the most recent epoch of the ice age began, named the Pleistocene Epoch.

The start of the Pleistocene created a new phase shift for the climate on Earth. With the renewed glaciation on Earth, everything changed. Glaciation has a big effect on all life, especially during the harsher time, when around a million years ago the glaciation endured longer, with the intervals thereby becoming longer in duration.

We see the last 500,000 years expanded

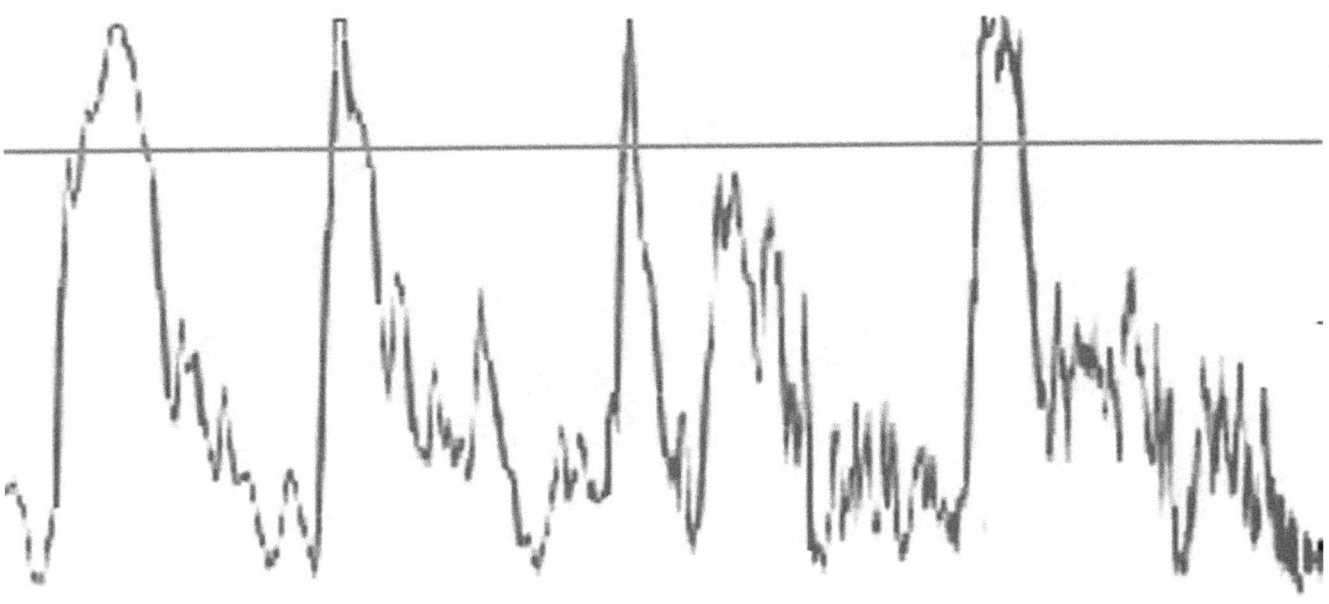

That's what we see in the plot to the right, where we see the last 500,000 years expanded again.

We see the same large-spike climate oscillations

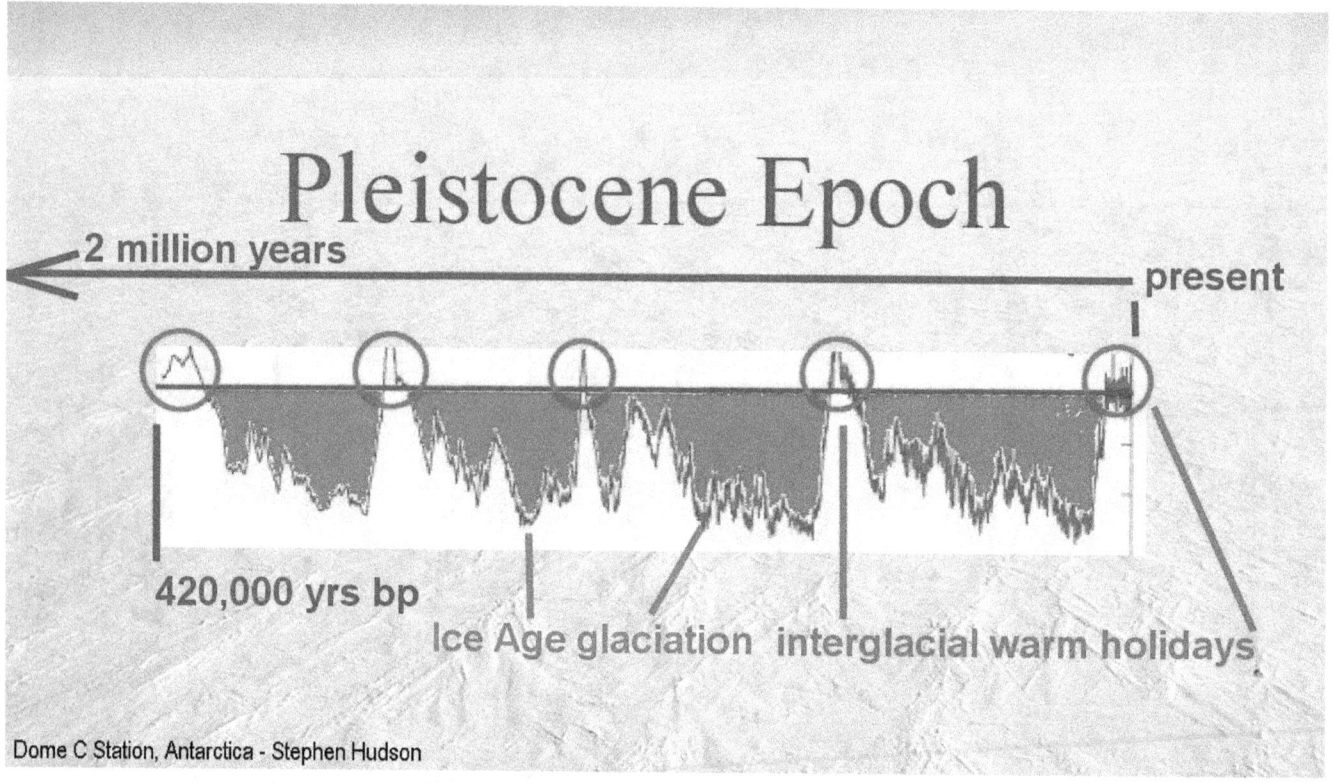

We see the same large-spike climate oscillations also reflected in the ice core samples of Antarctica, a sequence of glacial and interglacial periods.

The Pleistocene climate is of a type that affects us hugely

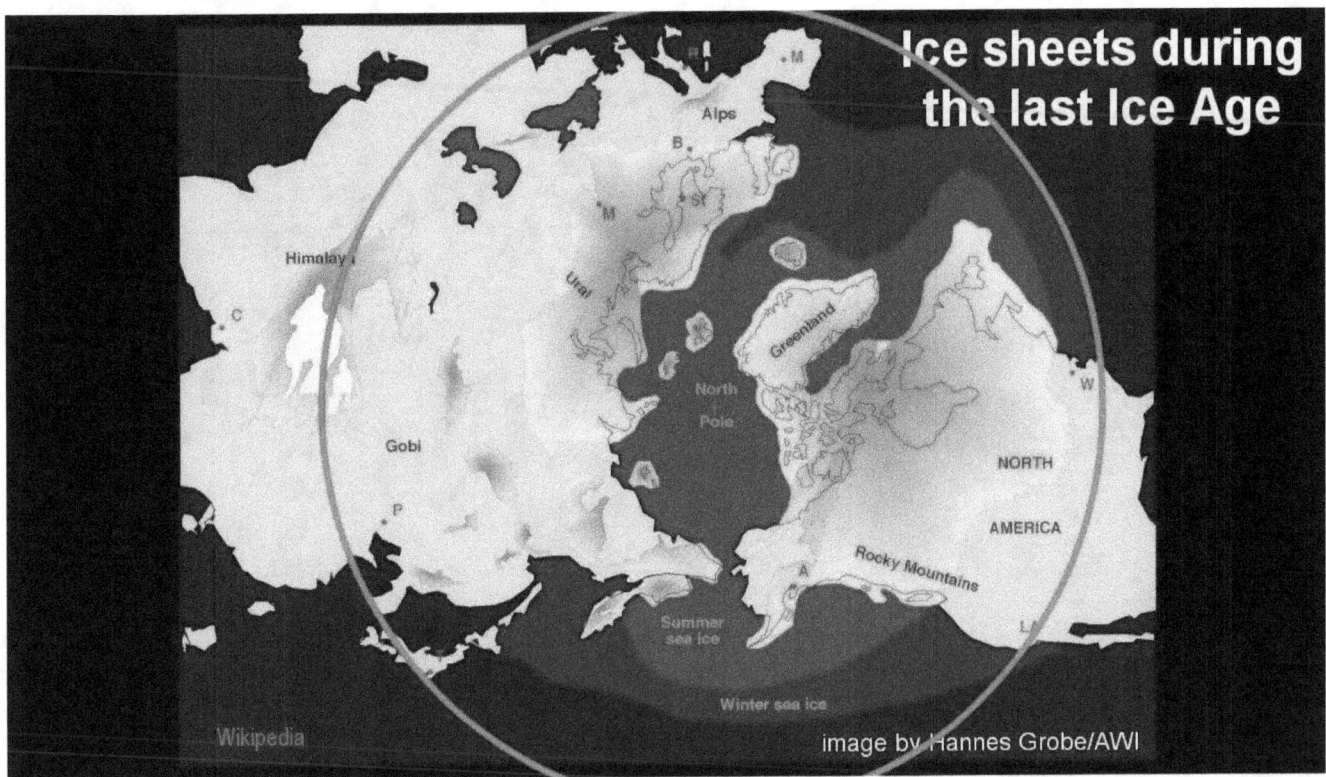

The Pleistocene epoch is an epoch of a type of 'macro' climate, in historic terms. While the Pleistocene is minuscule in comparison with the 'mega' fluctuations in Phanerozoic climate history, the Pleistocene climate is of a type that affects us hugely, because the current climate is functionally an extension of it, for the simple reason that the Pleistocene Ice Age cycles have not ended.

We call the current warm period, the Holocene epoch

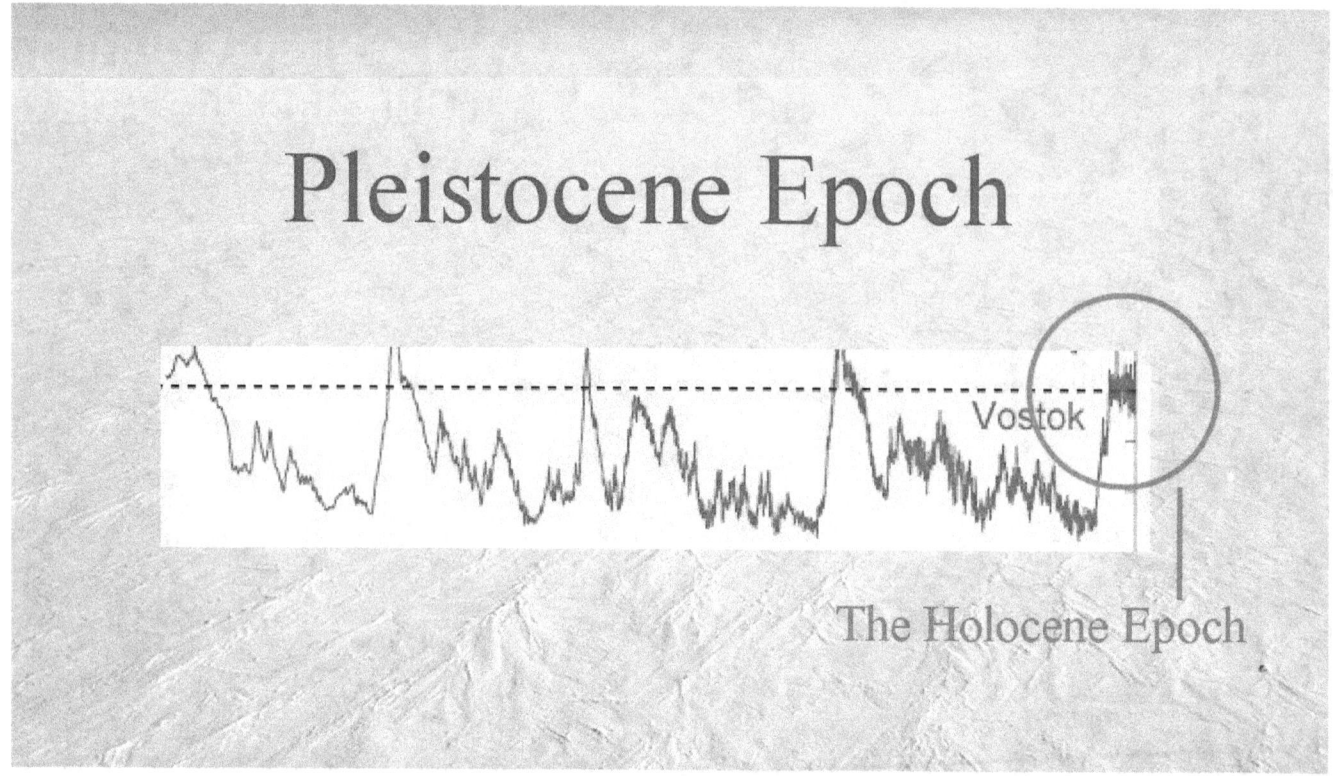

We call the current warm period, the Holocene epoch, a kind of holiday from the Pleistocene.

The Pleistocene Ice Age epoch will continue

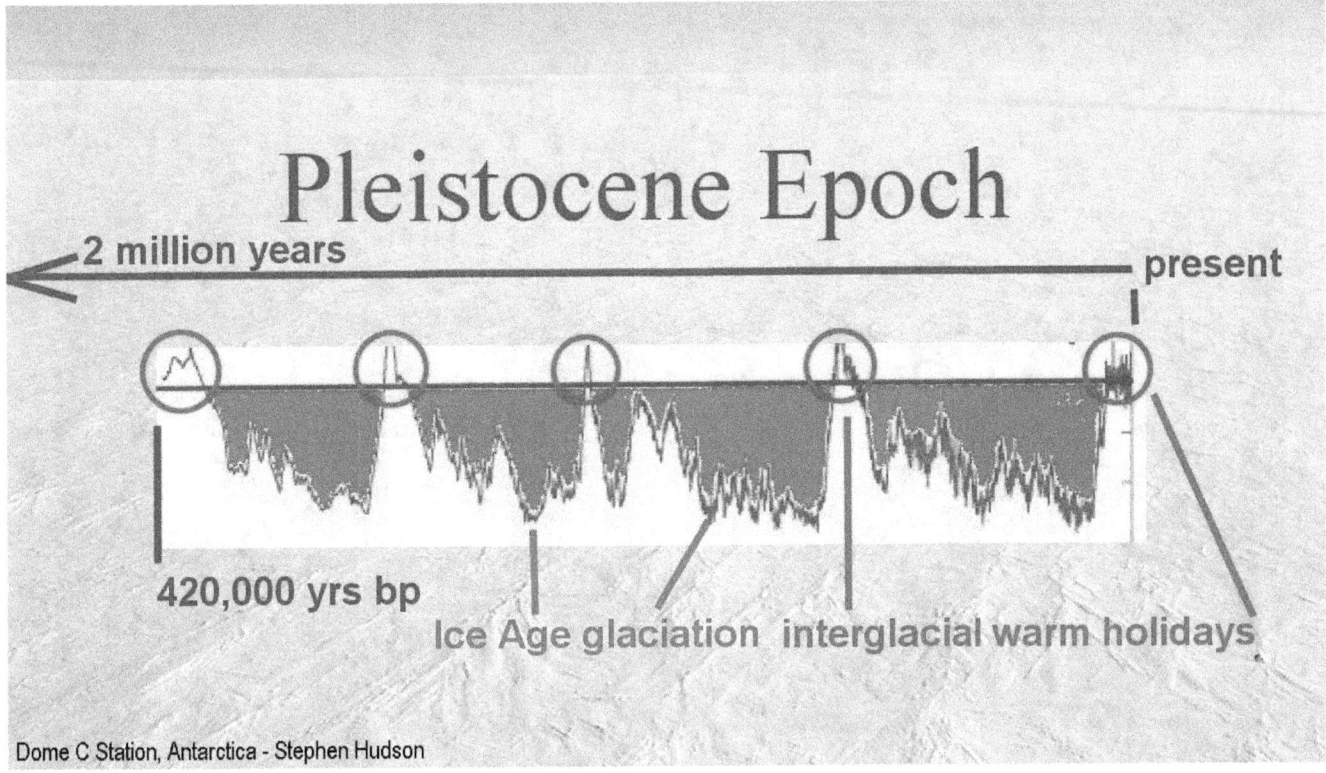

The Pleistocene Ice Age epoch will continue on for several more million years, after the Holocene interglacial ends and many more ice age cycles follow.

Part 2 - Climate-Change History in the last 5 million years

Part 2 Climate-Change History in 'Macro-Time' - the last 5 million years.

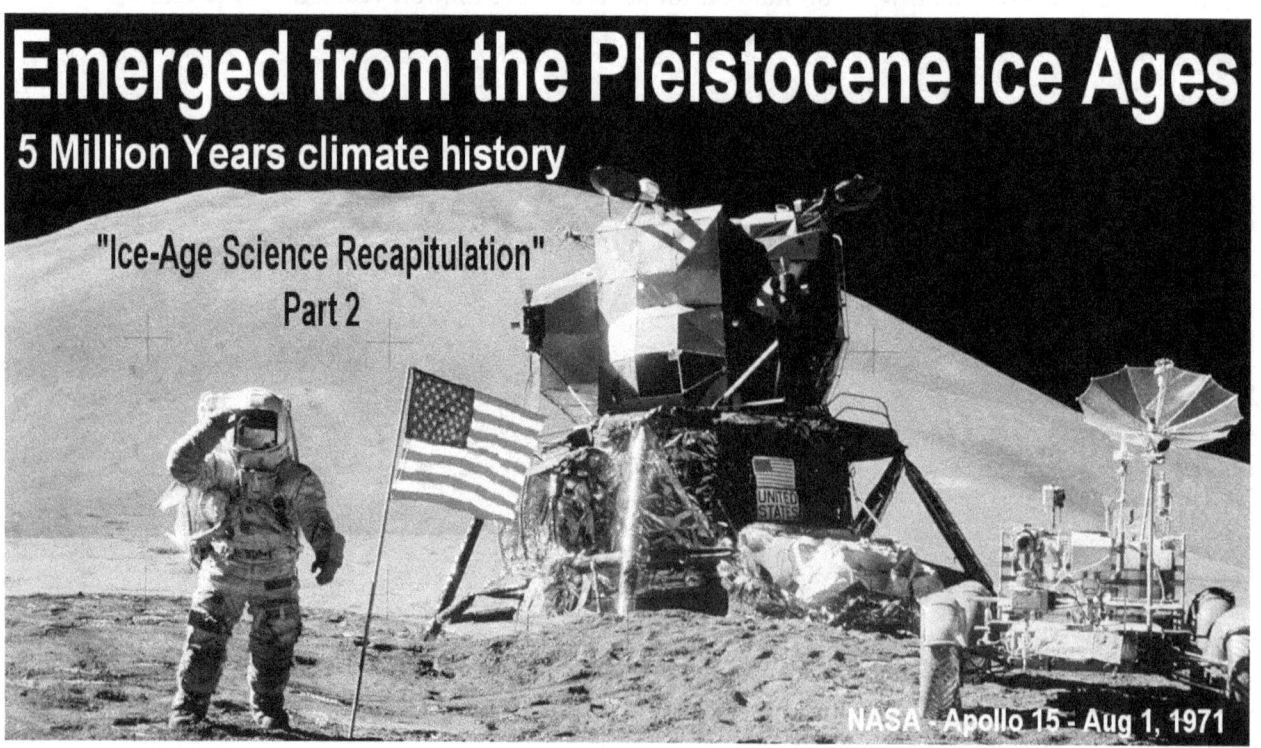

timeframe in which human development started

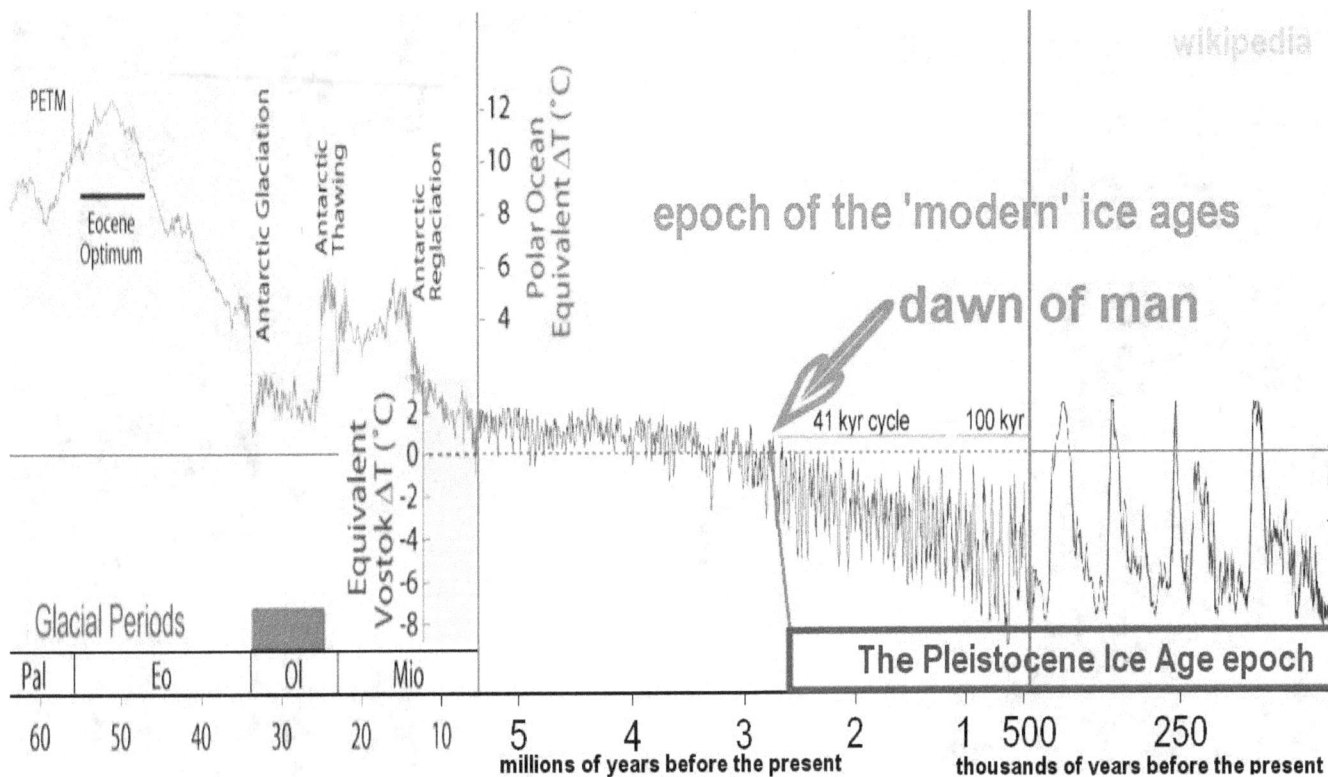

As I said before, in the middle of the last 5 million years, the climate on Earth dropped below the temperature level where glaciation begins, with which the cycles of the modern Ice Age epoch started, termed the Pleistocene Epoch. Archaeology tells us that this is also timeframe in which human development started. We are in broad terms, the children of the Pleistocene, a product of the ice ages.

Celebrating the near Ice Age

Celebrating the near Ice Age

Ice Ages periods of high volumes of cosmic-ray flux

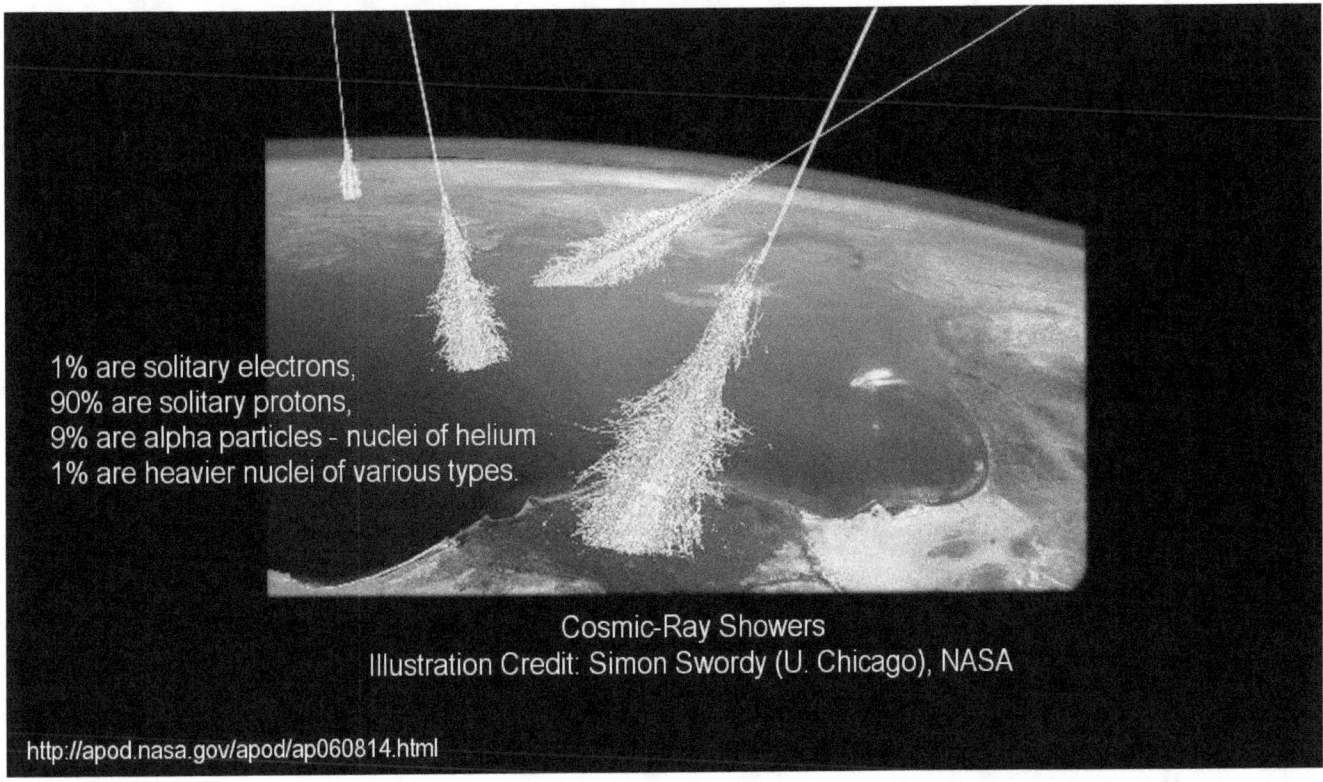

Ice Ages are not only cold periods, but are also periods of high volumes of cosmic-ray flux escaping from the plasma fusion cells on the surface of the Sun. Cosmic rays are singular events of highly energized plasma particles. While most of these are intercepted in the atmosphere, some penetrate to the surface, and some also penetrate the human body, by which they appear to have a beneficial effect.

When cosmic-ray particles pass through us

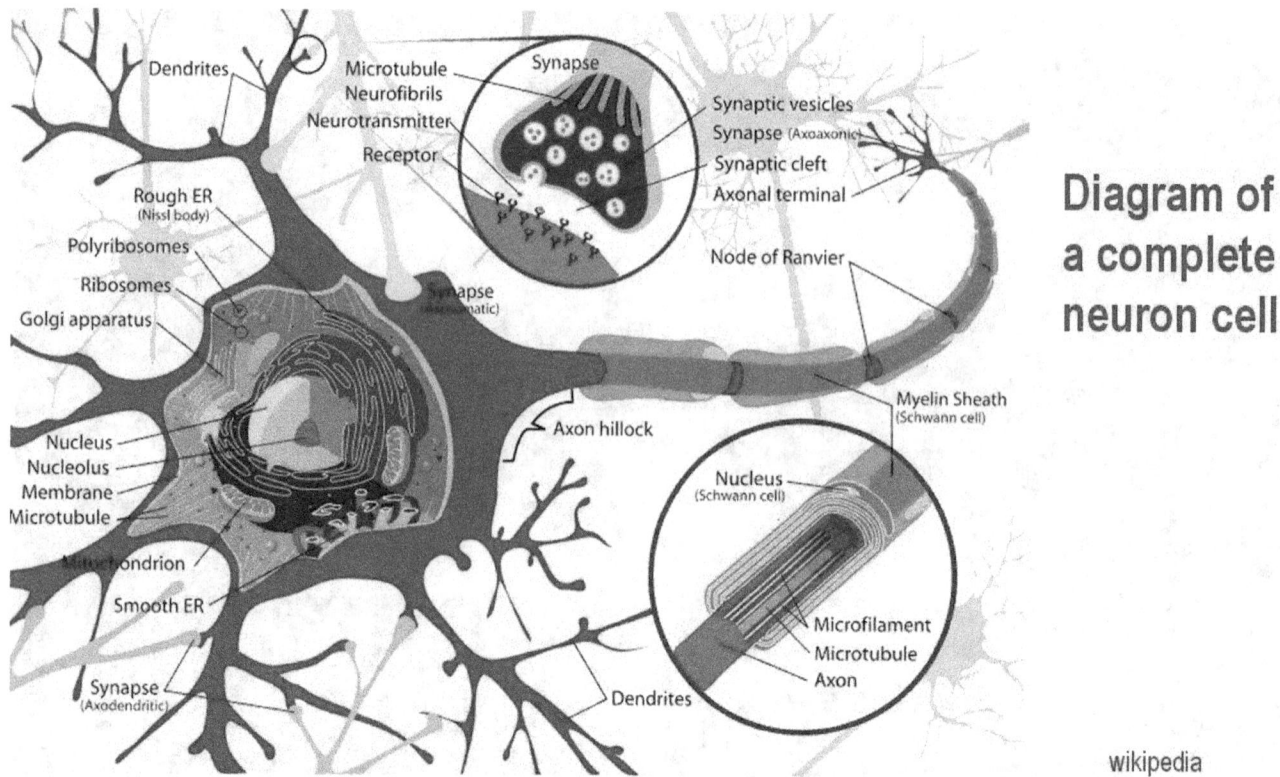

The complex neurological system of the human body, operates to a large extend electrically. When cosmic-ray particles pass through us, they impart minute electric currents by electromagnetic induction, which appear to aid neurological processes.

The evidence is found in cultural development

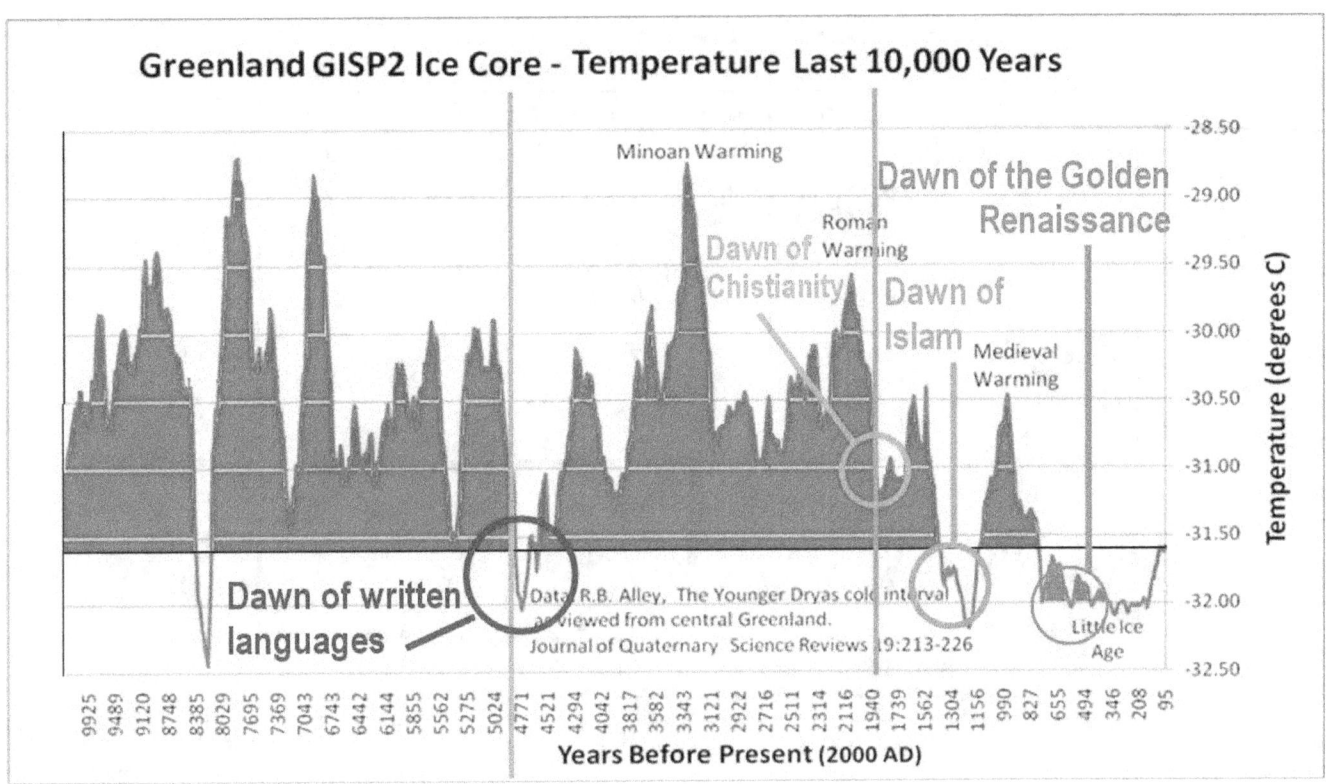

The evidence is found in cultural development. All the great cultural achievements in 'modern' time, such as the development of written languages, occurred in cold periods, that are periods of larger volumes of cosmic-ray flux.

Cosmic-ray flux larger during the last Ice Age

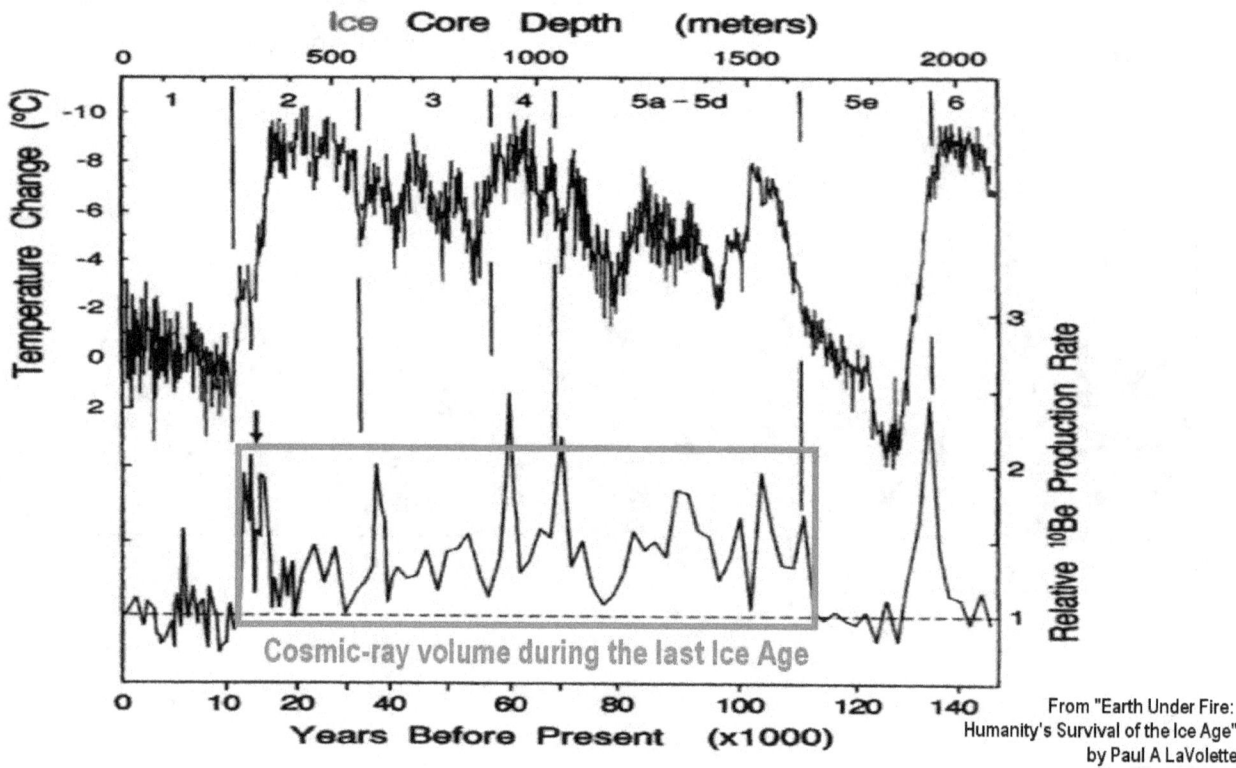

The cosmic-ray flux volume was measured to have been significantly larger during the last Ice Age, which evidently applies to all Ice Ages. The development of humanity occurred to a large extend in this cosmic-ray enriched background, during the glaciation timeframe. This enriched, electrically active background, aids our neurological and cognitive processes.

Around a million years ago

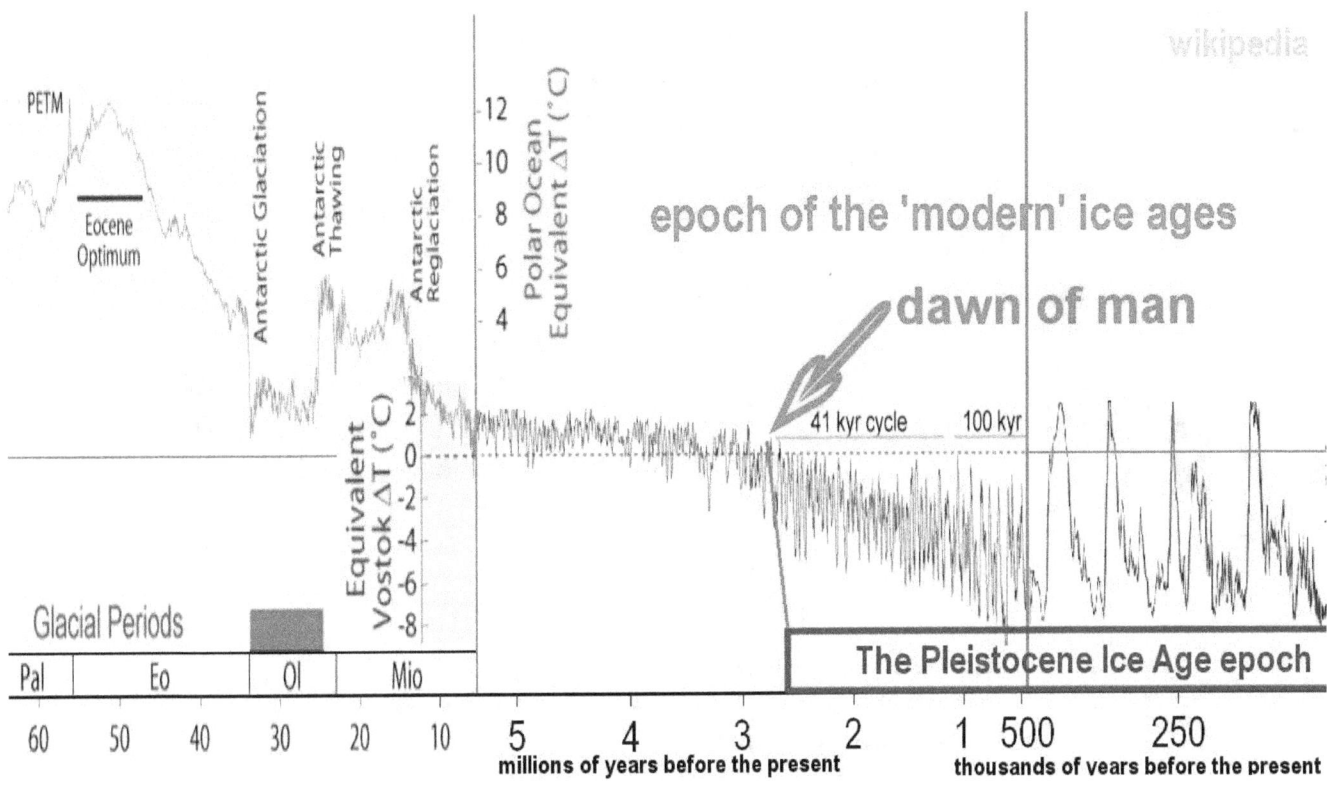

It is interesting to note here that as the Pleistocene Epoch became colder, around a million years ago, and the glaciation periods became longer.

The amazing proliferation of the human species

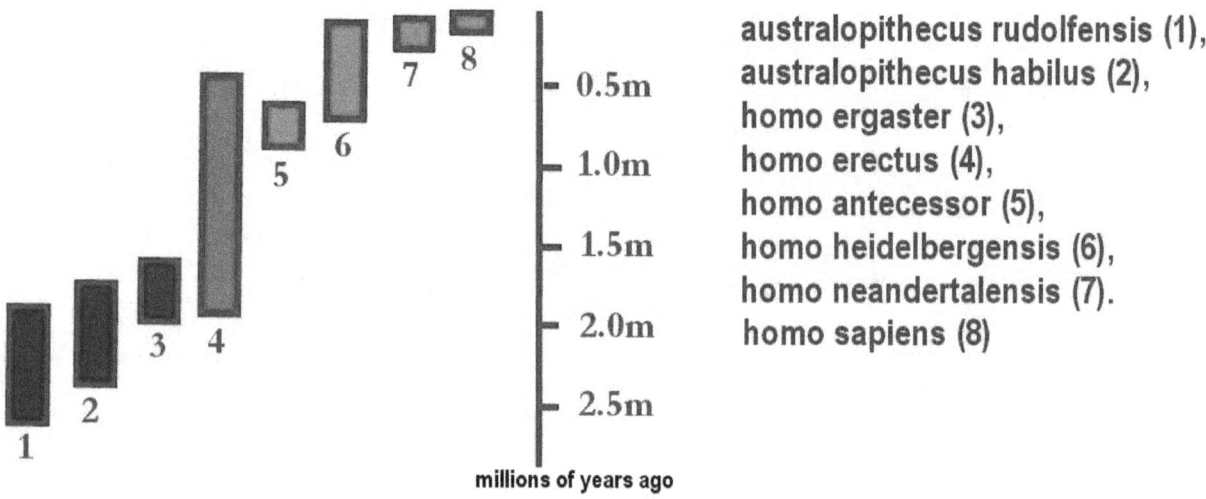

We, the homo sapiens (8), are the only surviving, and the shortest lived of all the the human species, at barely 200,000 years of age.

It is not surprising, therefore, that the resulting increase in cosmic-ray exposure coincides with the amazing proliferation of the human species that occurred at this time, a million years ago.

It may well be, that as the children of the Pleistocene, we owe what we became, to the cosmic dynamics that cause the ice ages and their effects.

The Pleistocene is our cradle

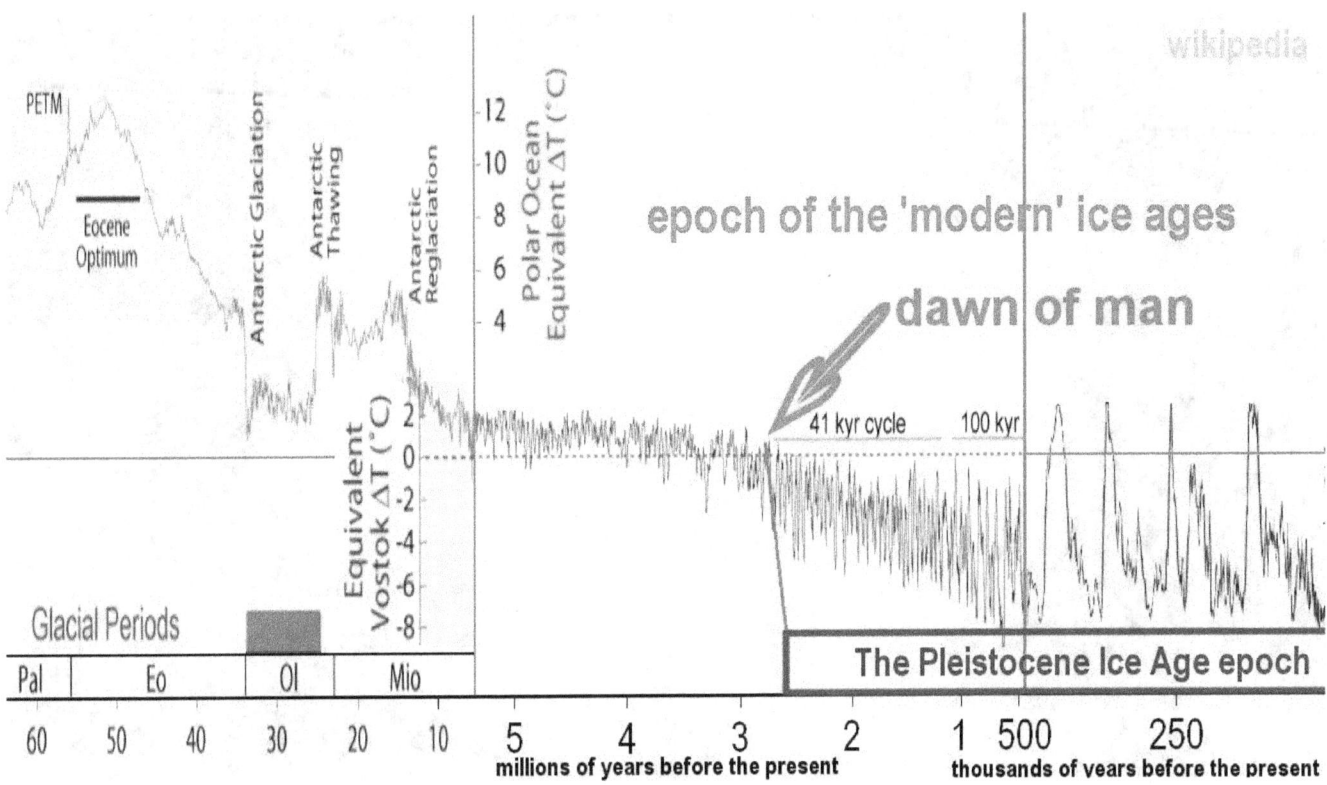

The Pleistocene is our cradle. While the Ice Age climates pose significant challenges, they are not an enemy that we should fear. They bring with them great opportunities.

That's something to celebrate and to move forward with

That's something to celebrate and to move forward with, as we are progressing towards the next glaciation period.

China, which had developed one of the earliest civilizations

China, which had developed one of the earliest civilizations may still have some cultural links in its historic background to the Pleistocene people and their humanist achievements that likely became echoed in some of the advanced ideologies that China became known for,

Echoed in China's Belt and Road initiative

which is being echoed again in China's Belt and Road initiative to foster worldwide economic development. Without an established foundation in worldwide economic cooperation, humanity stands no chance to master the Ice Age Challenge.

Chinese people, aided by their cultural background

It may be Chinese people, aided by their cultural background, who will grasp the great opportunities that the Ice Age Challenge inspires, to build a New World for human living that enables humanity to have a future on the Earth that is poised to become an ice planet to a large degree from the 2050s onward, potentially.

How the Sun generates Ice Ages

How the Sun generates Ice Ages

Expanded view of the last half-million years

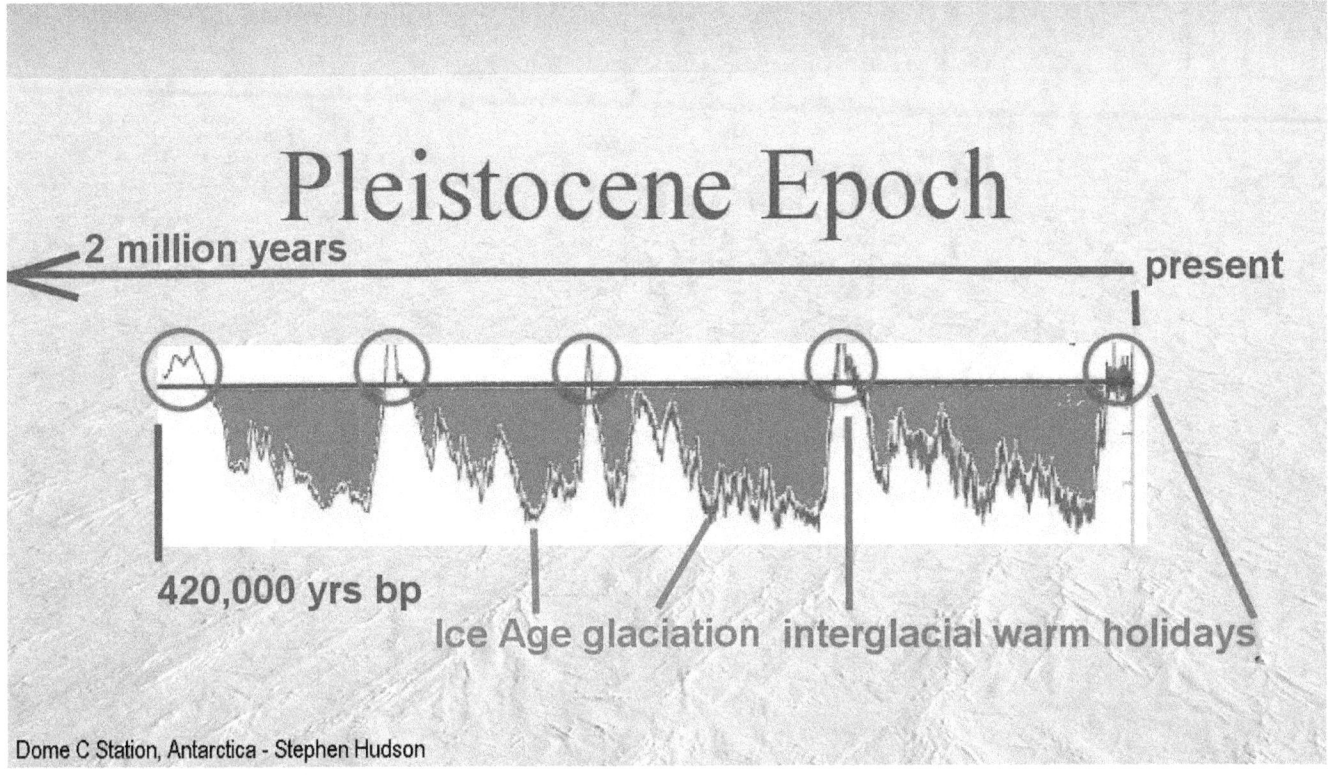

We see in this expanded view of the last half-million years of climate fluctuation in ice core sample,

For 85% of this period the climate was cold

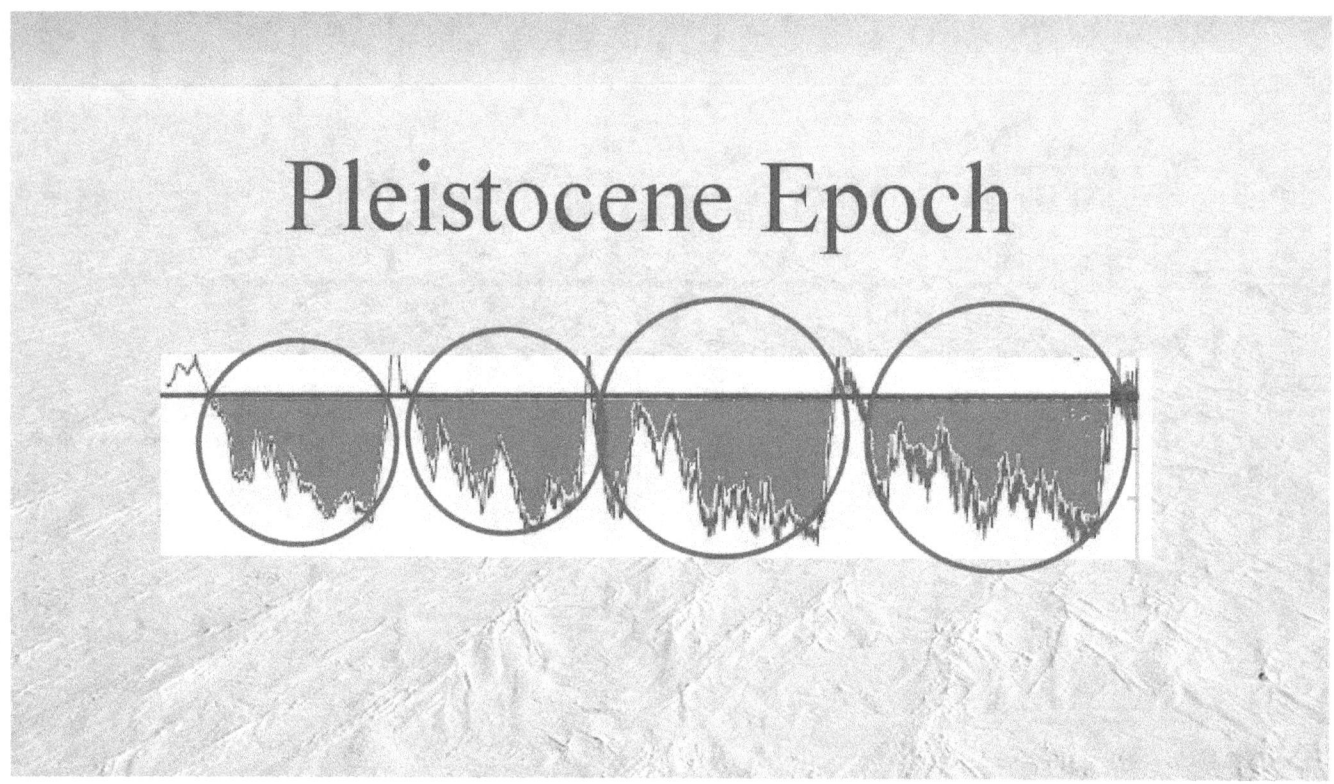

that for 85% of this period the climate was cold,

The remaining 15% spaced between the glacial times

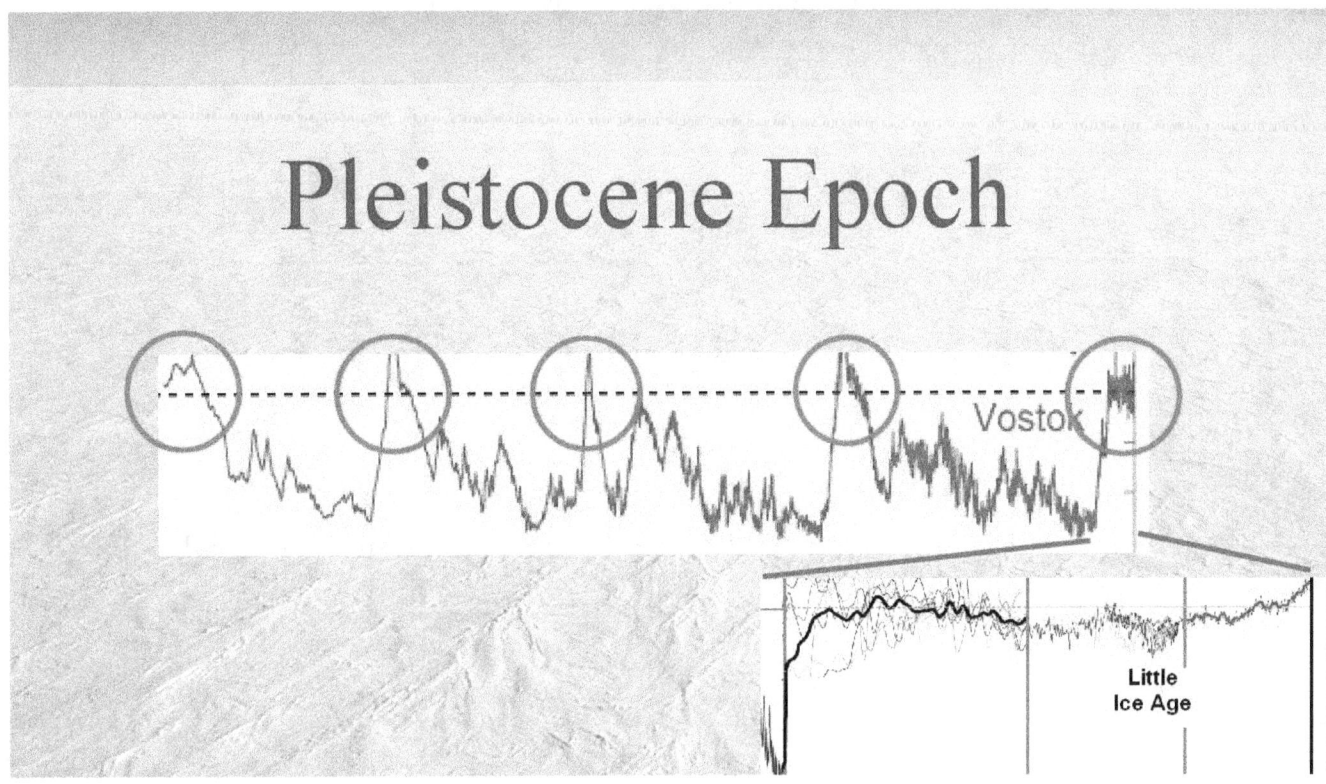

with the remaining 15% spaced between the deep cold glacial times in 100,000-year intervals, in the form of sharp interglacial spikes erupting.

The same interglacial spikes that we have measured in the ice layers

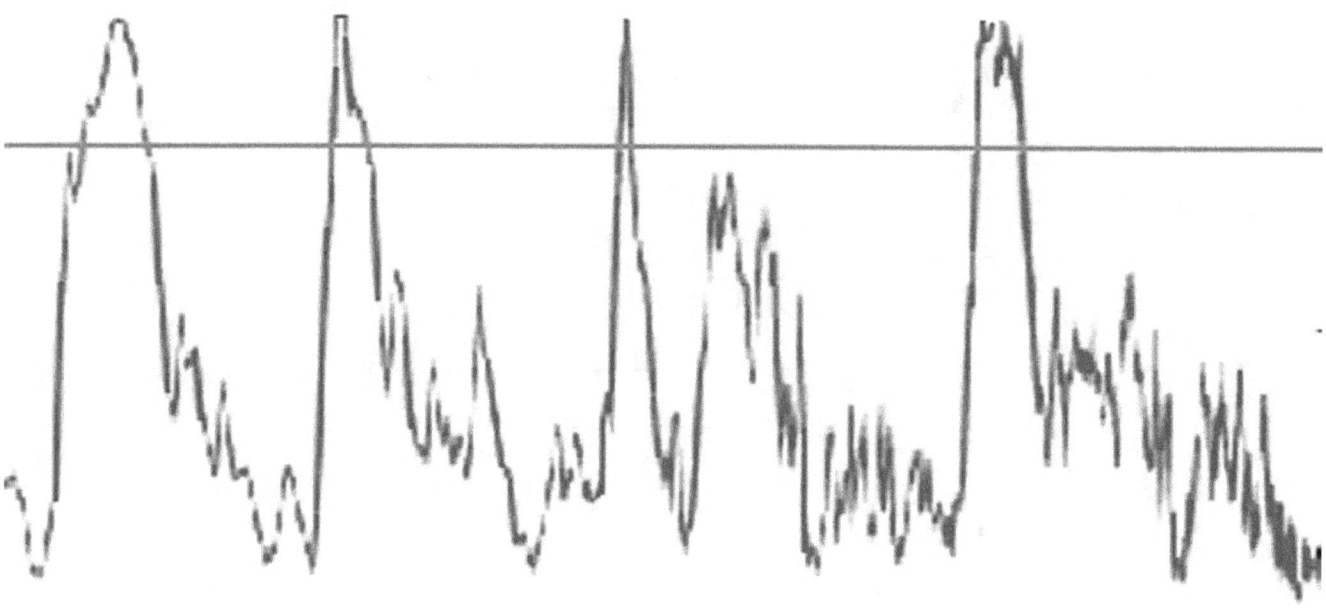

We see the same interglacial spikes that we have measured in the ice layers in Antarctica as ratios of heavy hydrogen, also measured in deep sea sediments as ratios of heavy oxygen, as is shown here.

But what is it that we see ?

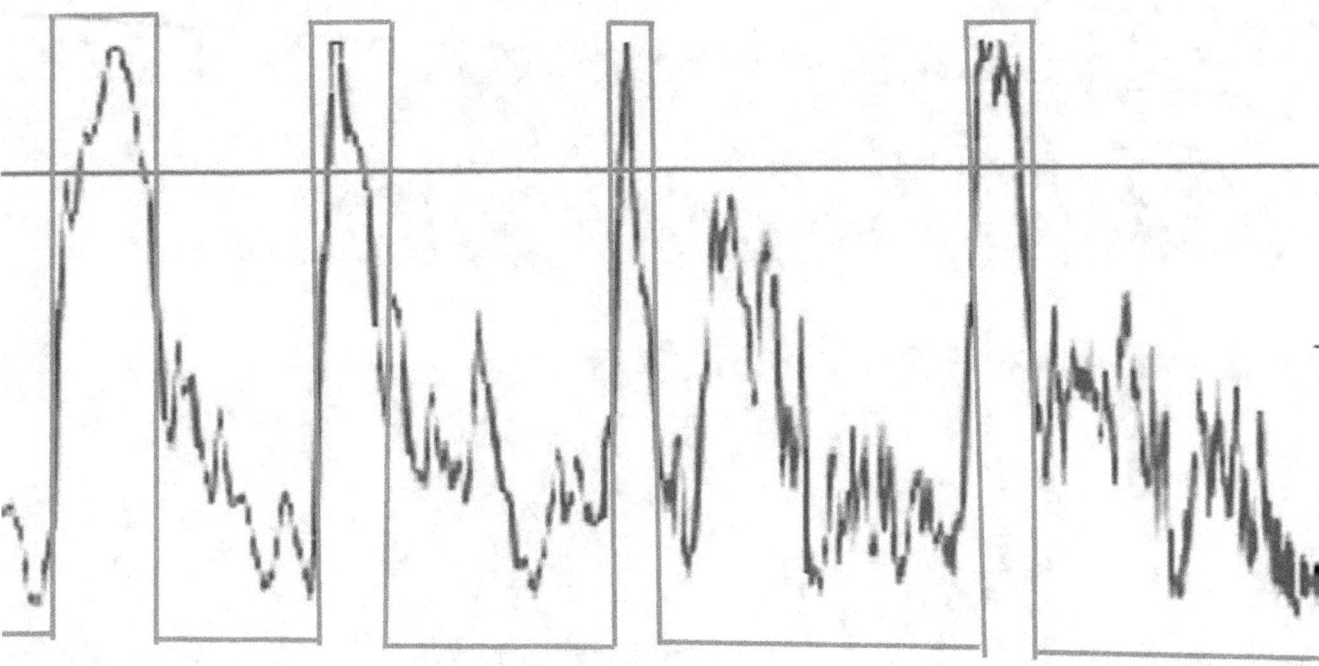

But what is it that we see when we look at these huge warming spikes, which we have measured in proxy, that have pulled the Earth repeatedly out of deep glaciation into interglacial climates, for brief periods?

The same type of dynamics that NASA has photographed

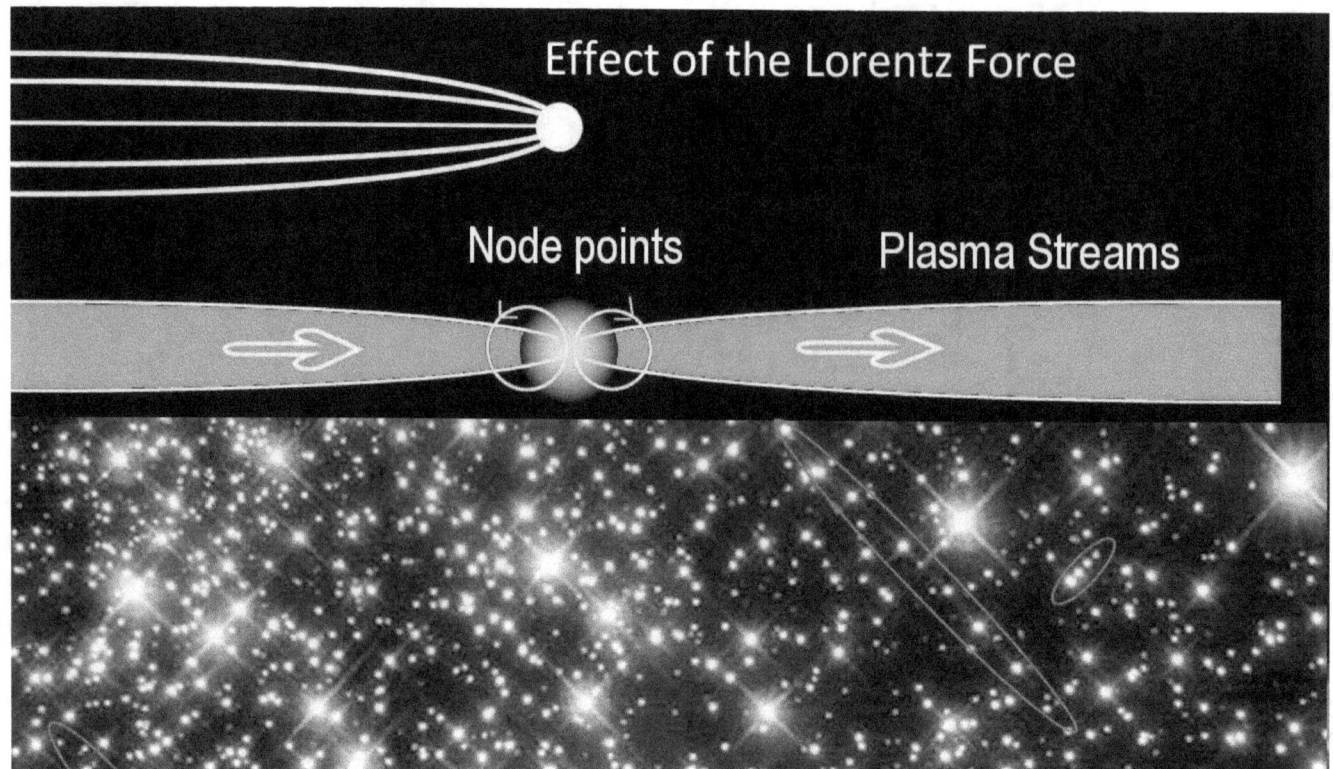

Surprisingly, we see the result here of the same type of dynamics that NASA has photographed on the galactic scale. Except we see the result happening here on the small scale of the solar system, where it is less-visibly apparent, but has nevertheless been measured.

In these measurements it becomes apparent that the solar system is powered by the same type of electromagnetic primer fields that power the galaxy, though with the plasma-streams flowing into them being radically smaller in scale.

The evidence that the Sun is located at a node point

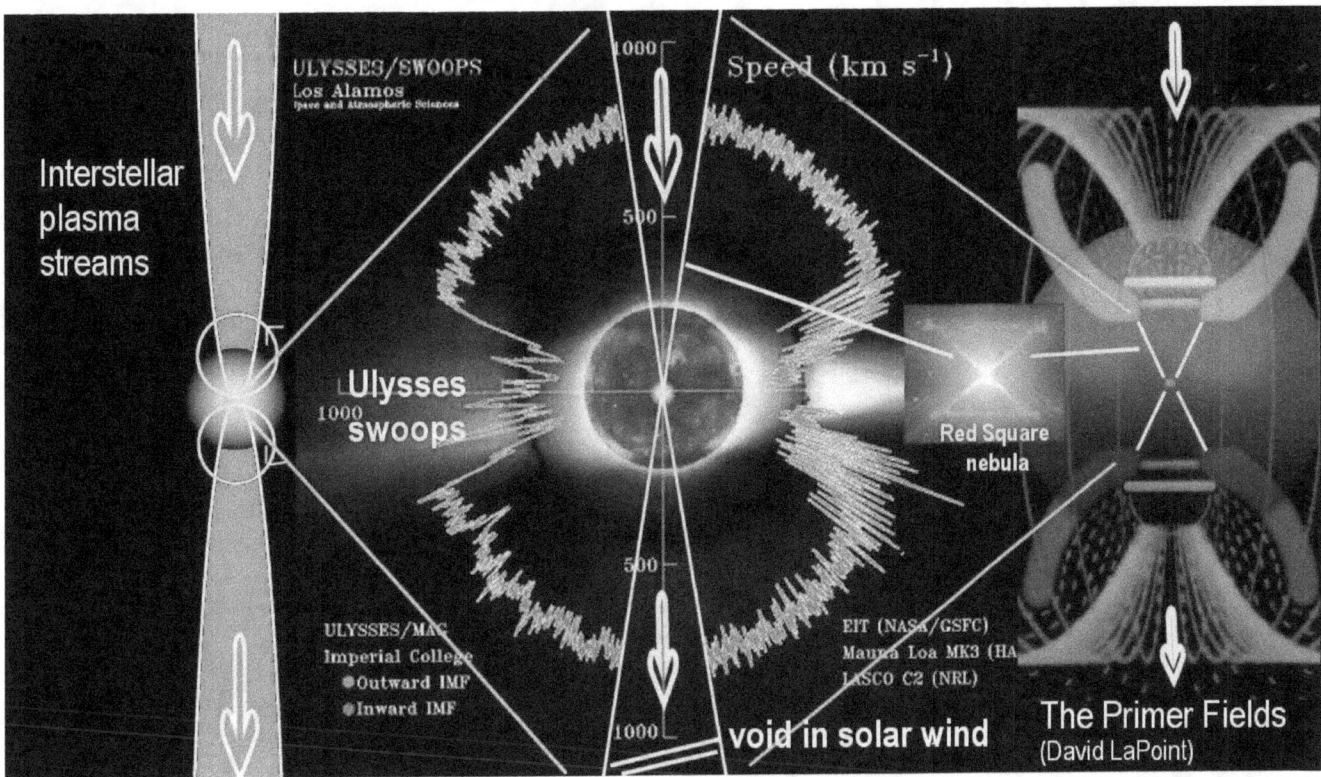

The evidence that the Sun is located at a node point between a set of primer fields, has been repeatedly measured by the Ulysses spacecraft during its three polar orbits around the Sun. The spacecraft measured a void in the solar-wind pattern over each of the Sun's polar region where the plasma connection with the primer fields are necessarily located, according to the discovered principles.

Between two long interstellar plasma streams

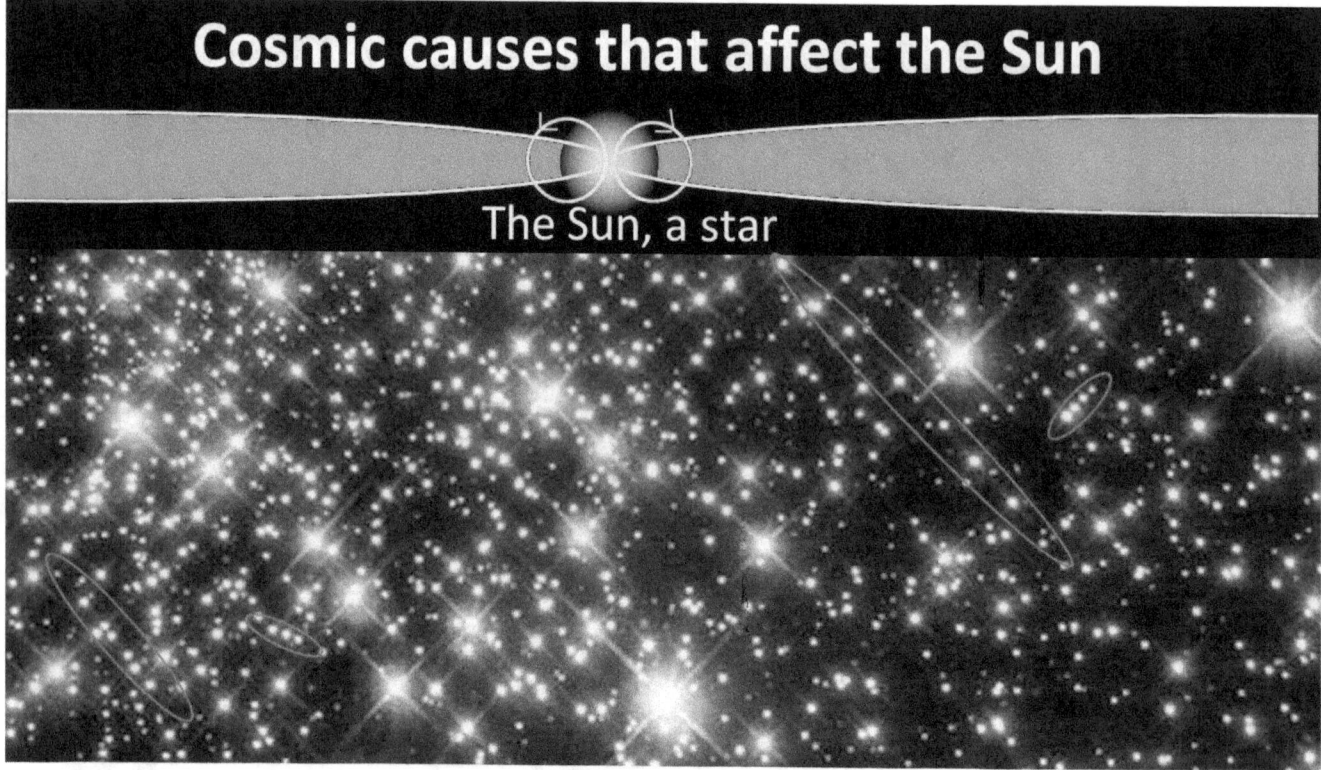

In this manner we see our solar system located between two long interstellar plasma streams, which likewise have their resonance characteristics.

The principle is the same

The principle is the same as that which powers the galaxies.

NASA measured the evidence

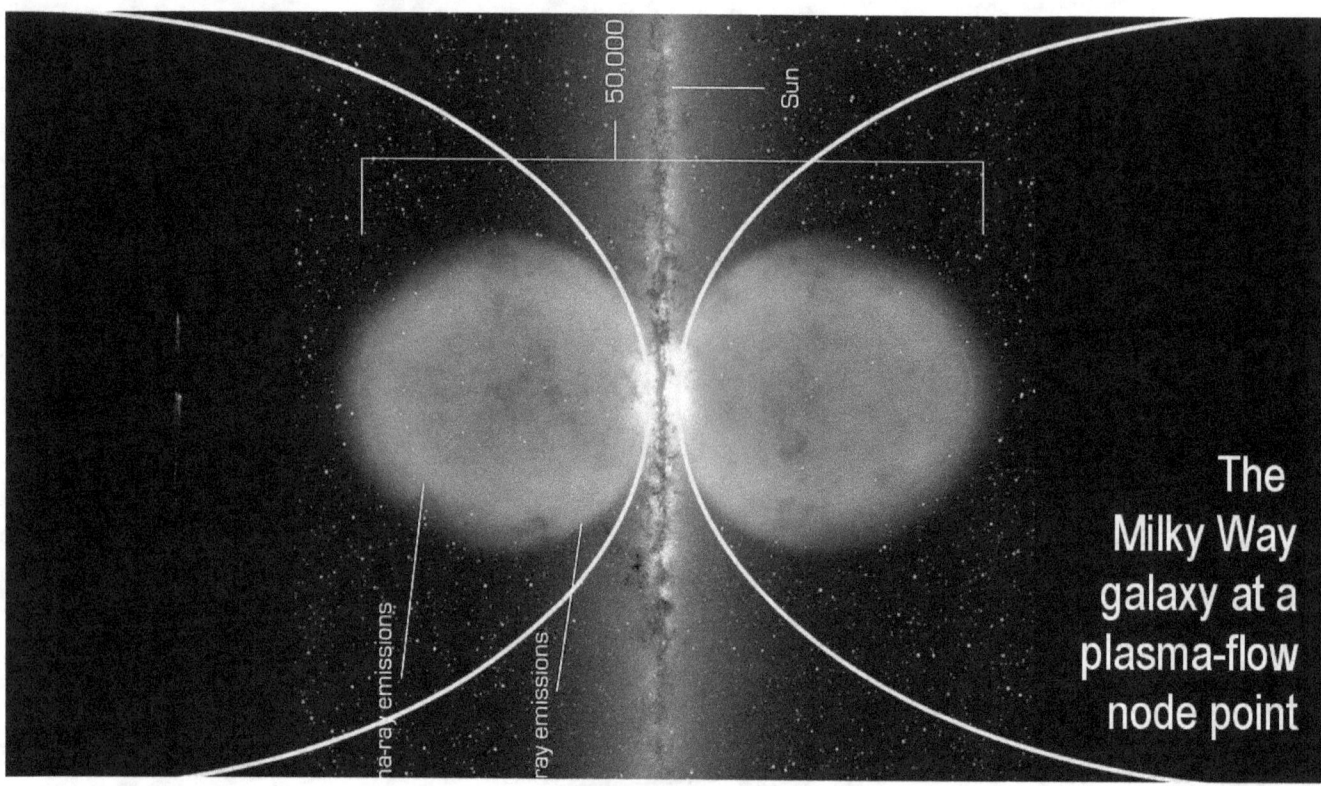

NASA measured the evidence in form of 'visible' plasma structures.

Ulysses measured a void in the solar-wind pattern

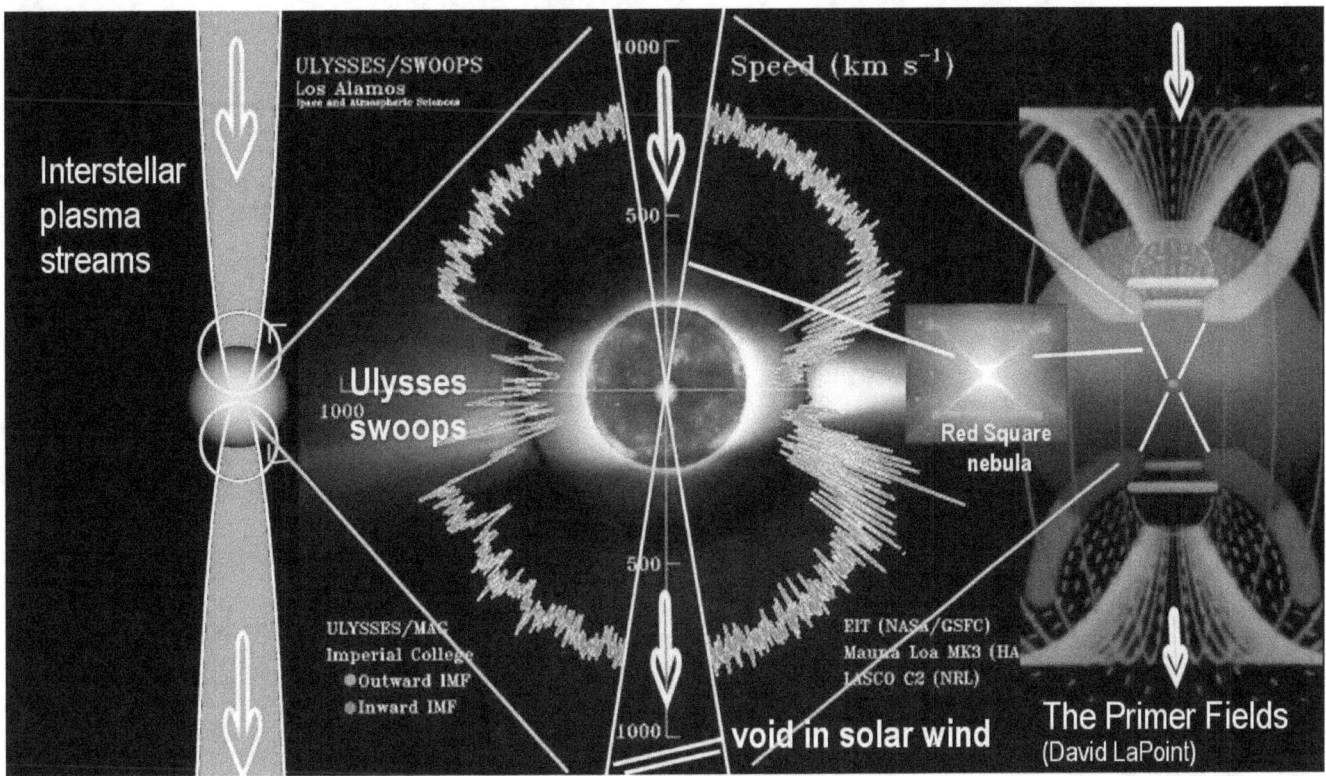

Ulysses measured the same as a void in the solar-wind pattern by a different effect of the same principle.

In the case of the solar system, the resonance in the plasma streams has been measured to be in the range of 100,000 years by its effects.

The sharp climate pulses

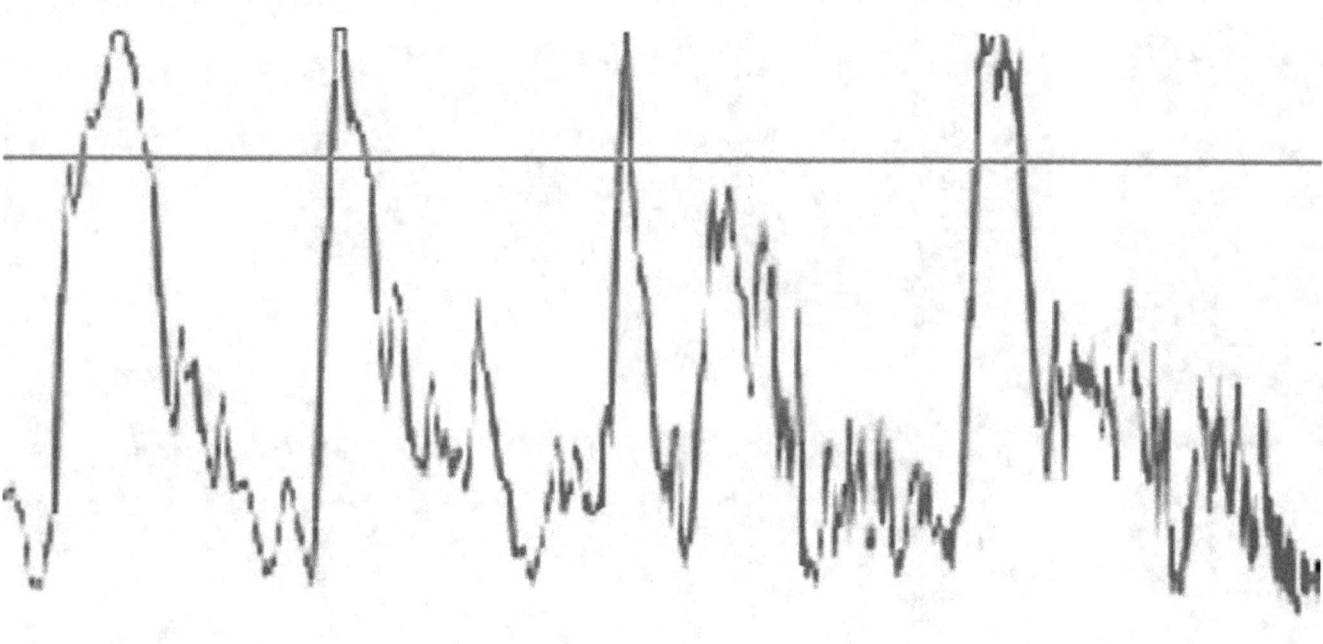

The sharp climate pulses that we see in both the sediment data and the ice core data, are climate fluctuations derived from the 100,000 years resonance in interstellar plasma streams.

Pulses represent the period in which the Sun is in a high-powered mode

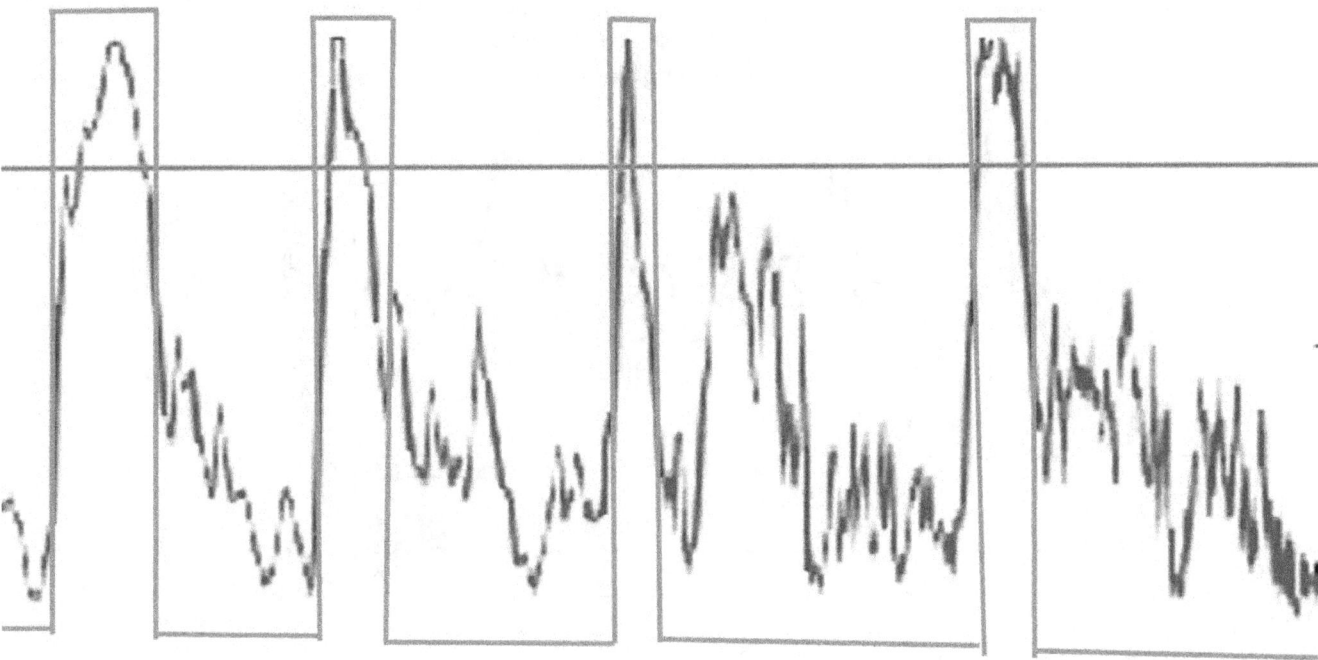

The sharp pulses represent the period in which the Sun is in a high-powered mode. The pulses are of short duration, because, with the galactic 'mega climate' being at the extremely low level that it presently is at, there is only enough plasma density in the galaxy for our Sun's primer fields to be active for a mere 15% of the time at the peak of the resonance effect in the interstellar plasma streams.

If the galactic plasma density wasn't at the deep low level

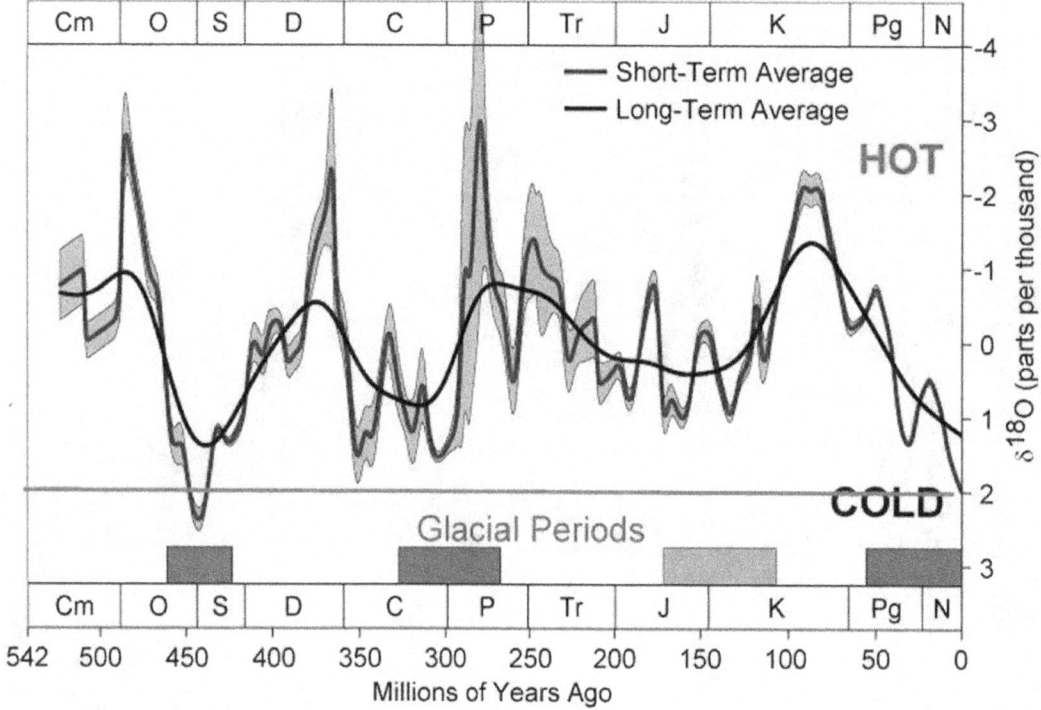

If the galactic plasma density wasn't at the deep low level that it presently is at, and has been for the last few million years, which is the lowest in 440 million years, there would have been enough plasma density in interstellar space for our Sun to be fully powered all of the time. In this case, no ice ages would have occurred.

We are the product of the ice ages

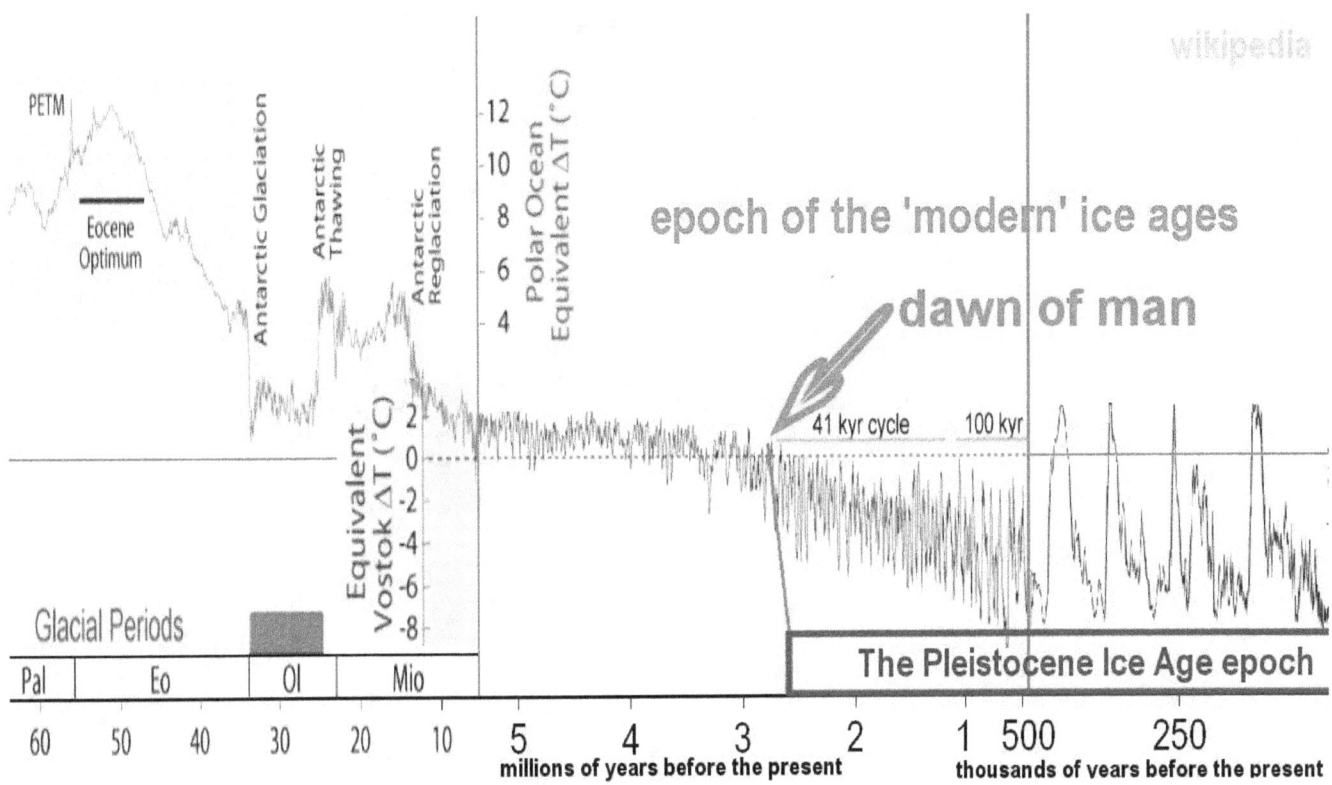

Neither would the conditions have occurred that have enabled humanity to be. We are the product of the ice ages and their high rates of cosmic-ray flux during the long glaciation periods.

False and true Ice Age theories

False and true Ice Age theories

False and true Ice Age theories

It was only theorized 150 years ago that Ice Ages have occurred. And it was only 100 years ago theories were developed as to what caused them.

The Milankovitch Cycles Theory

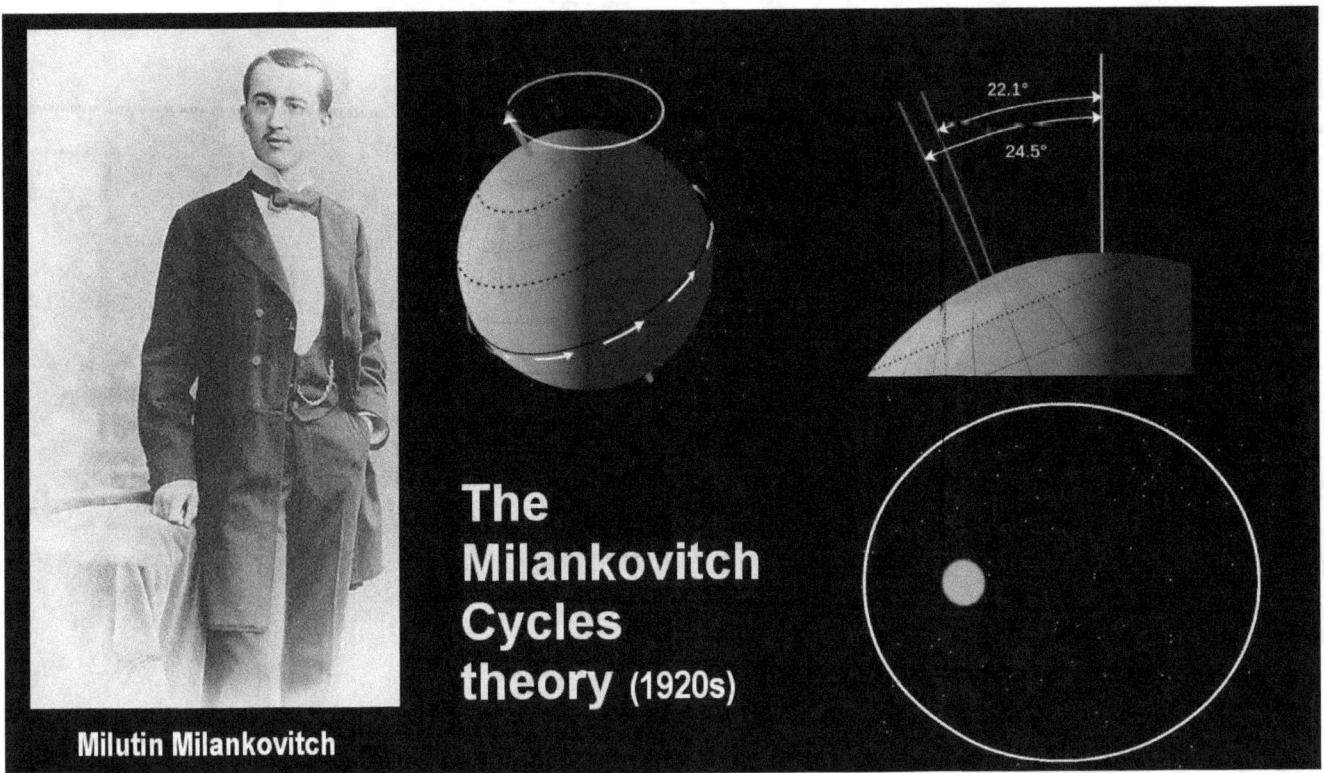

The most prominent of the early theories is the Milankovitch Cycles Theory. It proposes that ice ages are caused by the combined effect of long-term cyclical fluctuations of the spin-axis of the Earth and the eccentricity of its orbit around the Sun.

Exposure to the Sun at 65 degrees North

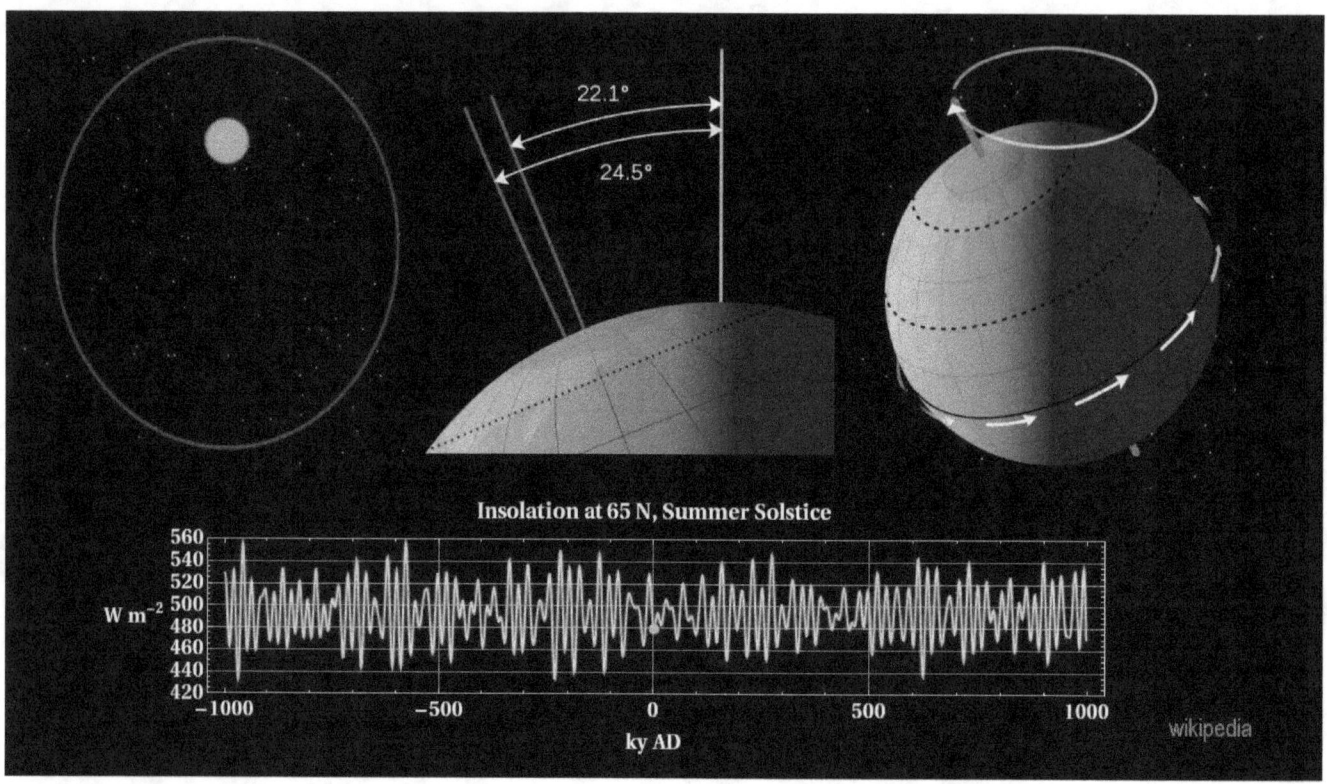

The climate effects were computed by their combined influence on the exposure to the Sun in high latitude regions, at 65 degrees North, which were deemed to cause sufficient variations to cause the Ice Age glaciation, even while the total exposure of the Earth to the Sun remains always the same.

This was the best theory that science could come up with 100 years ago, with the Sun being regarded as a constant factor. In more recent times Ice Age theories still continue this train, with theorized effects based on potential ocean-current fluctuations, or volcanism, and similar causes.

Ice Age glaciation phenomenon into the court of the Sun

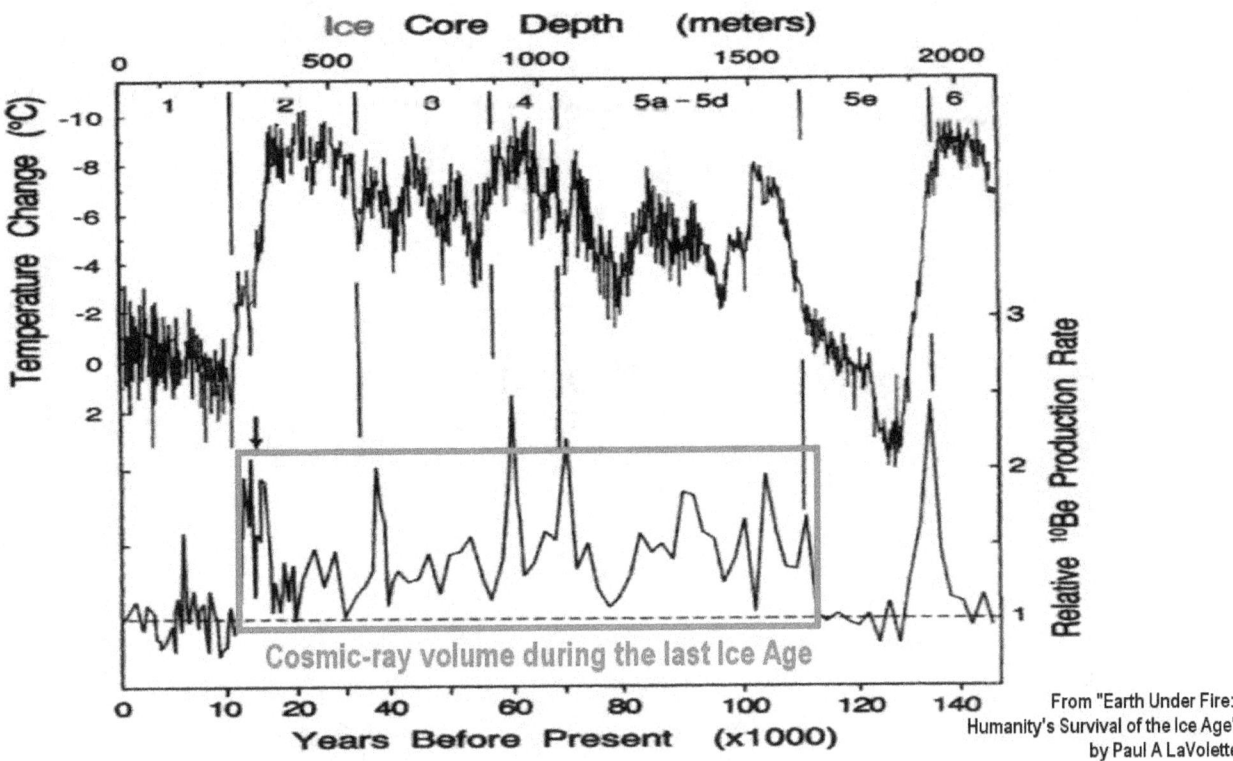

From "Earth Under Fire: Humanity's Survival of the Ice Age" by Paul A LaVolette

All of these mechanistic type theories have been rendered as simply false, by the measured physical evidence that places the Ice Age glaciation phenomenon into the court of the Sun.

Solar hibernation measured

Solar hibernation measured

Solar hibernation measured

Dramatically increased solar cosmic-ray flux

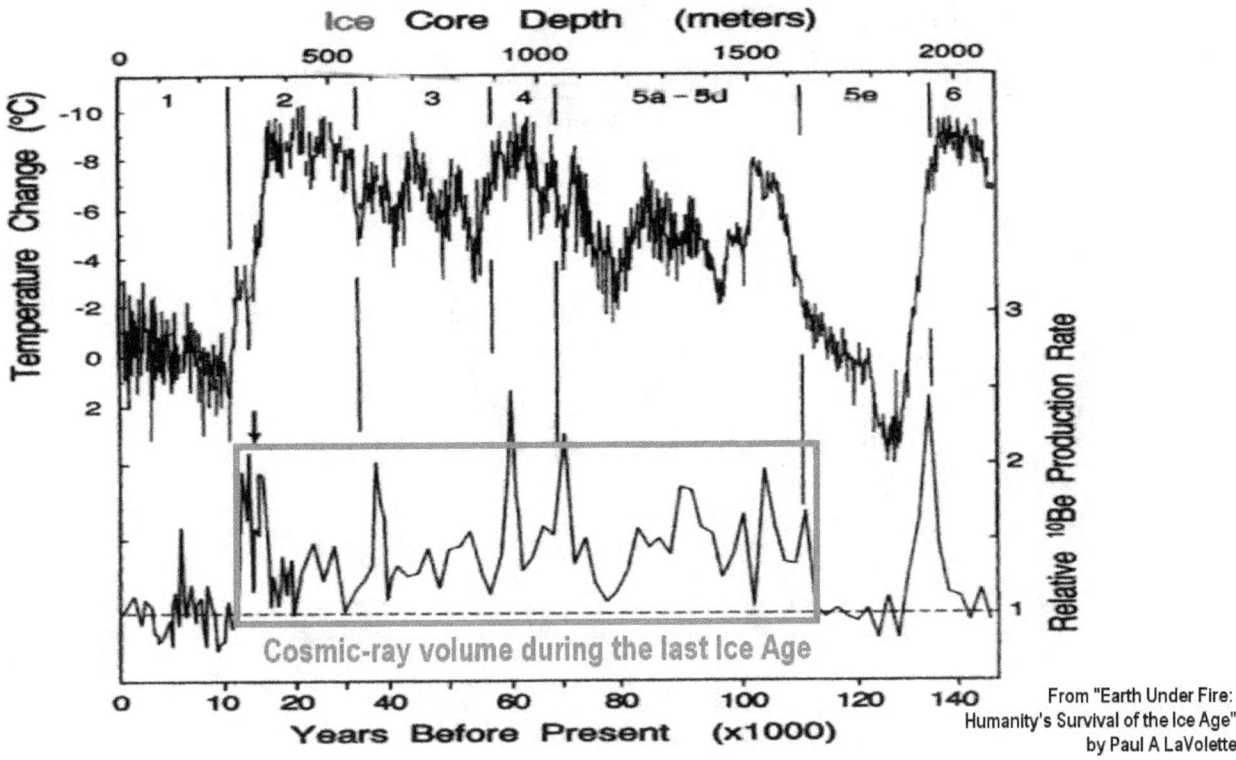

The measured evidence exists in the form of dramatically increased solar cosmic-ray flux during the entire period of the last Ice Age glaciation. The cosmic-ray flux is measured in ratios of the radio-isotope Beryllium-10 that is exclusively produced by cosmic-ray interaction with the atmosphere of the Earth. The high Beryllium ratio begins and ends with the glacial period. Nothing but the Sun operating in hibernation mode, without a plasma shield around it, is able to cause such high isotope ratios.

The measured high isotope ratio throughout the glaciation cycles disprove all mechanistic Ice Age theories, as their assumed causes have no effect on the Sun.

The bottom line is that Ice Ages are solar caused events

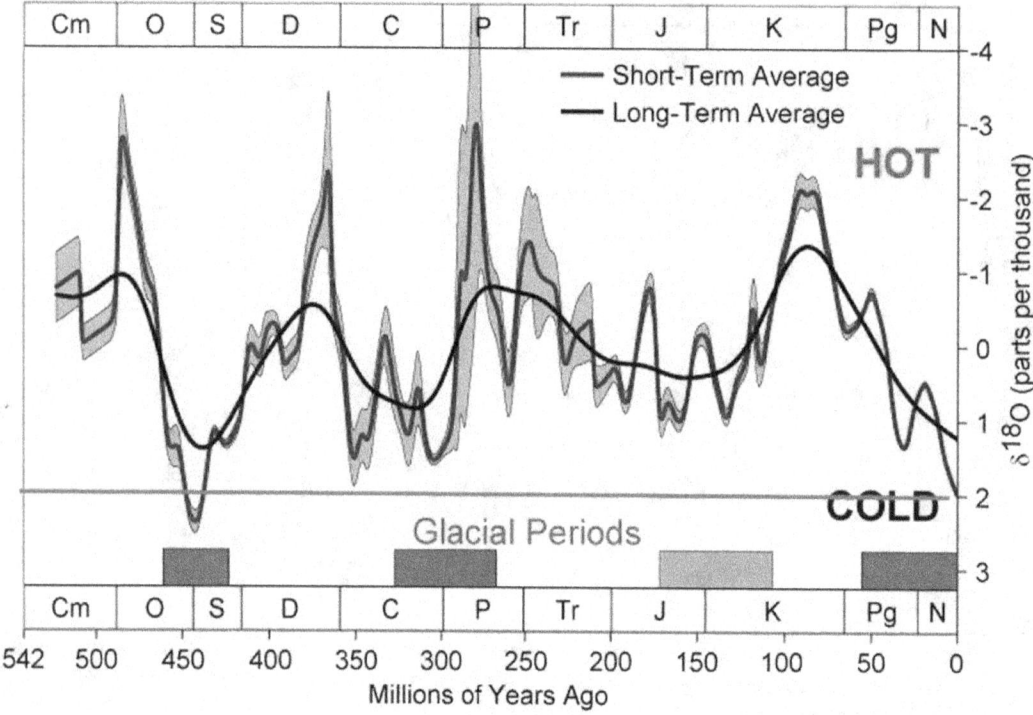

The bottom line is that Ice Ages are solar caused events, which occur only during the rare, extremely weak galactic epochs, as the one we are presently in.

Still enough density for our Sun to be fully powered

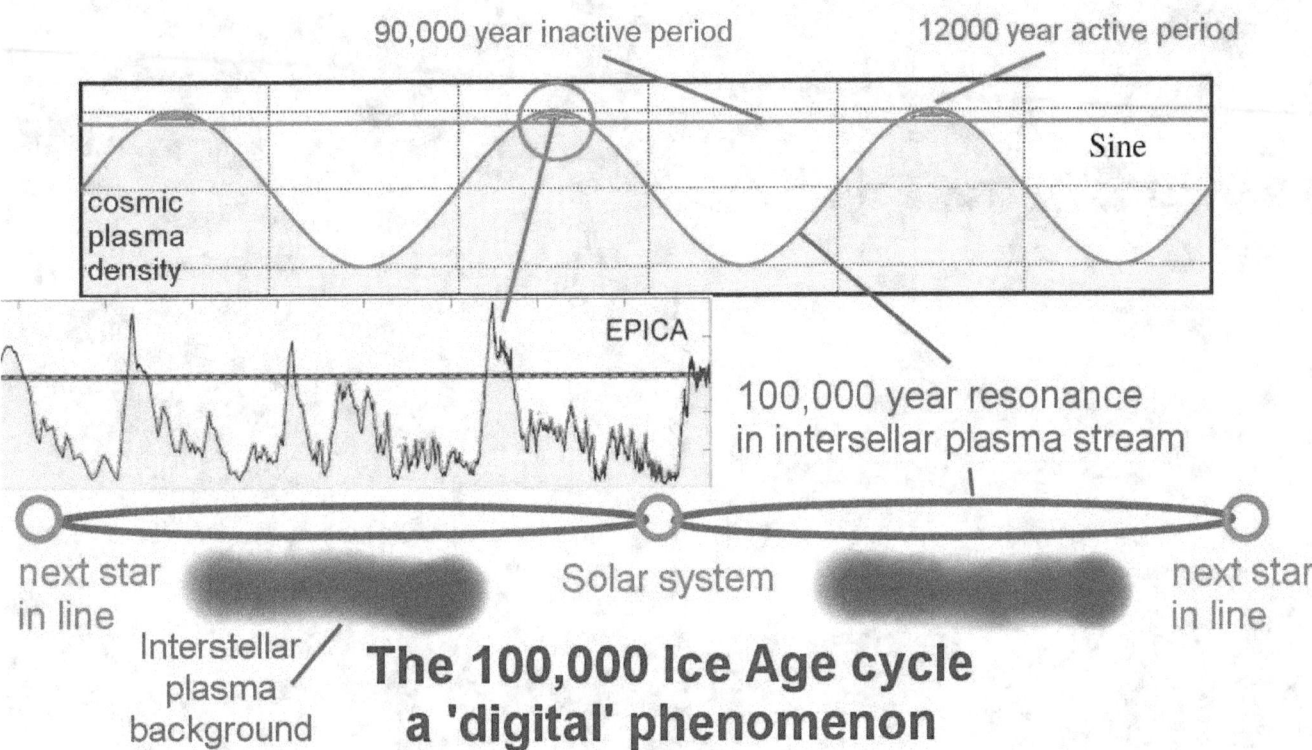

Fortunately for us, there is still enough density in the galactic system for our Sun to be fully powered for those 15% of the interstellar resonance that give us warm climates, when the interstellar plasma streams exceed the minimal threshold for the primer fields to form. For the remaining 85% of the time, when the primer fields are not active, the Sun operates at a low-level default state, a type of inactive state, or hibernation state, with 70% less energy being radiated by the Sun and extensive glaciation occurring on Earth.

This doesn't mean that the Sun is actually dying

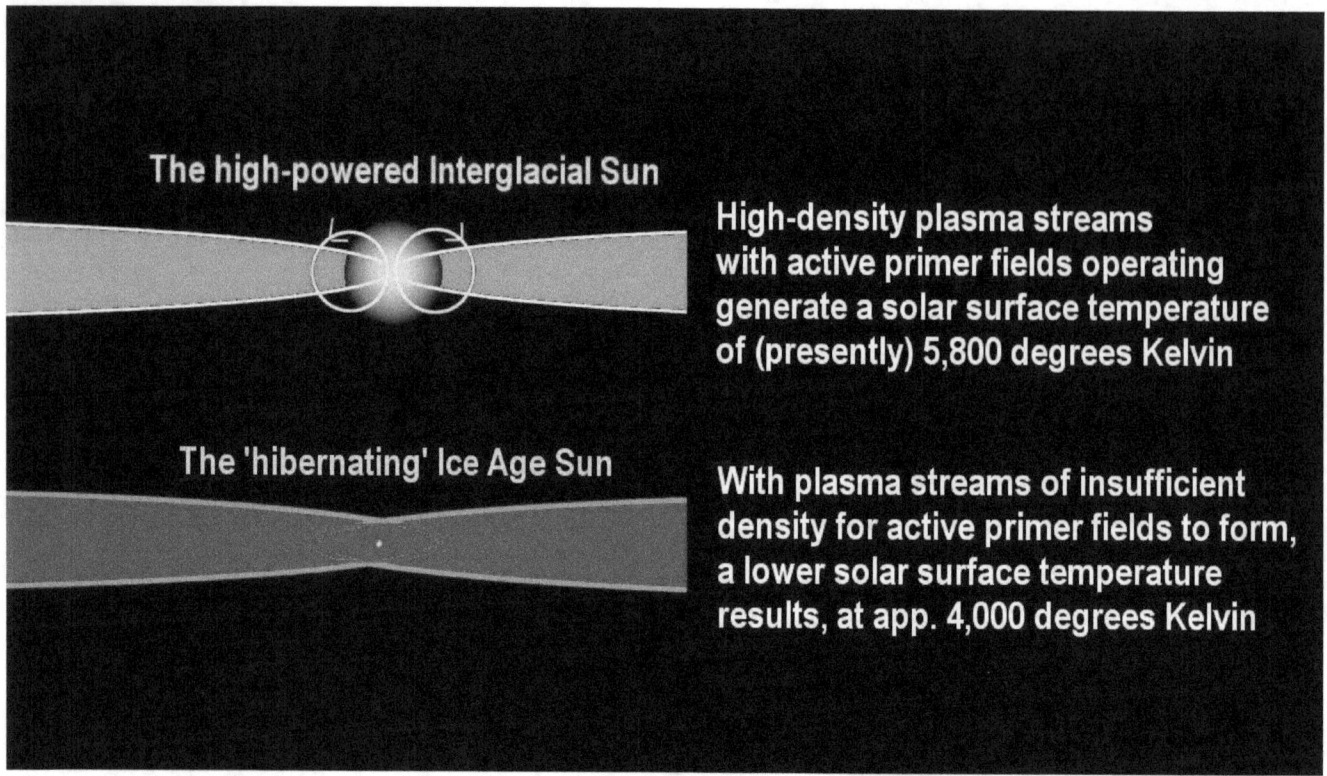

This doesn't mean that the Sun is actually dying. It is merely dying back in solar activity. The plasma streams that are powering the Sun are getting progressively weaker until there is not enough plasma density left in the system to maintain the primer fields that presently keep the Sun operating in its high-powered mode. When this happens, that is when the primer fields collapse, the Sun will revert back to a low-power default state, with which the Ice Age begins on Earth that typically lasts for 90,000 years.

During its low-powered hibernation mode

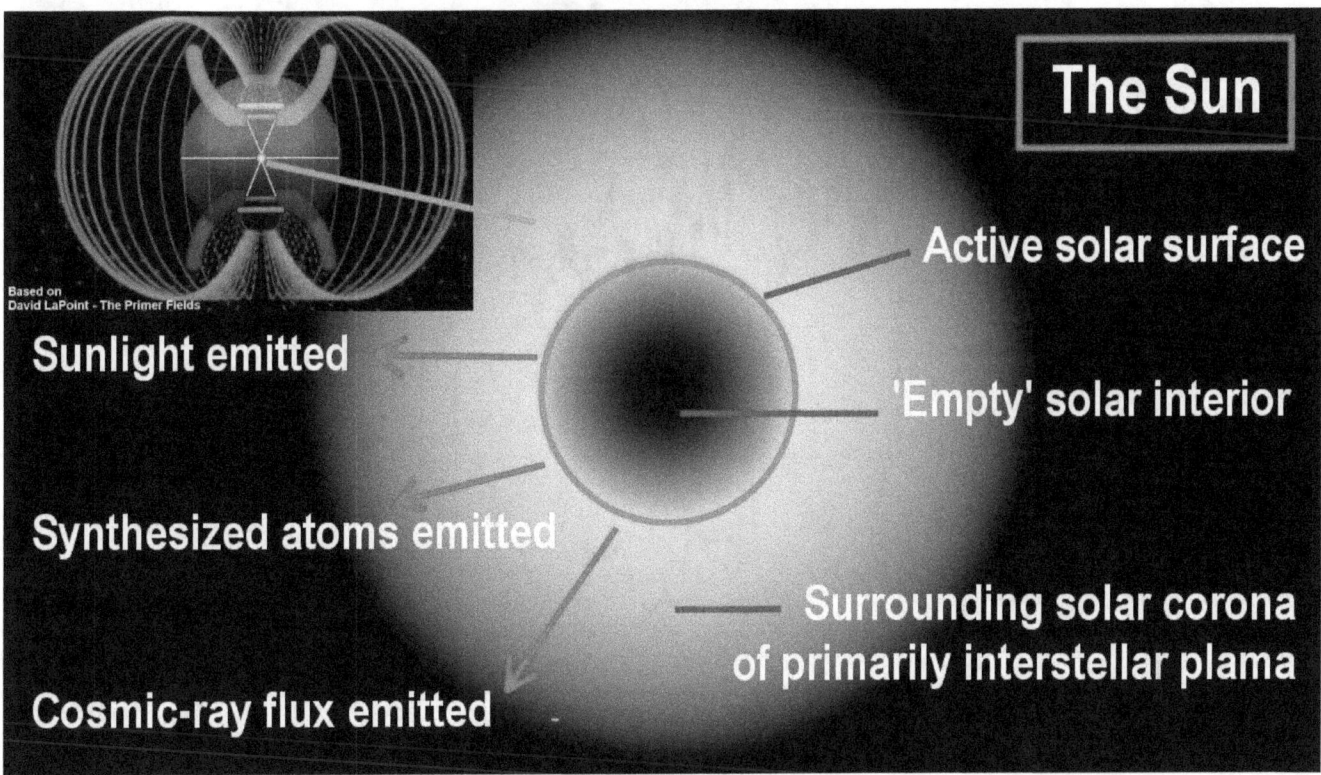

During its low-powered hibernation mode, the dense plasma mantle that the primer fields generate, no longer exists.

Glaciation is caused by the reduced energy

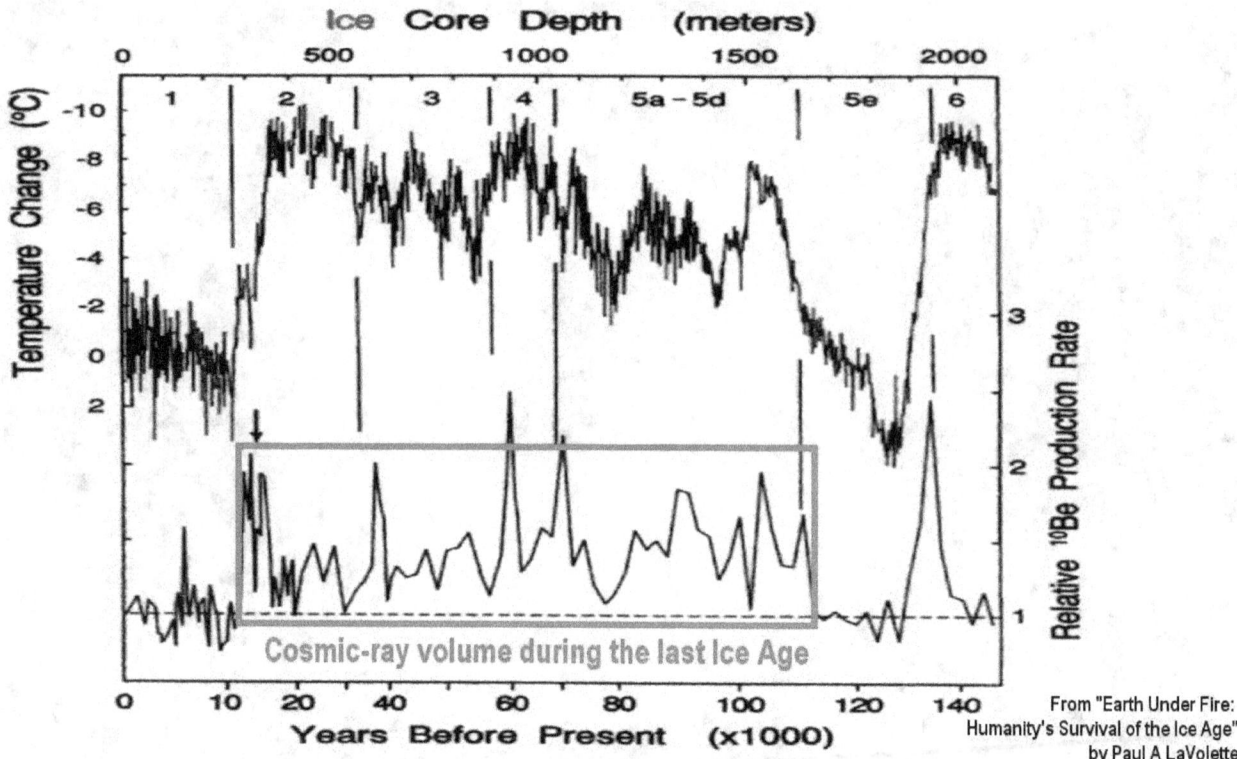

From "Earth Under Fire: Humanity's Survival of the Ice Age" by Paul A LaVolette

This means that larger volumes of cosmic-ray flux can now reach the Earth, which is reflected in increased isotope ratios. However, it is not the increased cosmic-ray flux that causes the glaciation on Earth. The glaciation is caused by the reduced energy that the Sun radiates in hibernation mode. The reduction in solar surface temperature from 5,800 degrees, to around 4,000 degrees, amounts to a 70% reduction in radiated energy. This is big, big enough to cause massive glaciation on Earth.

The glaciation will last until the next resonance effect

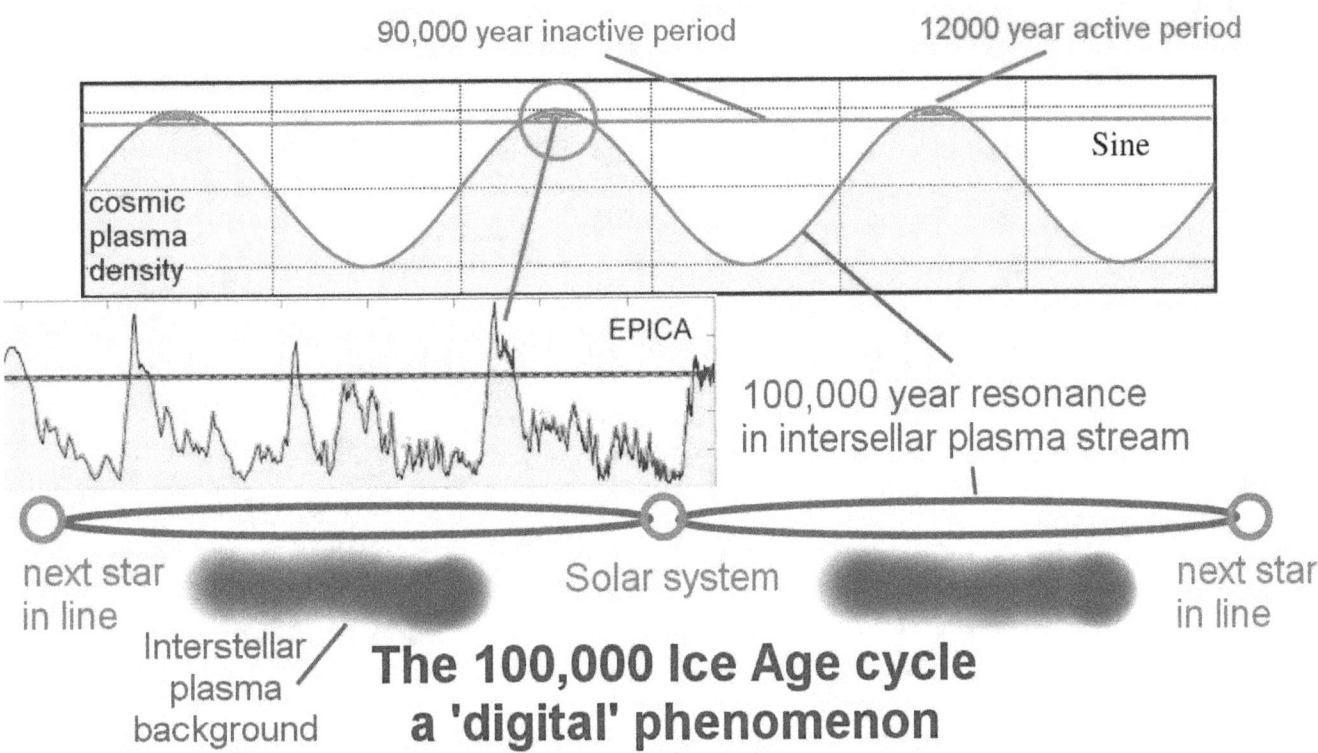

The glaciation will last until the next resonance effect in the interstellar plasma streams form the solar primer fields anew that create the conditions for the high-powered Sun that generates the next interglacial warm period for the typical 12,000 years till the fields collapse again.

The Asymmetric Interglacial

The Asymmetric Interglacial

The Asymmetric Interglacial

We are 4,400 years past the point of symmetry

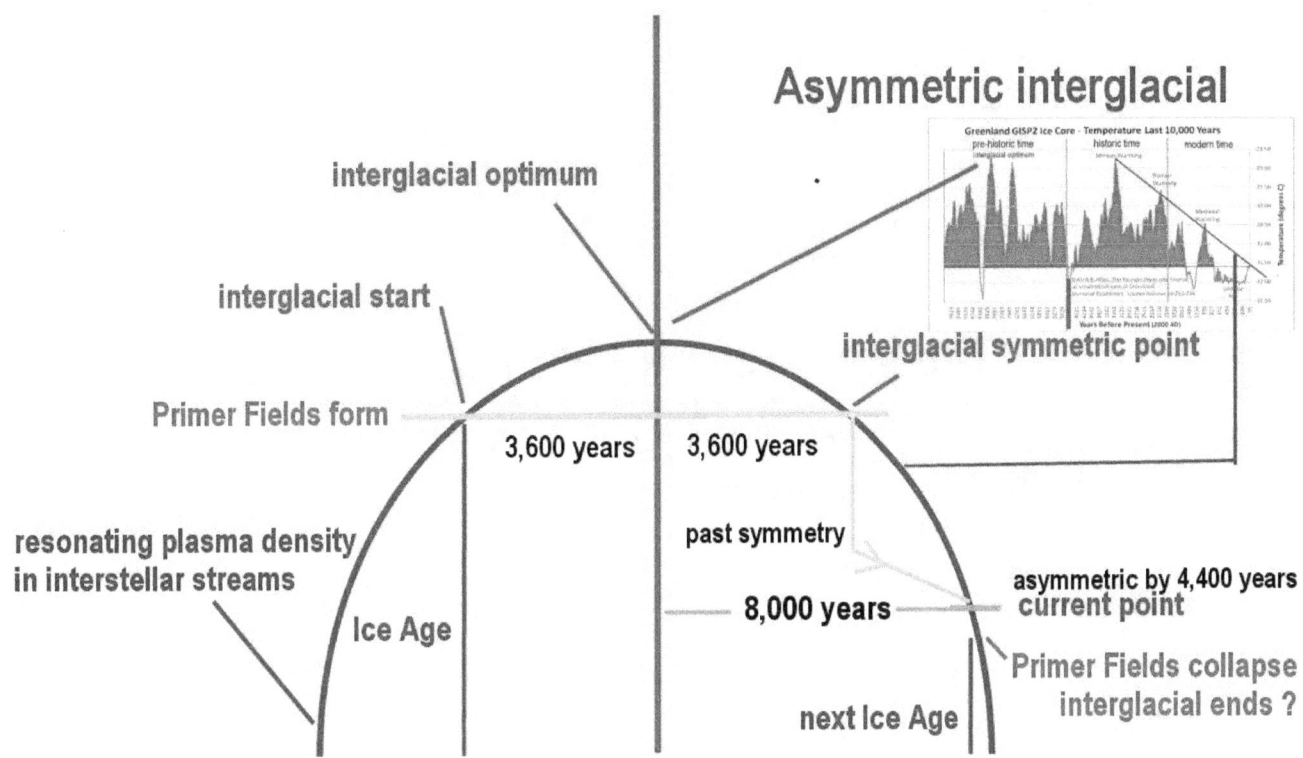

The dynamics are such that the turn-on and the collapse of the primer fields do not occur symmetric in time. The 'measured' evidence indicates that it takes a greater flow density in the plasma stream for the primer fields to form, than is required to maintain them. If the fields would collapse at the same level at which they were formed, they would have collapsed 4,400 years ago, before civilization began to develop. Fortunately for us, this didn't happen. The operation of the fields generate a positive feedback that keeps them operating at lower levels, while the solar activity keeps diminishing with the weakening primer fields..

We are presently 8,000 years past the interglacial optimum that marks the center of the interstellar resonance pulse. We are a whopping 4,400 years past the point of symmetry. The stark asymmetry of the interglacial period, delivers an additional element of proof that our Sun is a plasma Sun, powered by interstellar streams of plasma, and that its high-intensity mode is activated by electromagnetic primer fields. The proof is valid, because the asymmetry reflects the characteristic dynamics of the primer fields.

The forming of the Primer Fields

The forming of the Primer Fields

The forming of the Primer Fields

Primer fields focus interstellar plasma unto a Sun

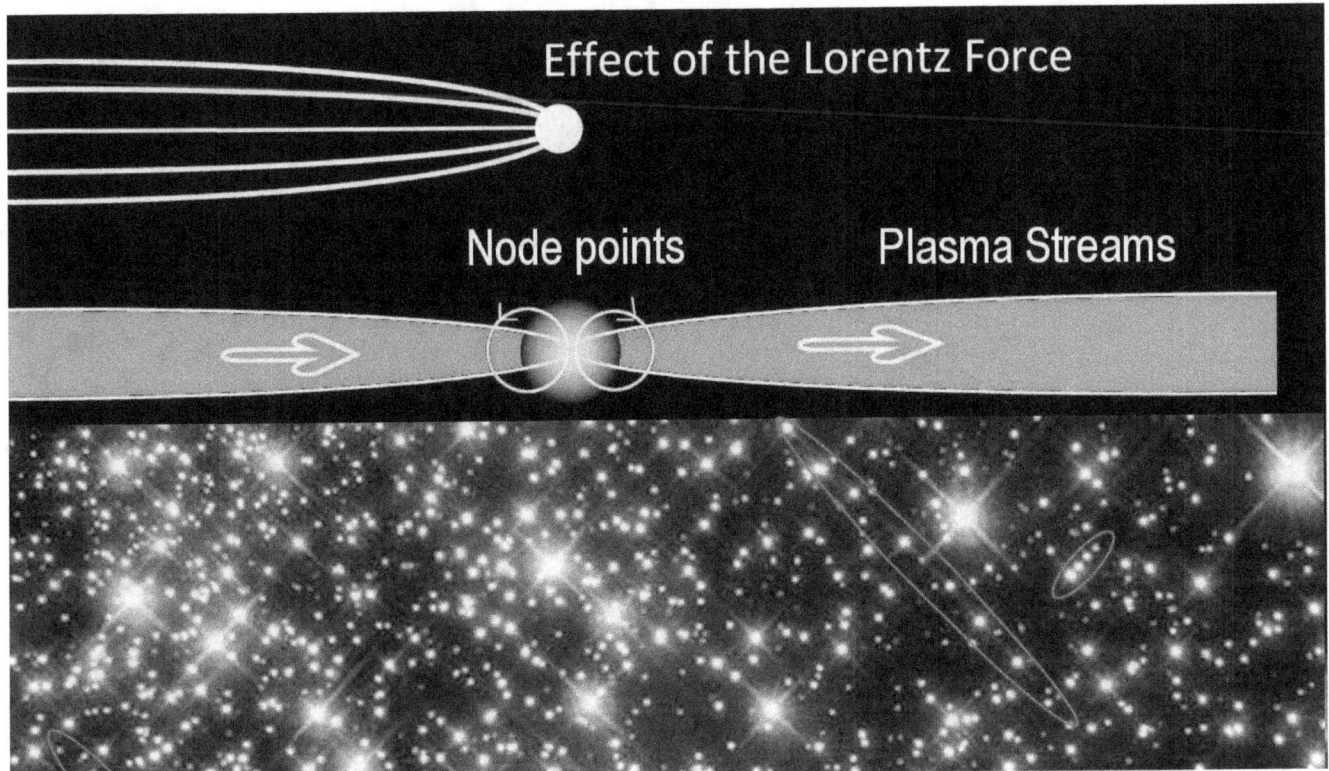

As I said before, the stellar primer fields focus interstellar plasma streams unto a Sun. This is accomplished by the pinching effects of magnetic fields that form as the result of flowing electrically charged particles moving in the same direction, such as in plasma streams in space.

David LaPoint explored what happens

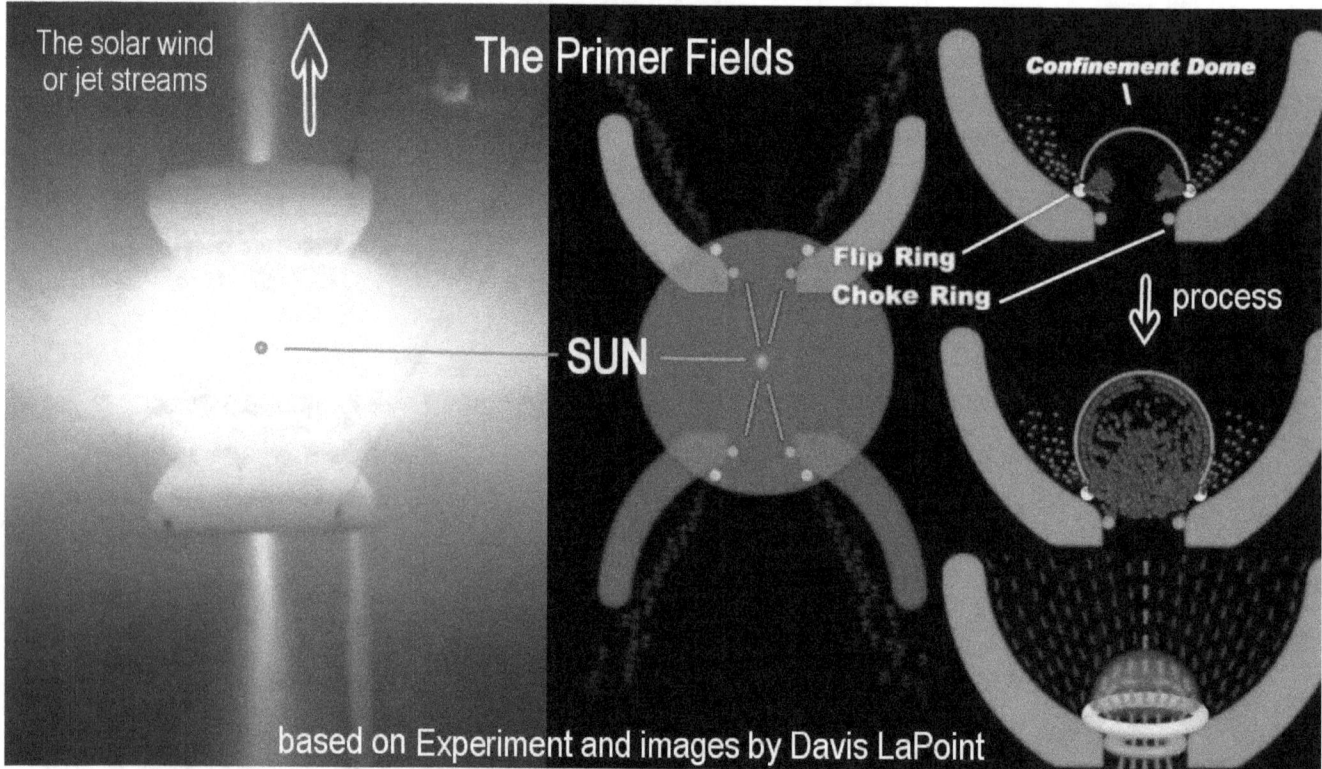

based on Experiment and images by Davis LaPoint

The researcher David LaPoint explored what happens at the extreme region of the plasma stream, where the pinch effect of the magnetic fields is so great that the pinched plasma flows backwards, and becomes trapped thereby and intensely concentrated under a magnetic confinement dome, which the back-flow of the plasma creates. If the back-flow is strong, some of the concentrated plasma escapes through the top of the dome, where the confining magmatic field is the weakest. On the Sun, the escaping plasma becomes the solar wind.

The highly concentrated plasma that accumulates under the confinement dome, is also magnetically focused onto a Sun.

The concentrated plasma surrounds the Sun like a mantle

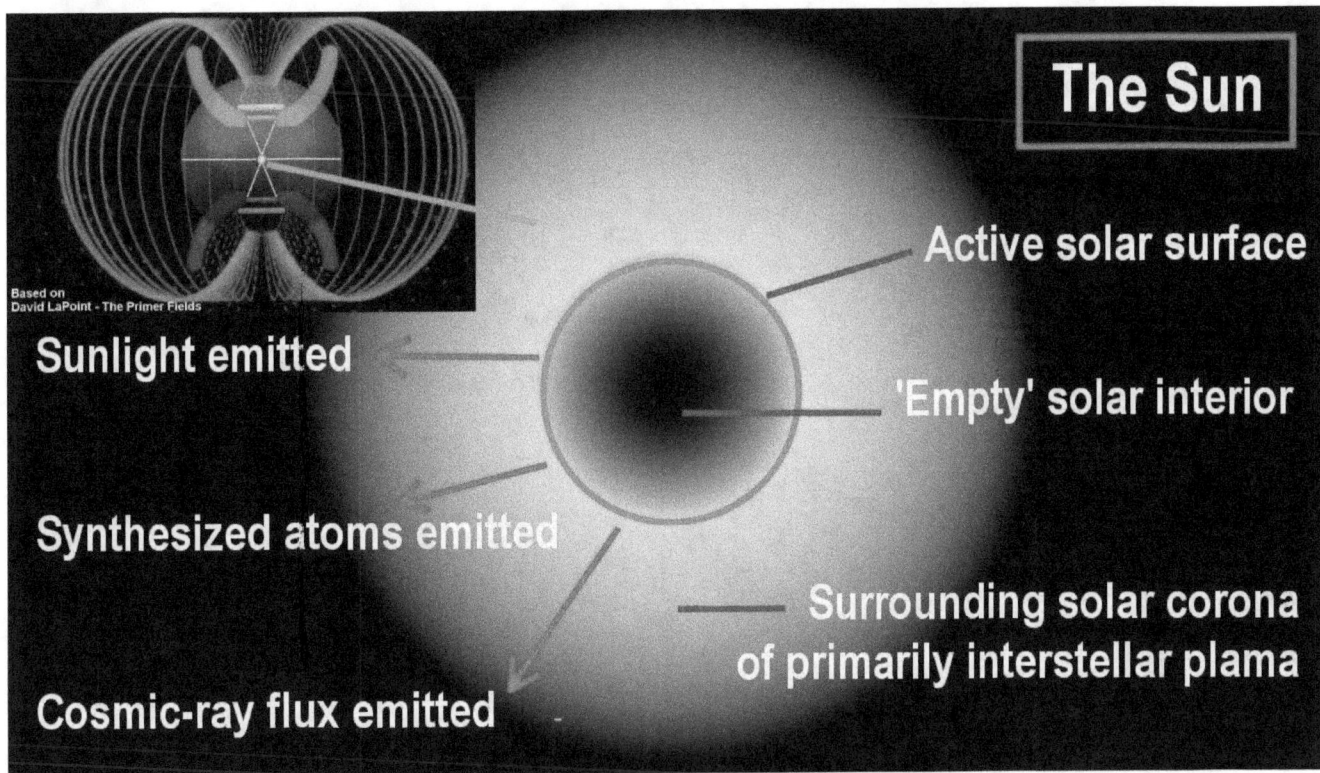

The concentrated plasma that is focused onto the Sun, is further drawn unto the Sun by the Sun's high-voltage electric attraction. The concentrated plasma surrounds the Sun like a mantle, and becomes accelerated towards it, whereby a portion of it is consumed by it in nuclear synthesis. The synthesis creates electrically neutral atoms. It thereby creates an electric sink that keeps the plasma streams flowing into the Sun forever and ever.

Whatever portion of the in-flowing plasma is not consumed

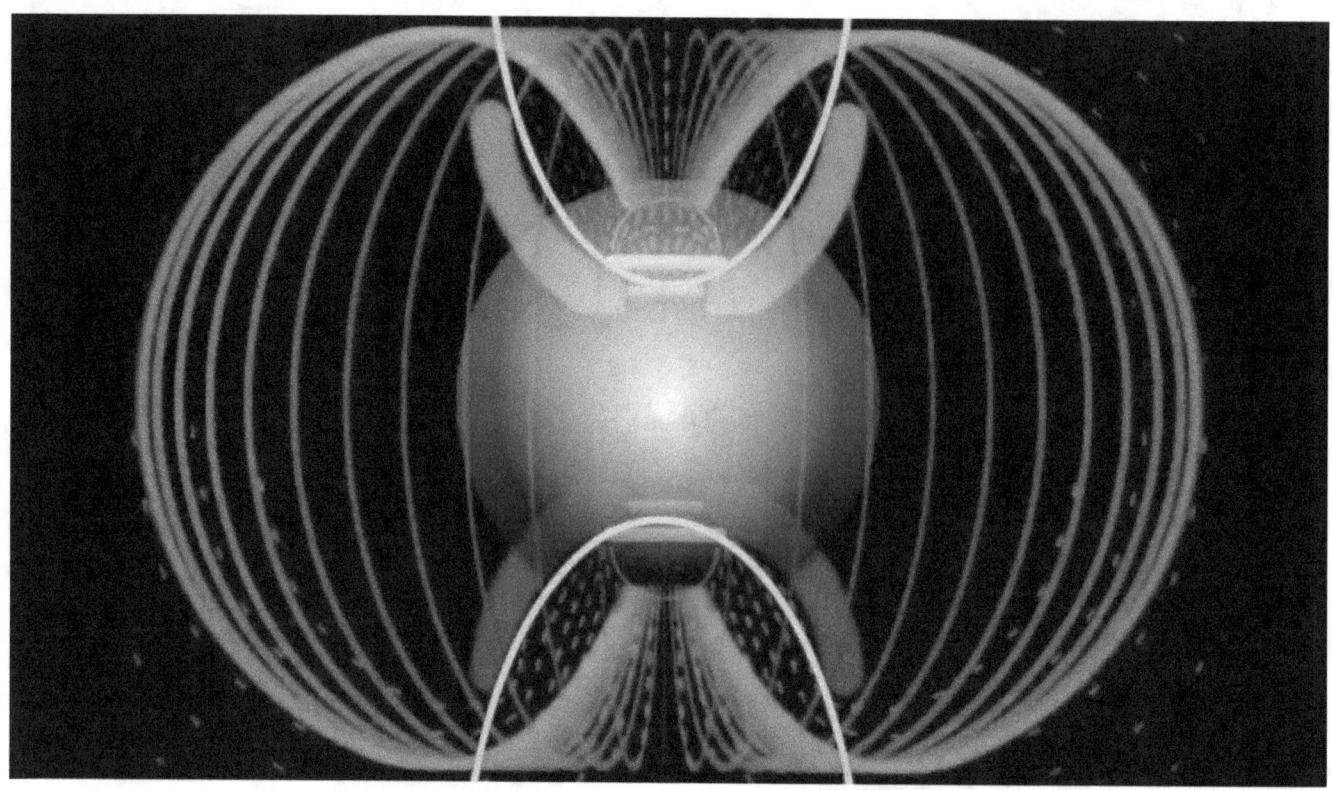

Whatever portion of the in-flowing plasma is not consumed by the solar process, flows on and away from the Sun by effects that create a reverse primer field with opposite characteristics.

The principle of the complimentary primer fields

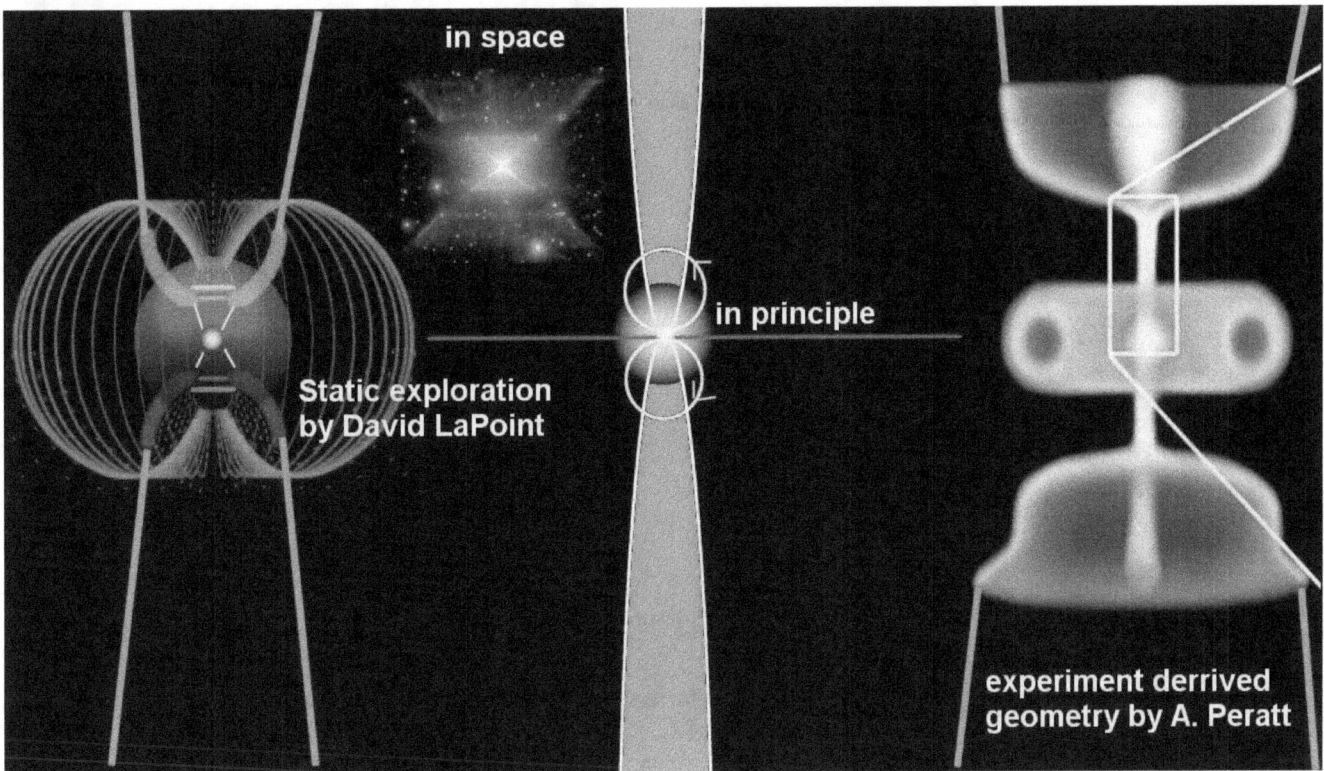

The principle of the complimentary primer fields has been replicated dynamically in high-energy discharge experiments at the Los Alamos National Laboratory, under the direction of Anthony Peratt, director of experiments.

The resulting geometry that is shown here in purple, was created exclusively by the magnetic effects of flowing plasma. The experiment illustrates a natural principle that is manifest in numerous ways in cosmic space. In some cases the effect is strong enough to be visible in space, as in the case of the Red Square Nebula.

In all cases, because the process that forms the primer fields, is dynamic, it depends on a minimal volume of flowing plasma for the magnetic fields to form, and to form with the type of complex geometry that focuses plasma onto a Sun. In such cases where the fields are formed, the effects of the fields increase the rate of flow onto a Sun, which keeps the fields active significantly longer in a weakening environment, such as our solar system is presently in.

Our Sun being a rather mediocre star among stars

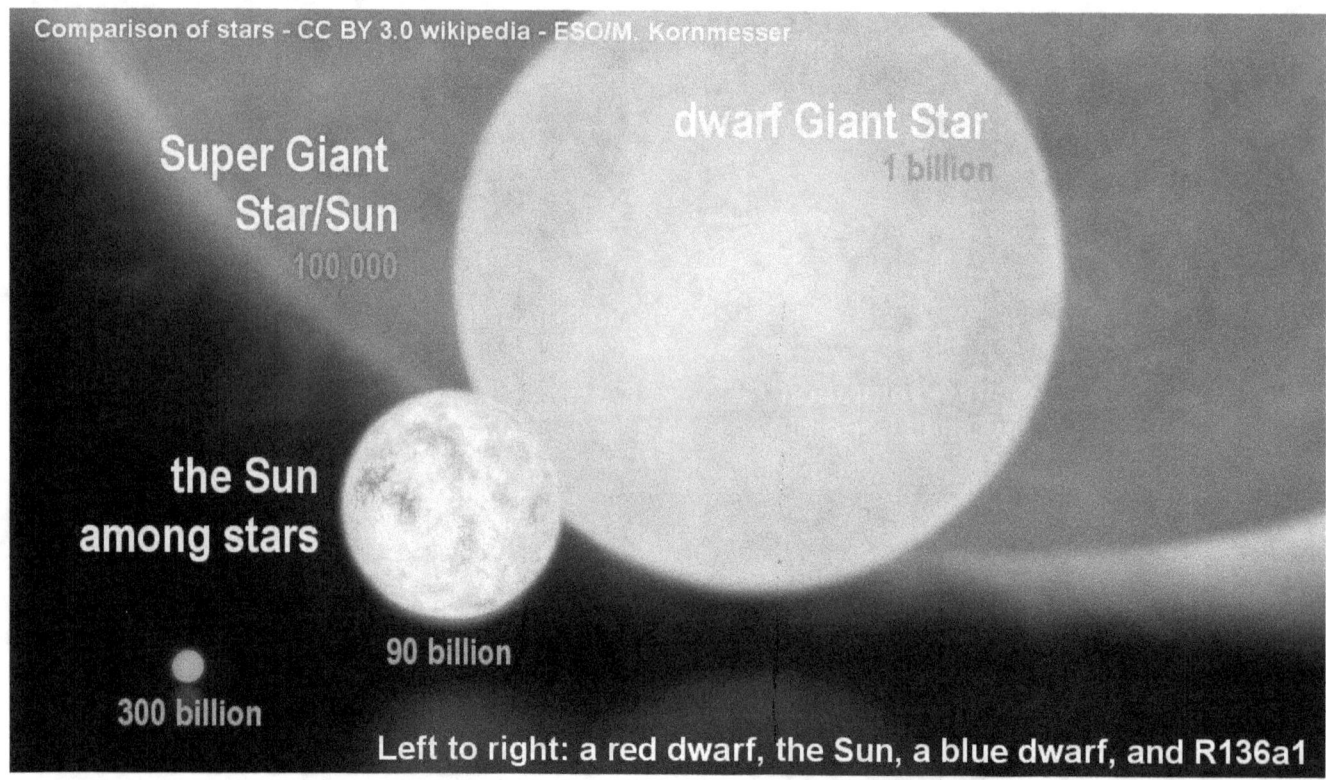

A part of the reason why our Sun is not constantly in its high-powered mode, in the presently weak galactic climate, is evidently due to our Sun being a rather mediocre star among stars, a type of yellow dwarf star.

The Sun is classified as a G class star

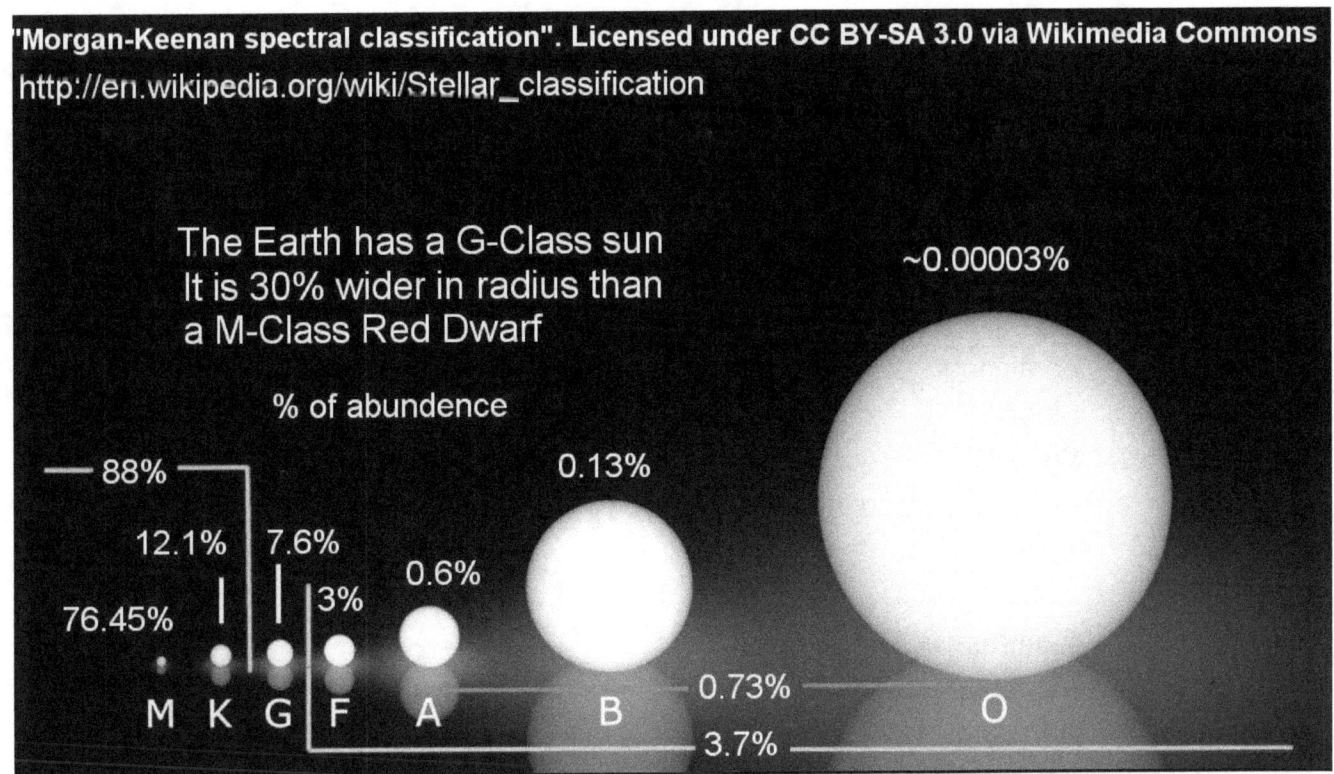

The Sun is classified as a G class star, which is the category between the low-powered red-dwarf stars, and the high-powered larger stars.

High-powered stars are larger in size

Size of active stars and their surface temperature

Star name	Times the Sun (diameter)	Surface temperature	
Sun		5,800 K	
Fomalhaut	1.8	8,500 K	
Sirius	2.0	9,900 K	
Spica B	3.64	18,500 K	
Antares B	5.2	18,500 K	
Spica A	7.4	22,400 K	
Arcturus	44	4,286 K	The universal default range for 'small' inactive stars
Aldebaran	65	3,910 K	

The high-powered stars are larger in size, and being larger, attract larger volumes of plasma, which results in higher solar surface temperatures. Evidently, these stars are presently unaffected by the low plasma density in the galaxy.

The very-much larger stars, however, appear to be affected by the low plasma density in the galaxy. It appears that there isn't enough density in the galaxy to form the correspondingly large primer fields. These stars operate at the apparent default level in the range of 4,000 degrees Kelvin.

When the primer fields for our Sun collapse, our Sun will likely revert back to what appears to be the universal default temperature that corresponds to the Ice Age climate on Earth.

Hanging by a fine thread

Hanging by a fine thread

Hanging by a fine thread

We are 4,400 years past the symmetry point

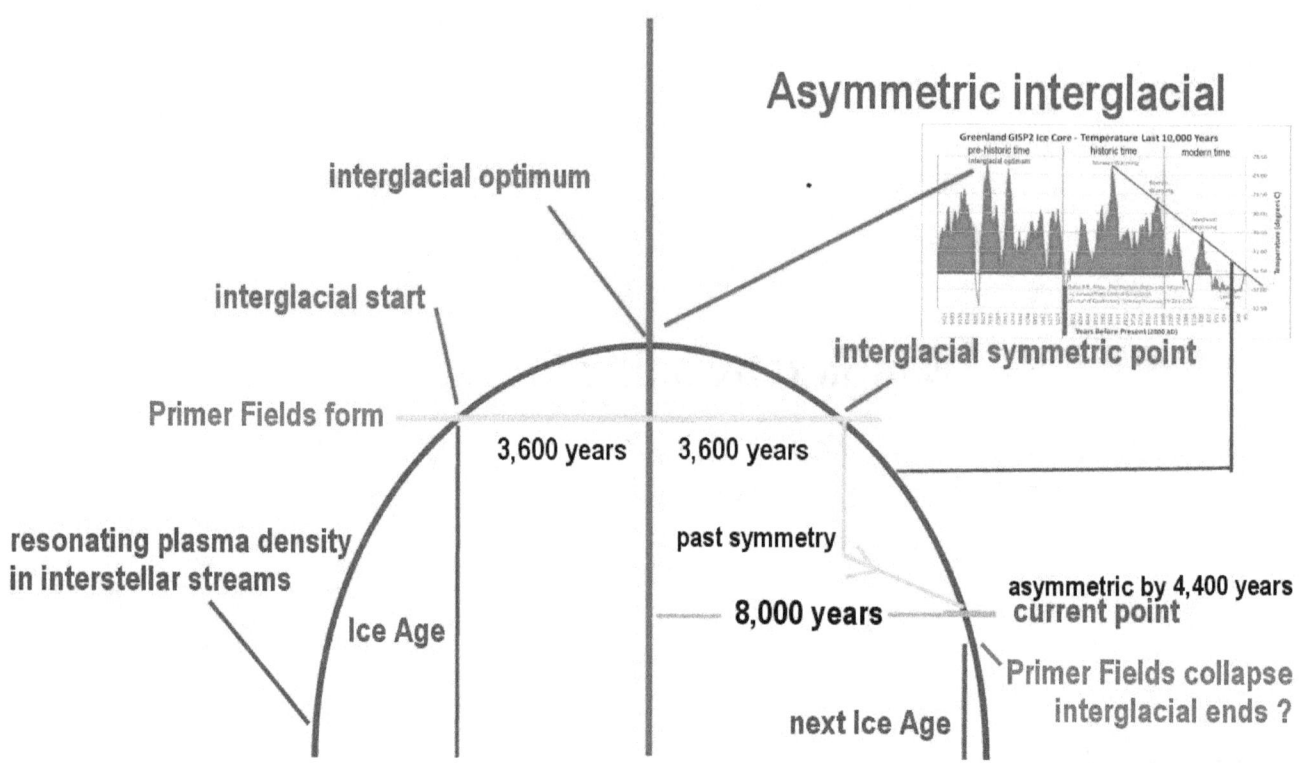

Since we are presently 4,400 years past the symmetry point of the current interglacial period, with the solar activity diminishing evermore rapidly, it seems reasonable to assume that the primer fields, and with them the interglacial period, which are both just barely hanging on, will collapse soon.

The primer fields will then collapse

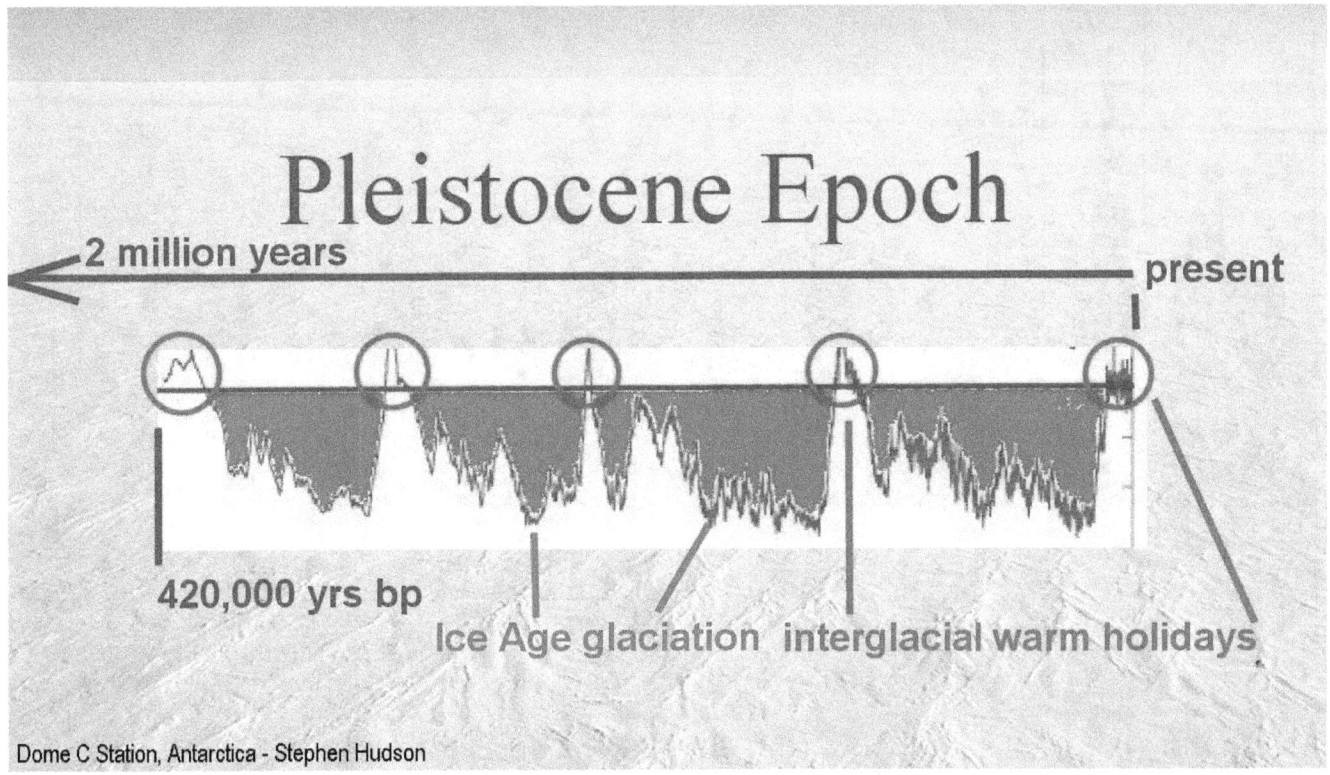

This means that the primer fields, and the interglacial climate with them, will then collapse as they always have. That's what we see in the ice core records. And the promise is not pleasant.

During the last 4,400 years

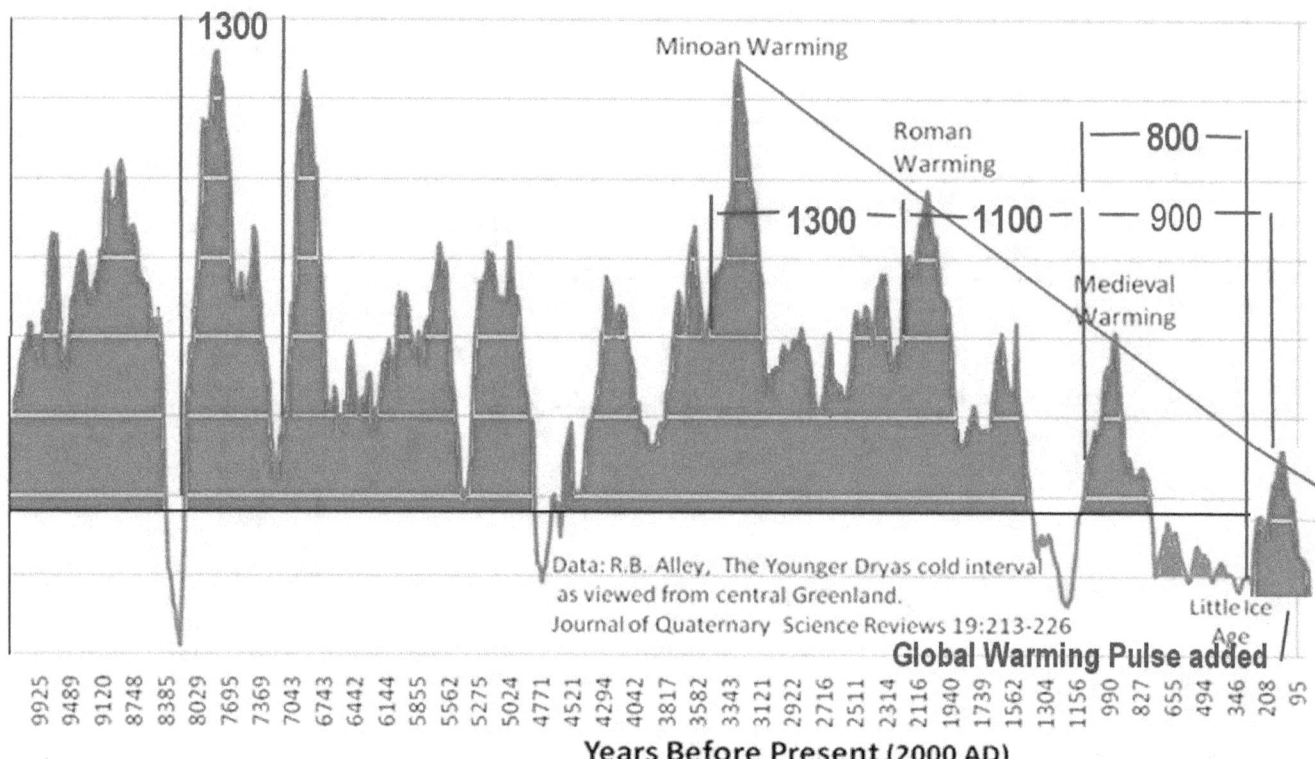

During the last 4,400 years the solar system has become progressively weaker, the big warming pulses progressively smaller, and the intervals between them shorter. The dramatic climate weakening that we see here, reflects the weakening primer fields that are barely hanging on as by a fine thread.

On the steepening slope of the last 1,000 years

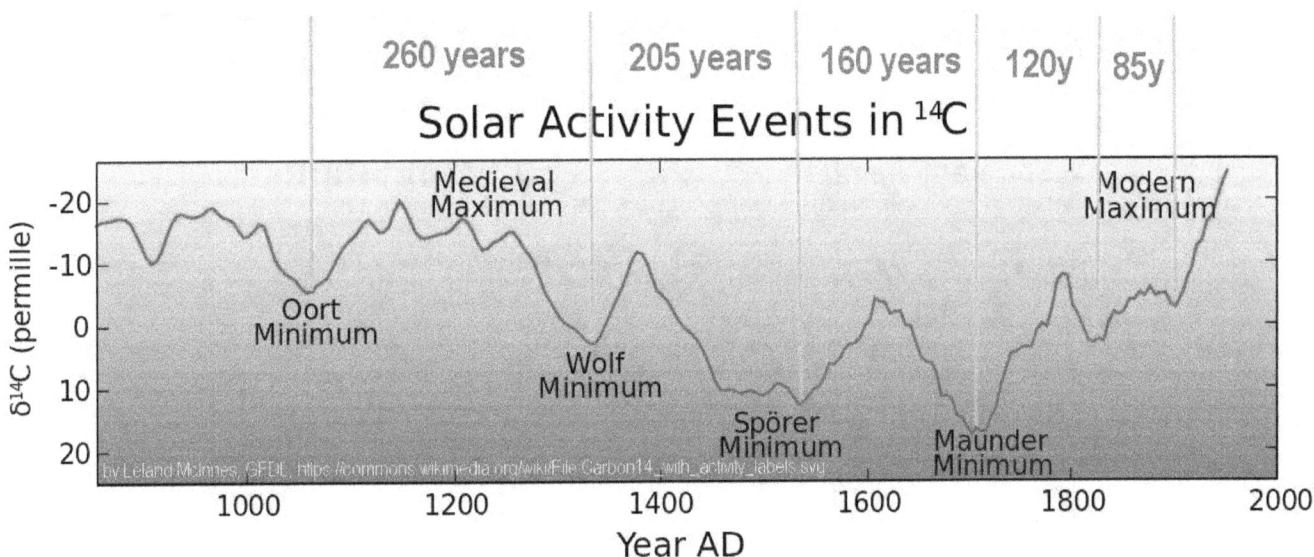

We have already seen a string of Little Ice Ages erupting on the steepening slope of the last 1,000 years. Harsh climate conditions have occurred in those Little Ice Age periods, between recovery periods. Now we face the fact that those cyclical events too, have become progressively smaller in amplitudes for recovery, and with ever shorter intervals between them.

Recognize why the cyclical events are diminishing

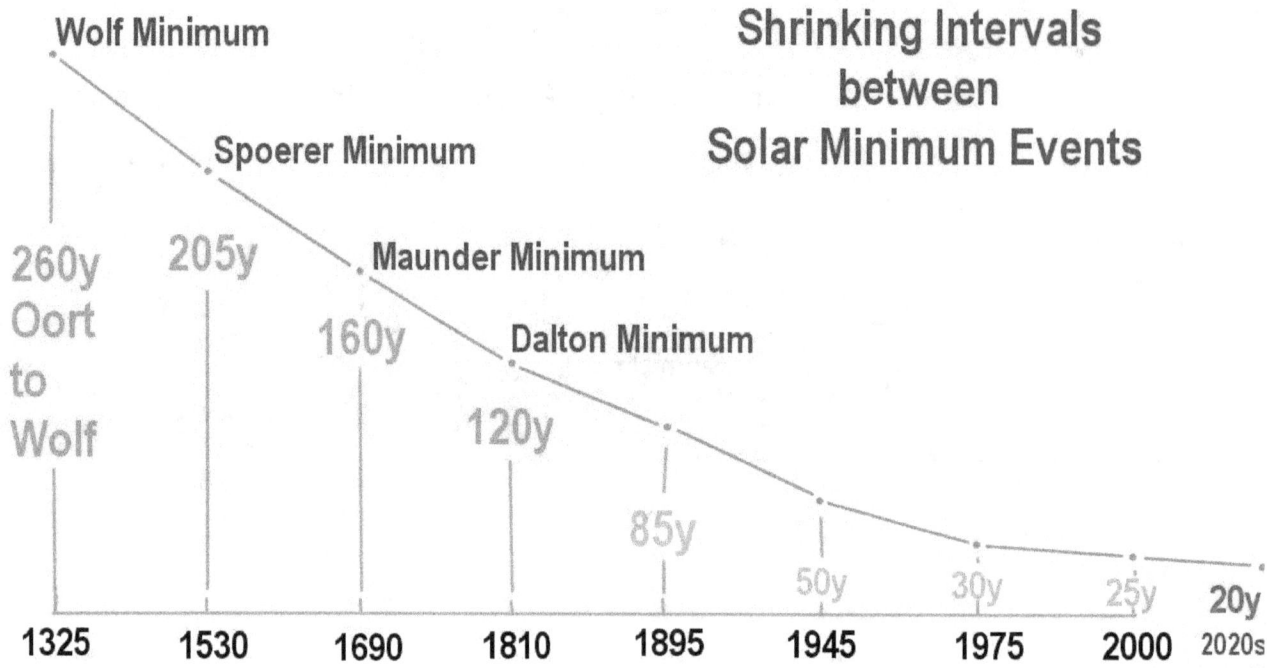

These cycles have collapsed so dramatically over the last 1,000 years that there isn't much left of them that we can count on for future climate recoveries.

It is critical to recognize why the cyclical events are diminishing, and the intervals between them are getting shorter.

Without this recognition it is tempting to assume

Without this recognition it is tempting to assume that the current climate collapse is but a replay of historic climate cycles. At the current stage, this is physically impossible. Thus, the imagined Grand Solar Minimum theories are basically dangerous illusions as they invite hope for a recovery that isn't possible, which closes the door to the Plan-B option that humanity's future depends on.

But why have the historic cycles fizzed out on both counts? The key may be found in their intervals getting shorter.

Warming pulses started with intervals of 1,300 years

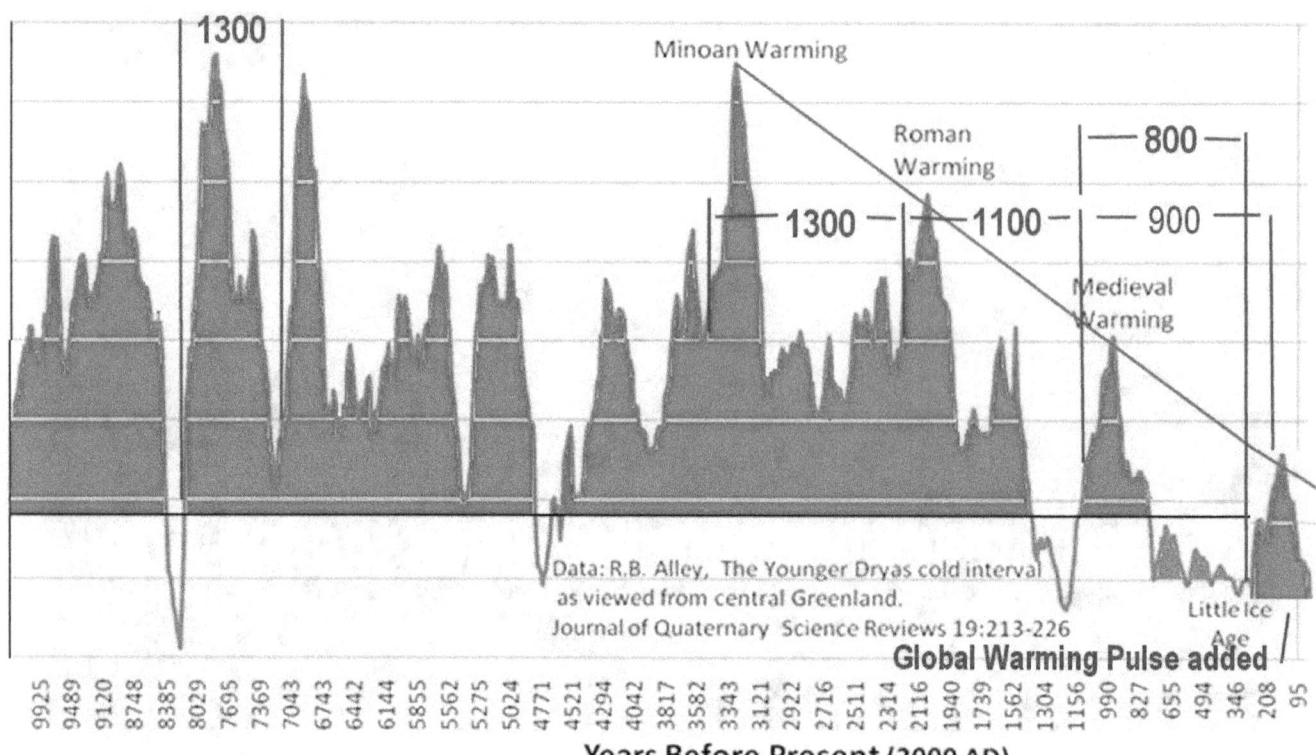

The big warming pulses started with intervals of 1,300 years between them.

Minimum cycles started with 260 years between them

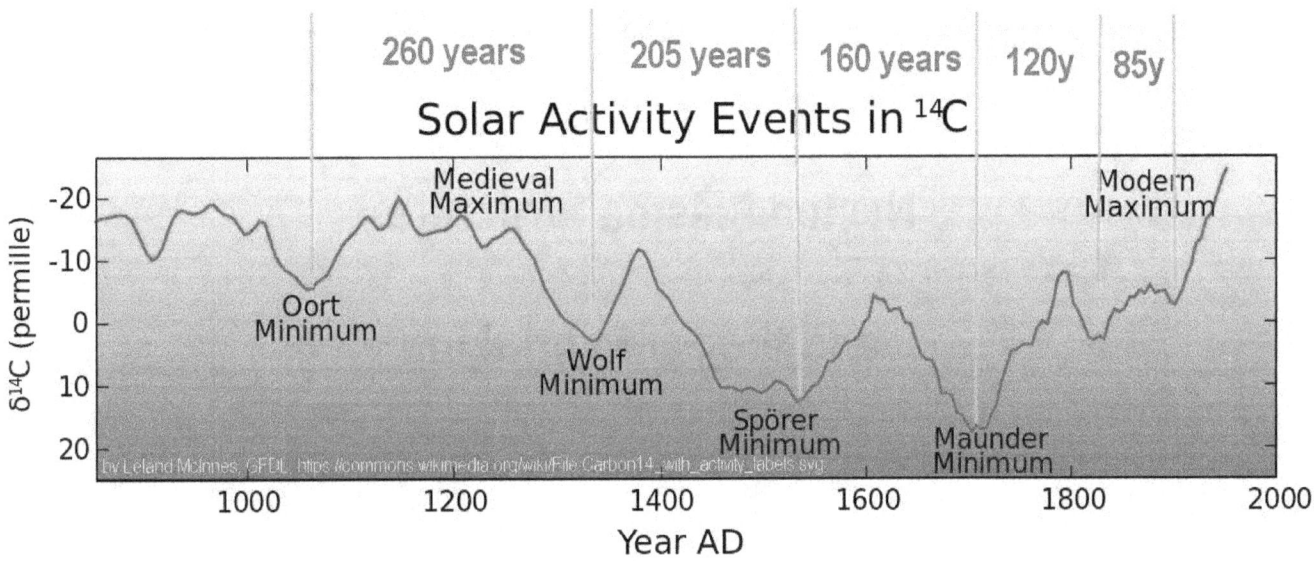

And the historic minimum cycles started with 260 years between them.

Since these repetitive pulses are evidently plasma resonance features in both cases, two separate large plasma structures must be involved, each with a size appropriate for its resonance time.

Nested Plasma Structures - and they are shrinking

Nested Plasma Structures

and they are shrinking

Nested Plasma Structures - and they are shrinking

The long-ago theorized Oort Cloud system

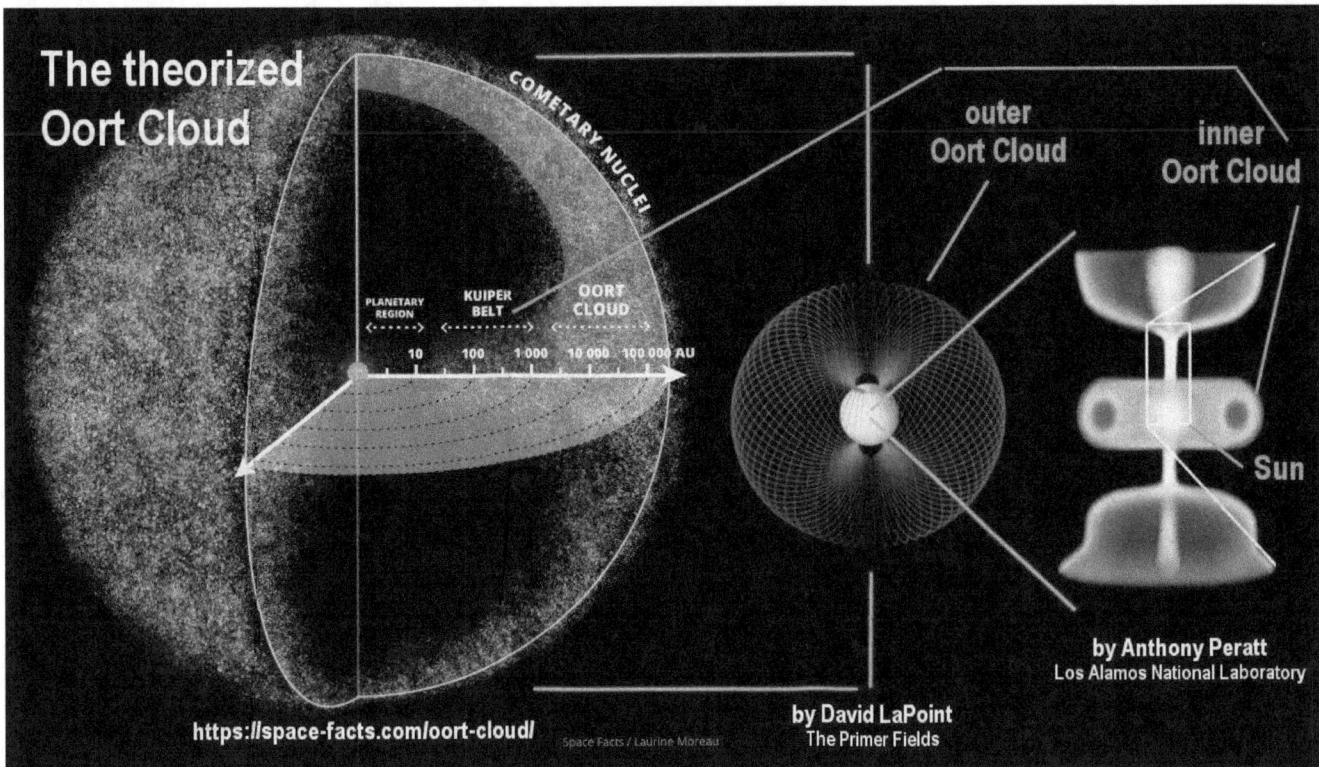

The long-ago theorized Oort Cloud system has the appropriate dimension to qualify as nested plasma structures surrounding the solar system. The Oort Cloud is theorized as two concentric clouds of asteroid-type objects, a donut-shaped inner cloud and a spherical outer cloud. Since no one as ever seen the theorized asteroid objects, it is highly probable that the Oort Cloud is merely a system of concentric plasma structures. The very large outer plasma sphere would then be the resonating driver for the big warming pulses at 1,300-year intervals that we see in the ice core records, and the smaller inner cloud would be the driver for the shorter 260 year cycles.

The nested plasma structures have both been shrinking

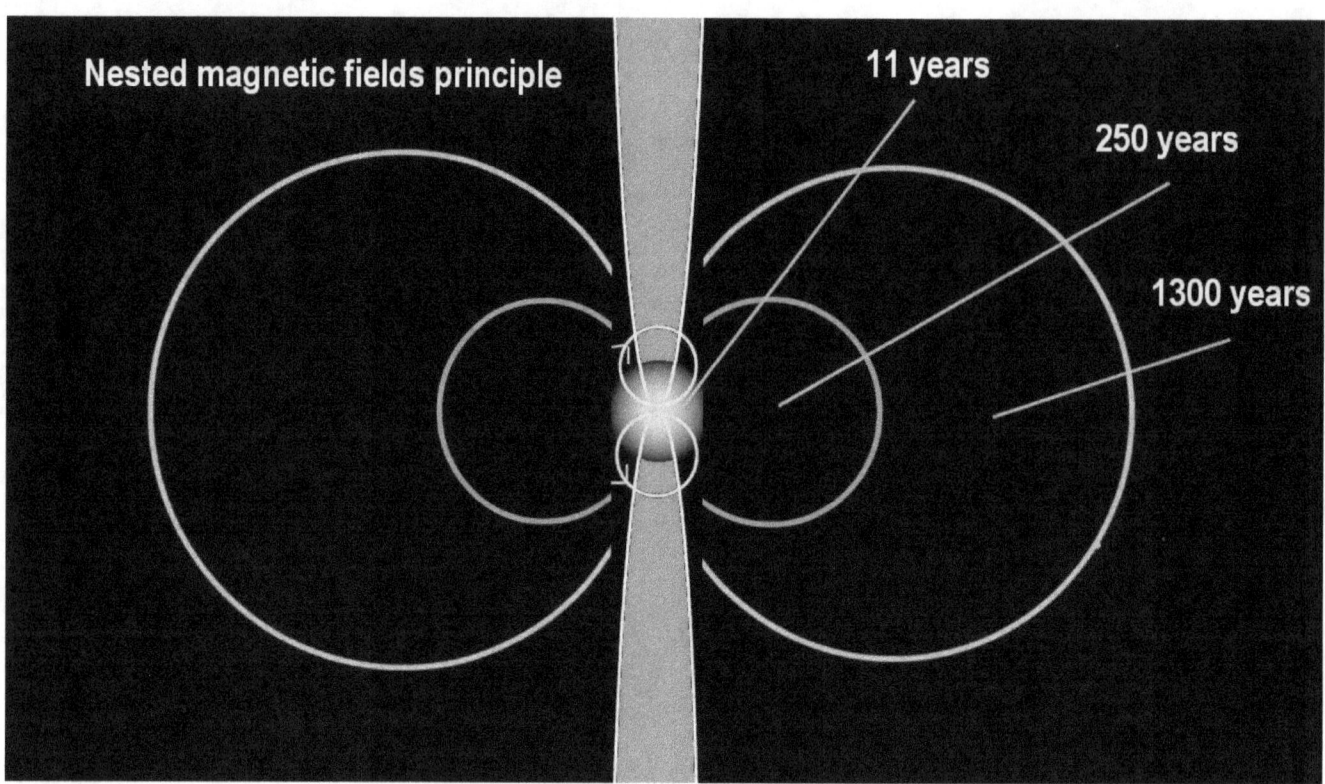

With the nested plasma structures flowing into the main plasma stream, each would modulate its density with its own periodicity. The fact that the periodicity has been getting increasingly shorter, suggests that the nested plasma structures have both been shrinking in size in accord with the weakening of the interstellar plasma streams, and have both been shrinking ever-faster.

Interval between the warming pulses has shrunk

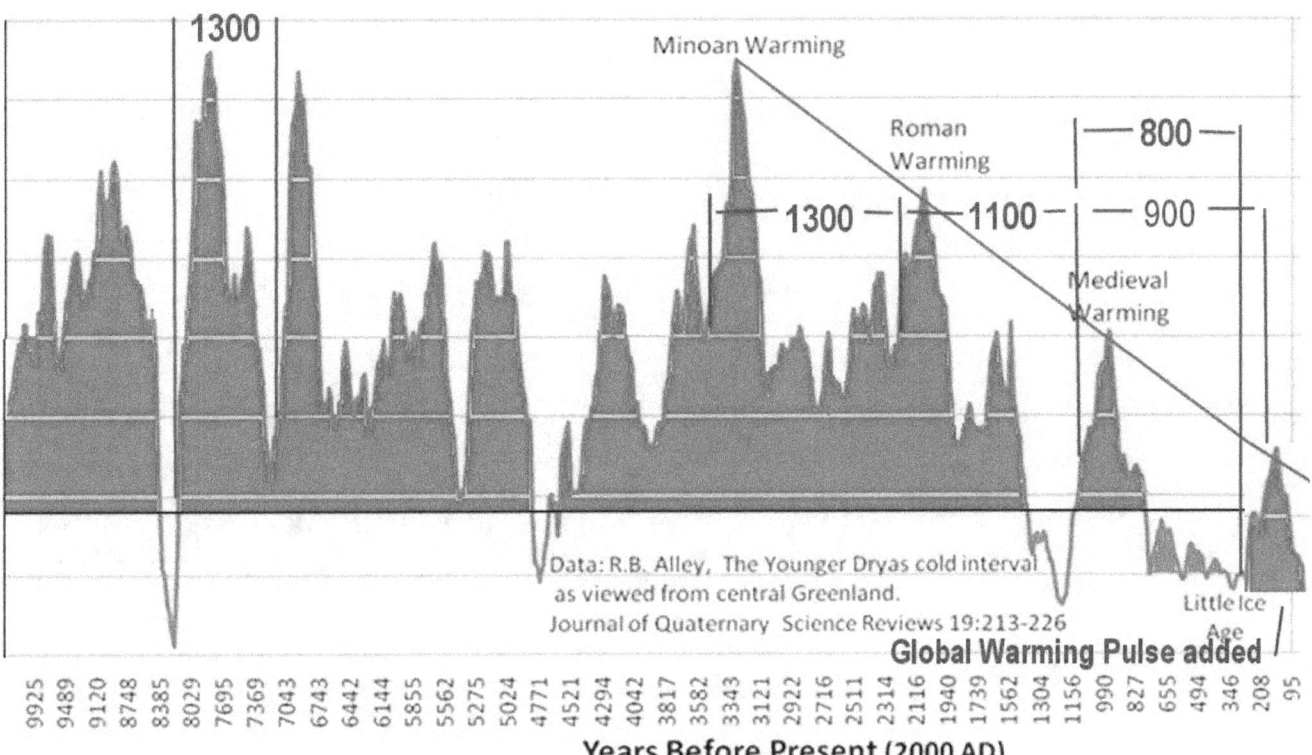

The 1,300-year interval between the warming pulses has shrunk to a mere 800 years, leading into the 1700s, while the pulses themselves have diminished.

Intervals between the minima been shrinking

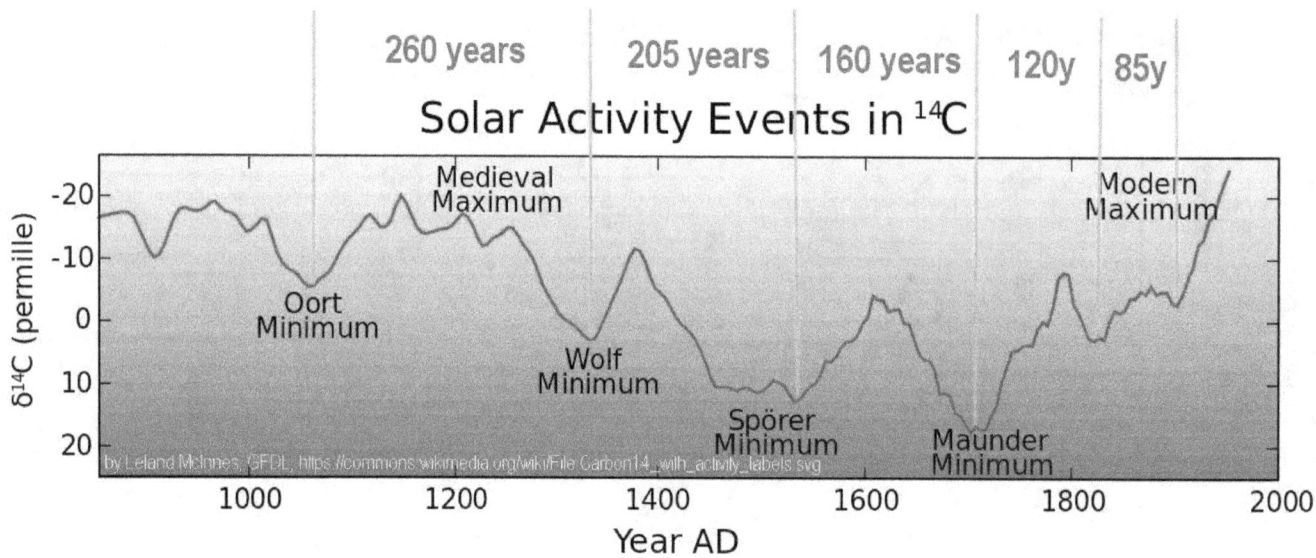

In a similar manner have intervals between the minima been shrinking from 260 years to 85 years, and so on, with the amplitude of their oscillations getting increasingly smaller.

Shorter intervals less-inflated, plasma structures

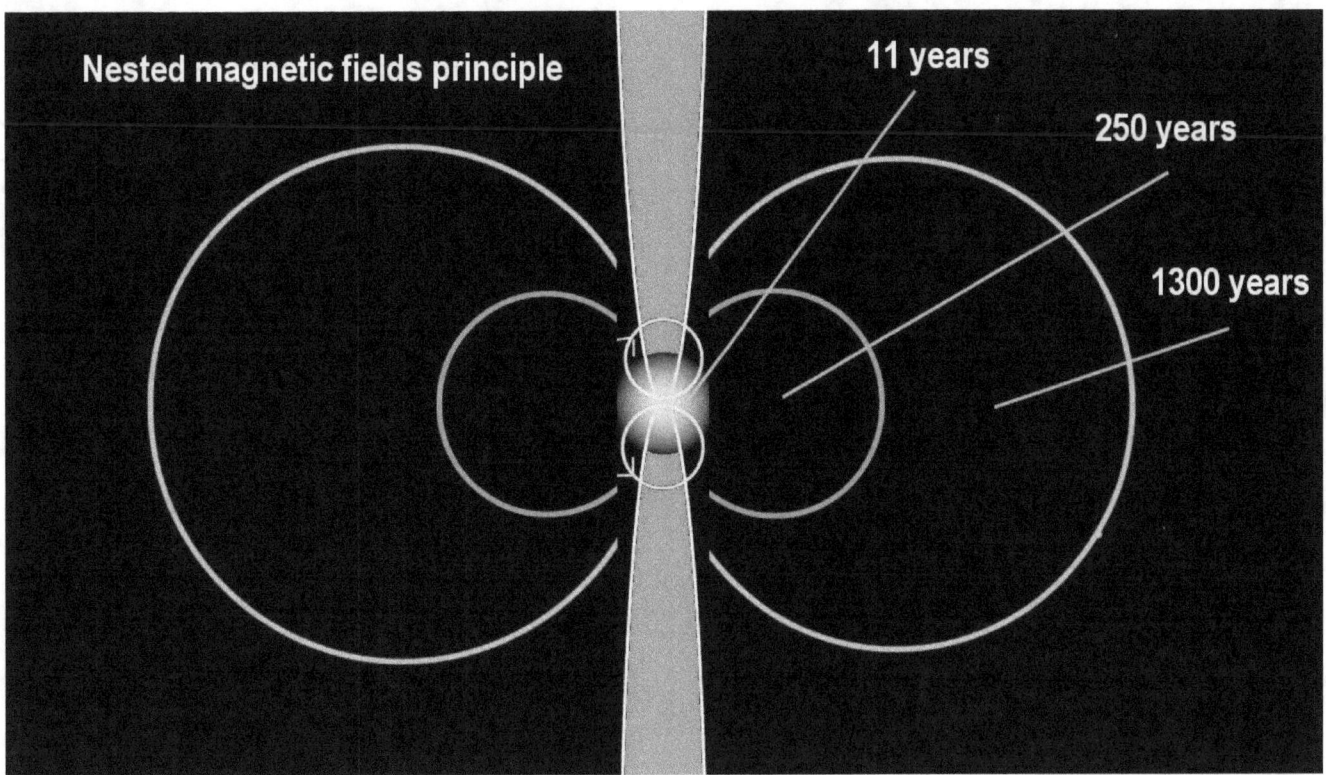

Since the shorter intervals represent smaller, or less-inflated, physical plasma structures, we cannot expect the historic cycles to be repeated. The physical structures that had once caused them no longer exist in the way they did. We are progressing towards the end of the line, so to speak, as the interstellar plasma streams that feed into the nested structures and inflate them have become dangerously weak themselves.

We are on a path that cannot reverse

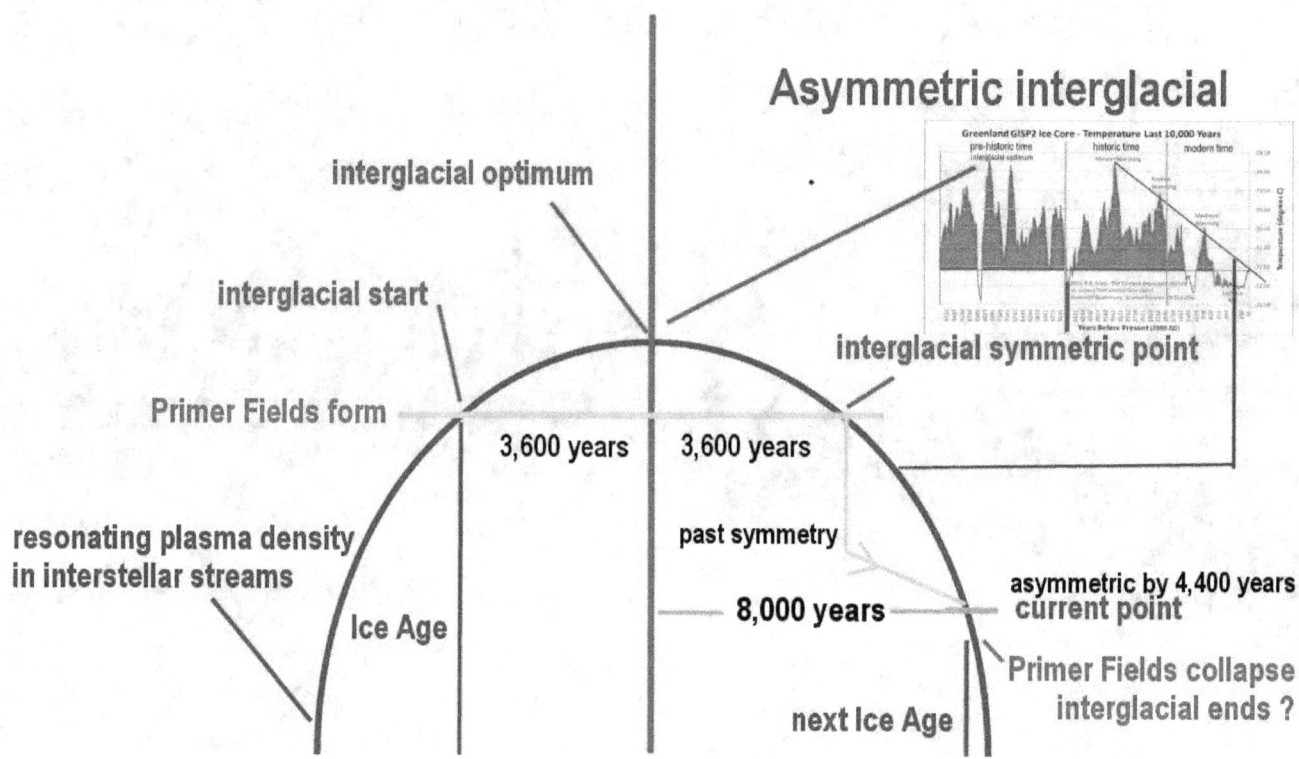

With the interstellar plasma streams diminishing evermore rapidly now, so that the primer fields are just barely hanging on, there is no strength left in the system for any type of significant recovery to happen. We are on a path that cannot reverse.

We face the full Ice Age at the end of the line

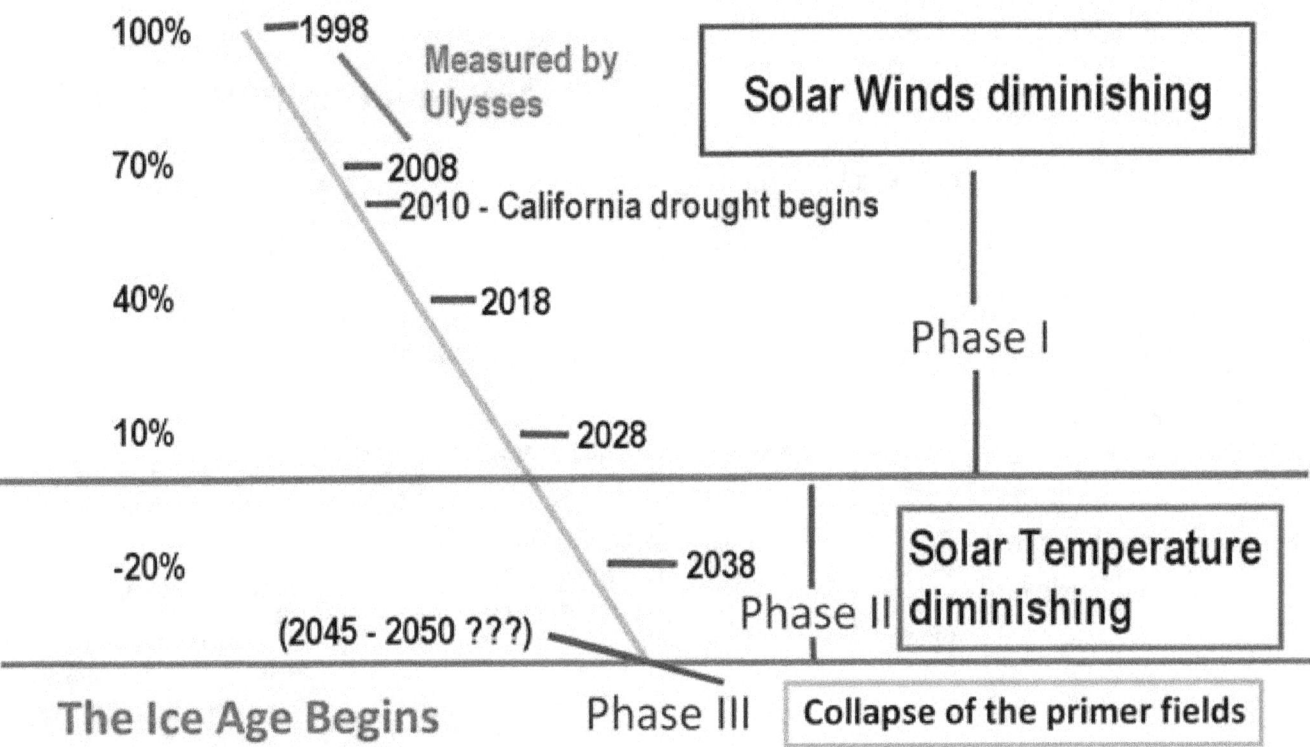

We face the full Ice Age at the end of the line, the Phase-3 stage that we cannot escape. We can only avoid the consequences by implementing the technological Plan-B option while we are in the boundary zone with a still functioning civilization. That's our only option to assure the continuity of civilization and defend humanity as a whole against the approaching Ice Age.

When the fine thread breaks

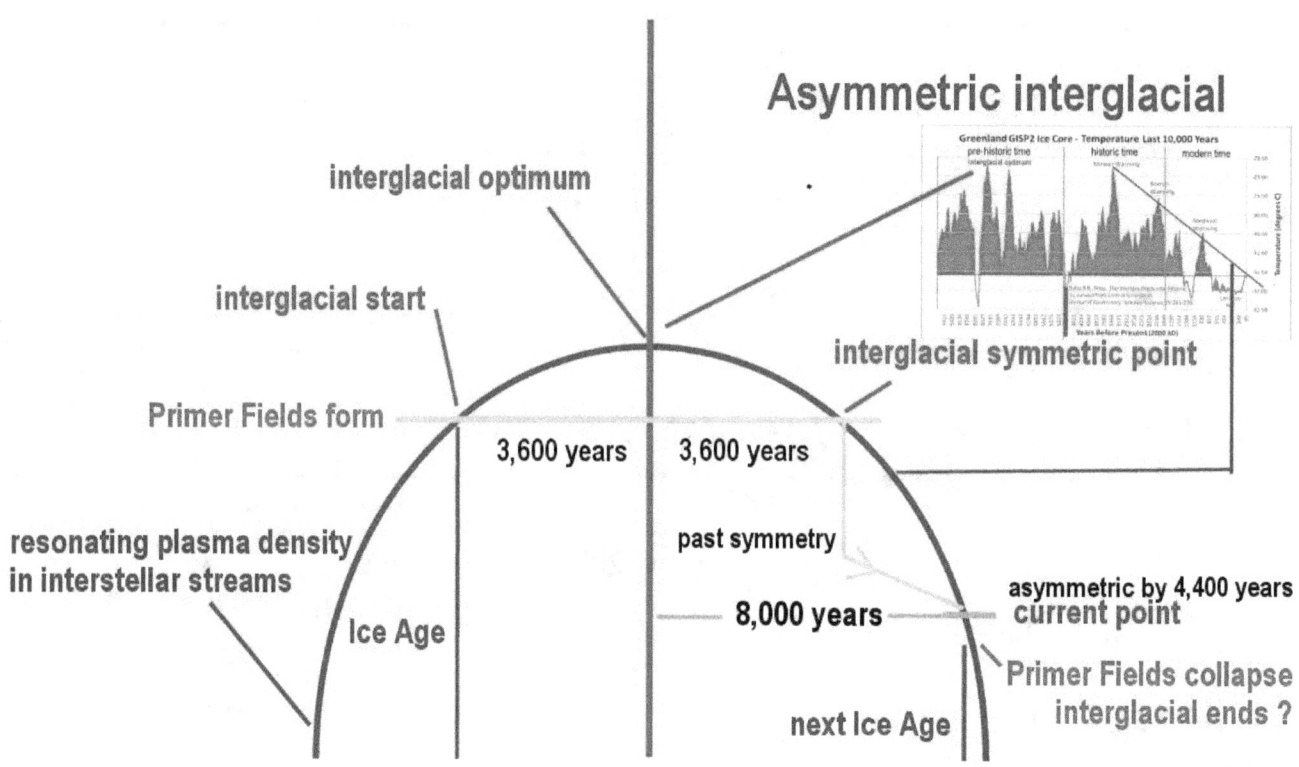

If we look at the current slope of the diminishing interglacial primer fields, we see that the slope is getting steeper, which brings us ever-faster to the turn-off point. The steepening down slope past the 8,000 year mark, where we are presently at, is also reflected in increasingly larger climate fringe effects that have already begun to affect agriculture in many areas of the world. When the fine thread breaks that keeps the primer fields still in operation, the full Ice Age will be upon us in short order, hopefully as late as the 2050s.

This is what we are not prepared for, the potential breaking of the fine thread.

At the present time, no Plan-B exists

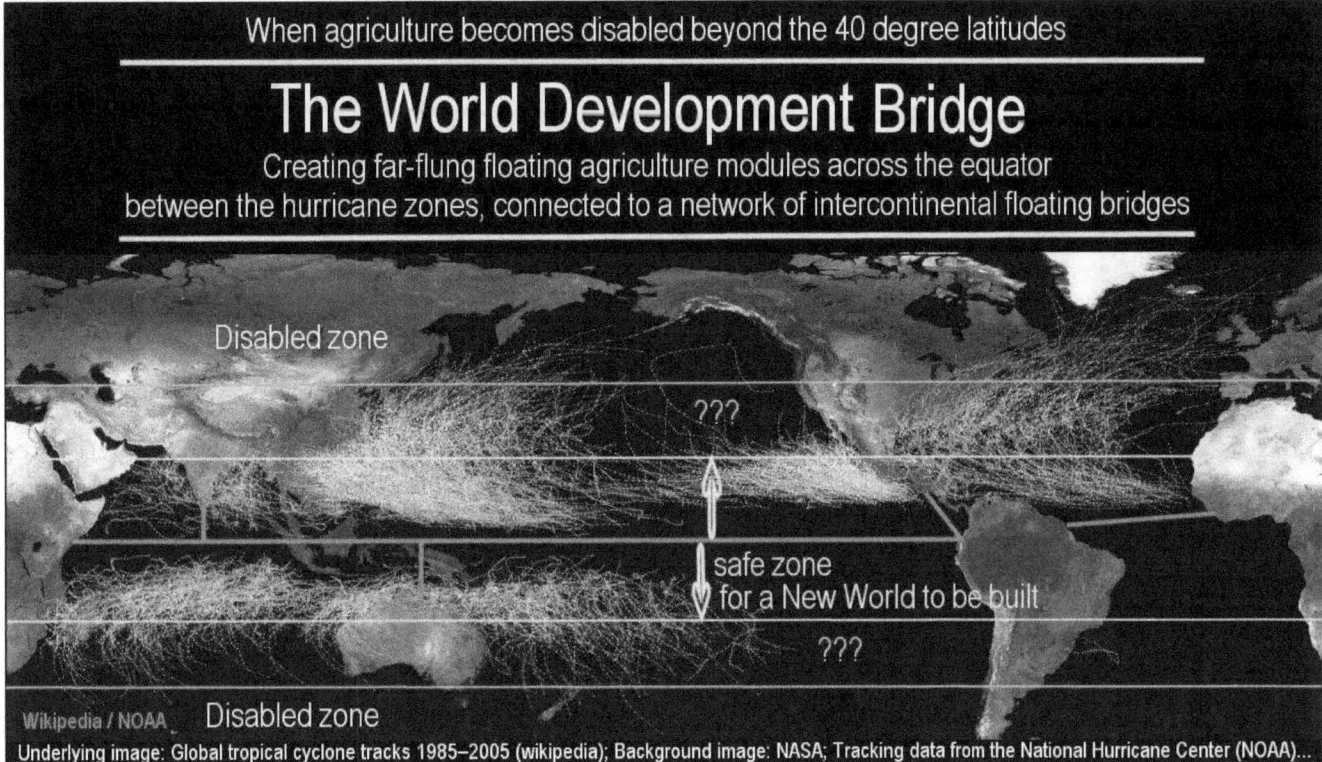

At the present time, no Plan-B exists to build us a New Word for our living under Ice Age conditions.

Part 3 The 50-years boundary zone

Part 3 - -Climate-Change History in 'Micro-Time' - The 50-years boundary zone

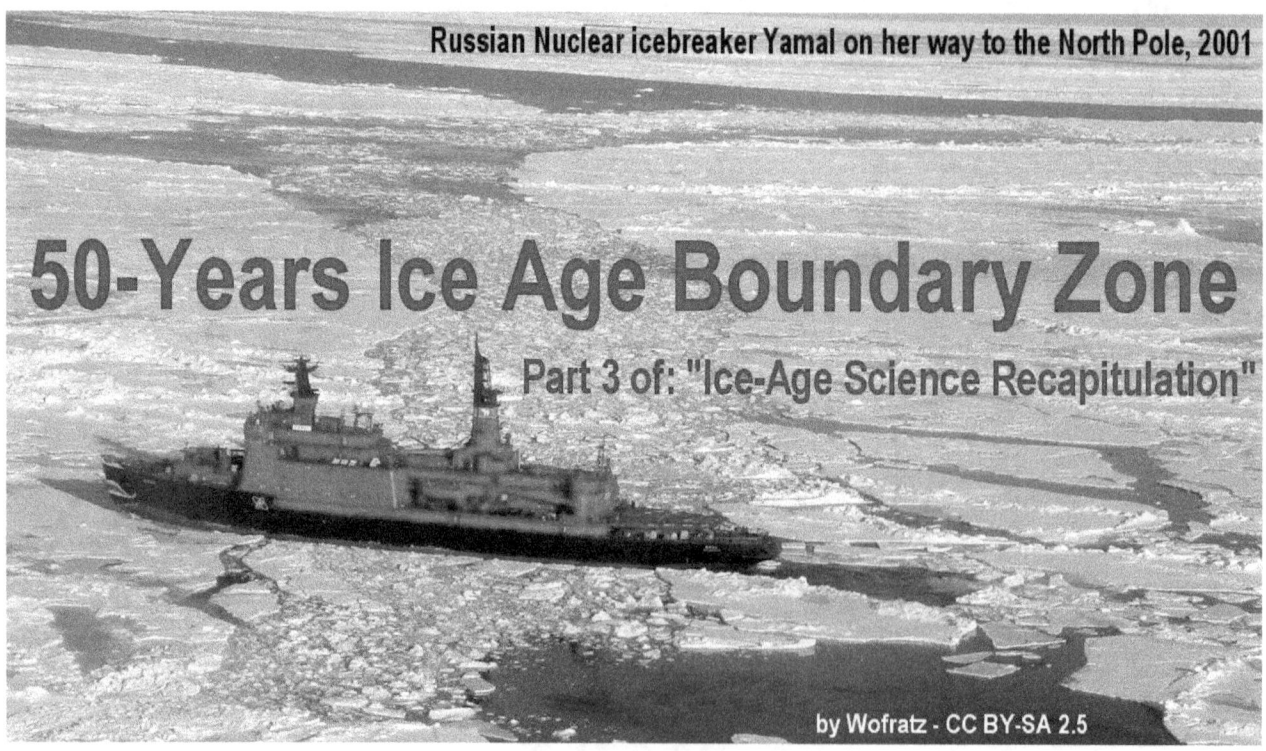

The boundary time-zone to the next Ice Age

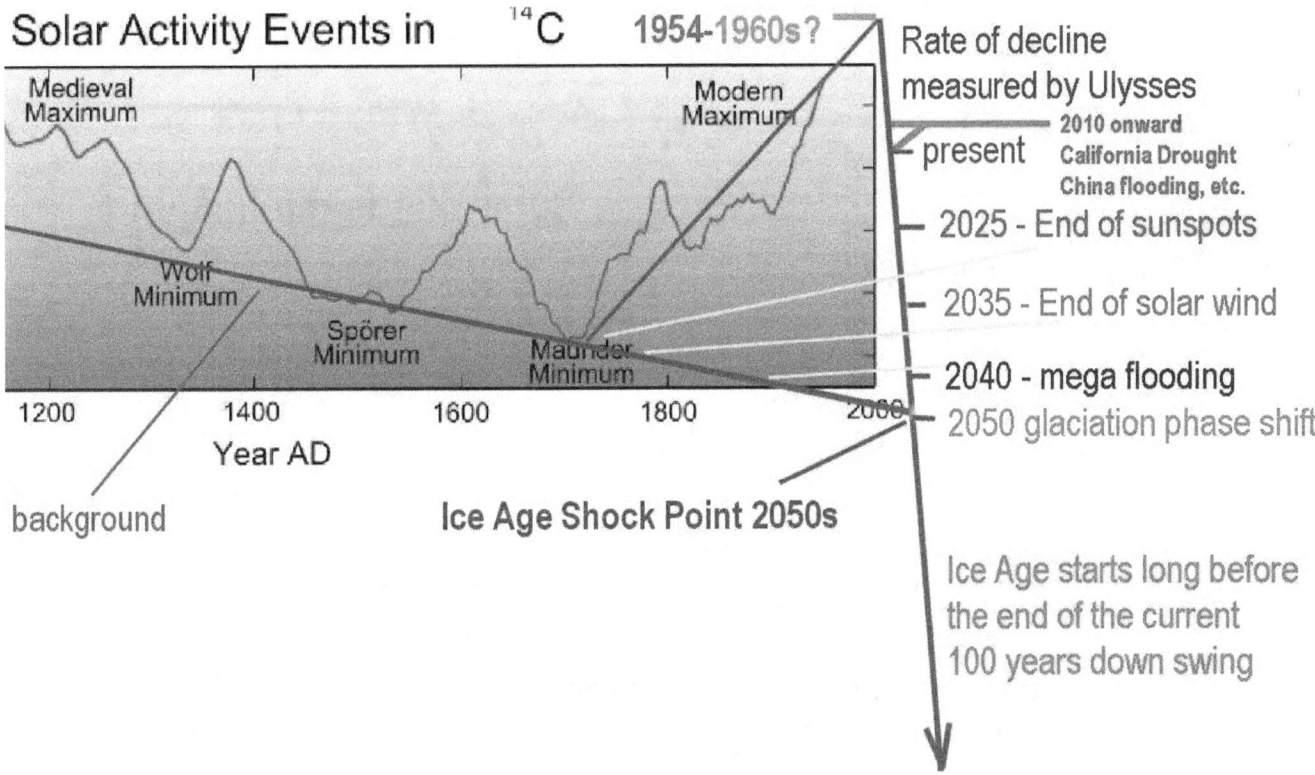

We are now in the boundary time-zone to the next Ice Age, and have been so for 20 years already. In this zone ever-greater climate events are the norm. Here traditional yard sticks no longer apply to predict the future, because we live in a different landscape now that is dominated by the Sun getting rapidly weaker.

The solar wind is diminishing

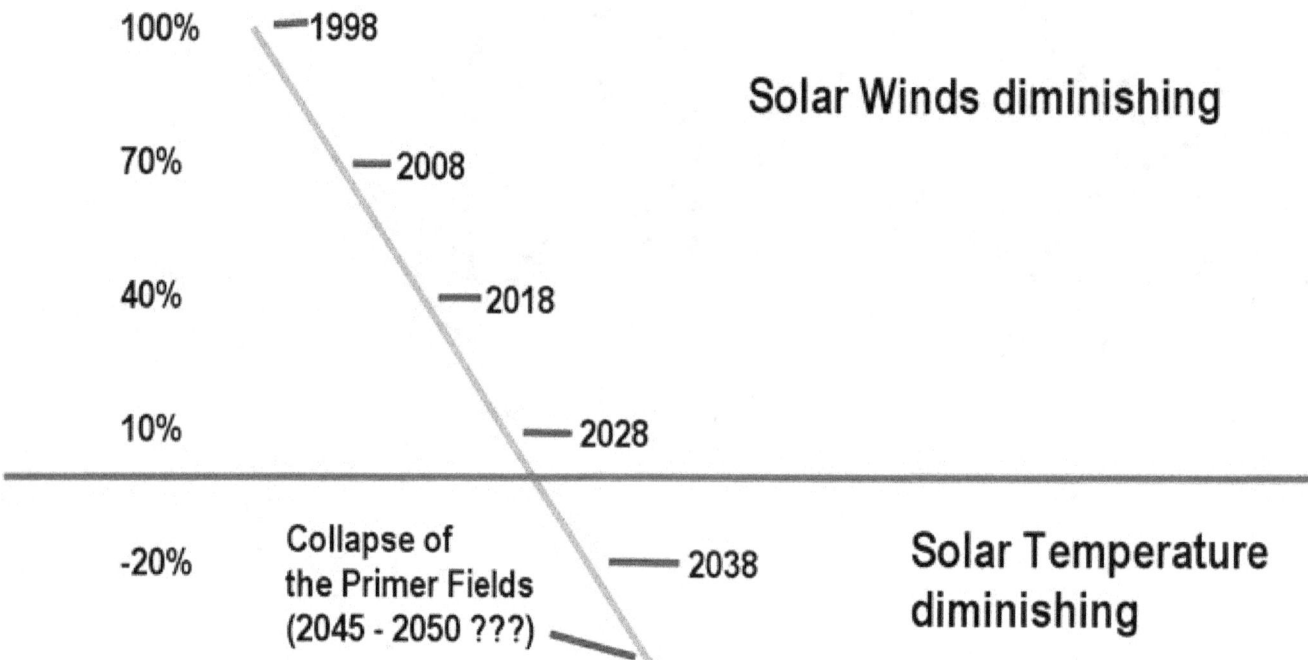

The solar wind that is an indicator of the strength of the Sun, is diminishing at a rate of 30% per decade, and will reach zero pressure in the 2030s. That's the end of the first phase of the collapse process. After that it gets worse.

Steam being boiled off from a heated kettle

The solar wind is comparable to steam being boiled off from a heated kettle. Water cannot be heated to more than 100 degrees at sea-level pressure. When more heat is applied, the excess heat converts water into steam. When the heat-input diminishes, the steam eventually stops flowing. After that, the water begins to cool. We know from this, that for as long as the kettle emits steam, the water in the kettle is 100 degrees hot, because it cannot get any hotter unless the pressure in the kettle is increased.

The solar dynamics are similar

http://www.zam.fme.vutbr.cz/~druck/Eclipse/ - an example of the amazing solar eclipse photography of Milloslav Druckmueller

The solar dynamics are similar. When solar wind is emitted by the Sun, there is more than enough plasma pressure surrounding it, to keep its plasma fusion process fully in operation. This means that for as long as solar wind is emitted, there is enough density in the system for the Sun's surface temperature to being maintained near the current temperature of 5,800 degrees. This happens all the way through the first phase of weakening of the solar system. The only effect on the Earth, that we see of the progressive weakening, is that larger volumes of solar cosmic-ray flux impact our atmosphere, which affects the cloud forming process..

When the solar system weakens

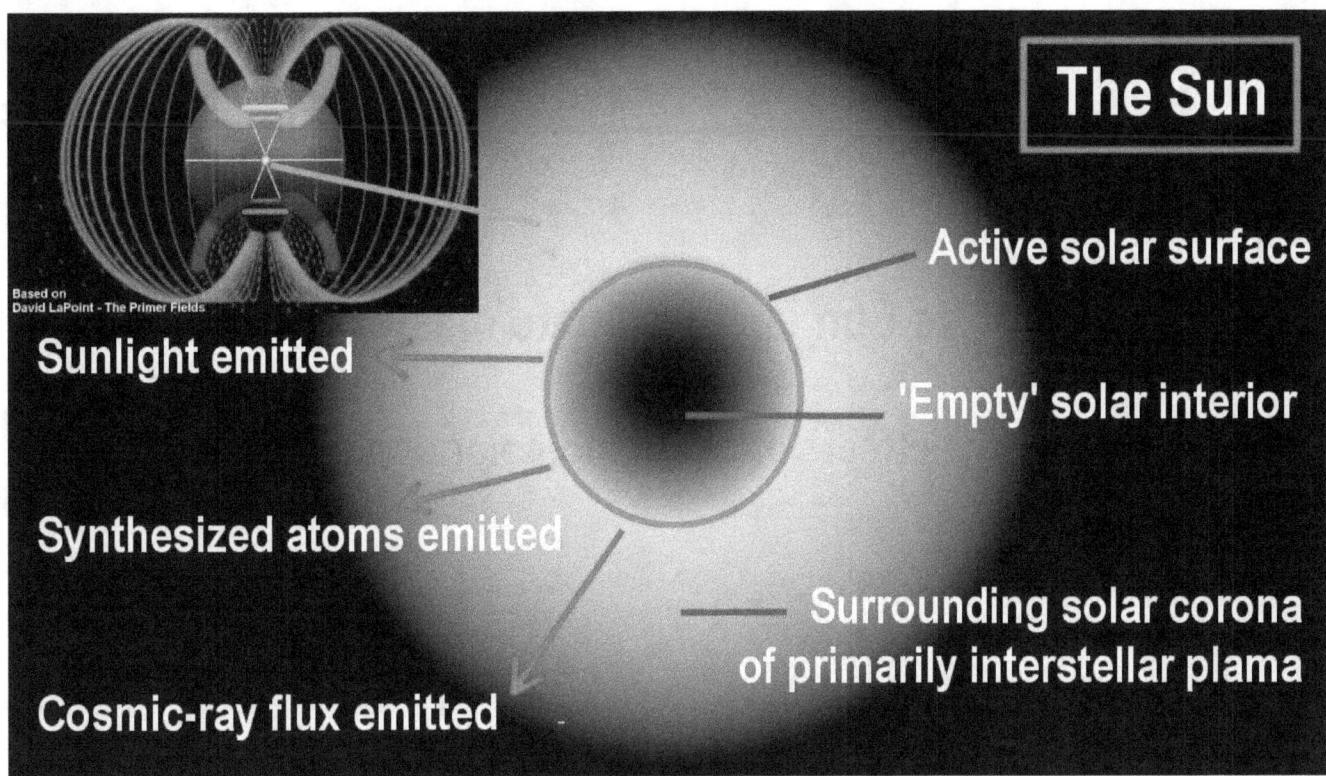

When the solar system weakens, the plasma-sphere around the Sun weakens likewise. This means that a larger volume of solar cosmic-ray flux can penetrate this barrier. The changing volume results in changing climate conditions on Earth.

The Sun is our climate master

The Sun is our climate master
largely by the effects of changing solar cosmic-ray flux

The Sun is our climate master - largely by the effects of changing solar cosmic-ray flux.

The cosmic-rays affect the Earth's climate enormously

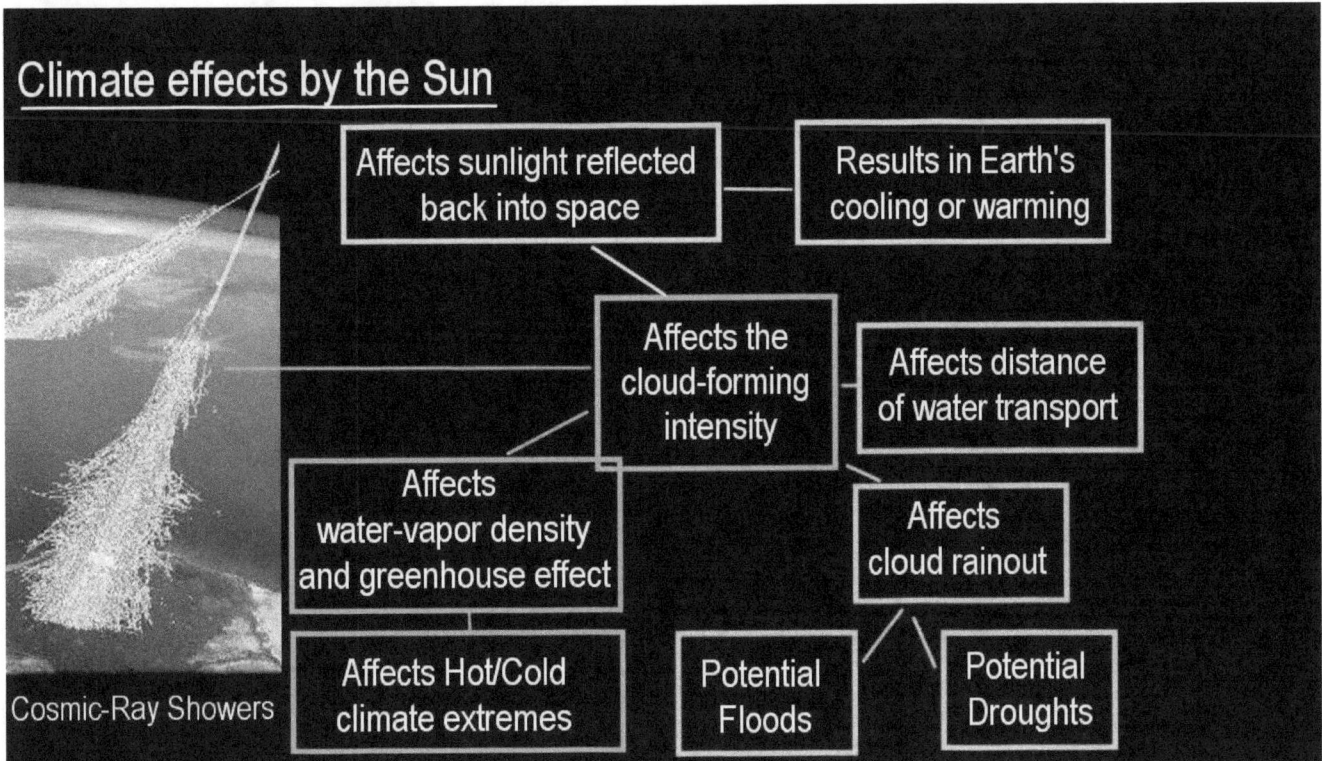

The cosmic-rays affect the Earth's climate enormously. Cosmic rays ionize the atmosphere, which boosts the cloud forming process and the condensation into rain. The effects of cosmic-ray increases include flooding, droughts, colder climates, and reduced greenhouse moderation. All of these effects increase throughout the first phase to 2030, while the Sun's surface temperature will remain the same throughout this period.

Cosmic-ray flux flows from the Sun

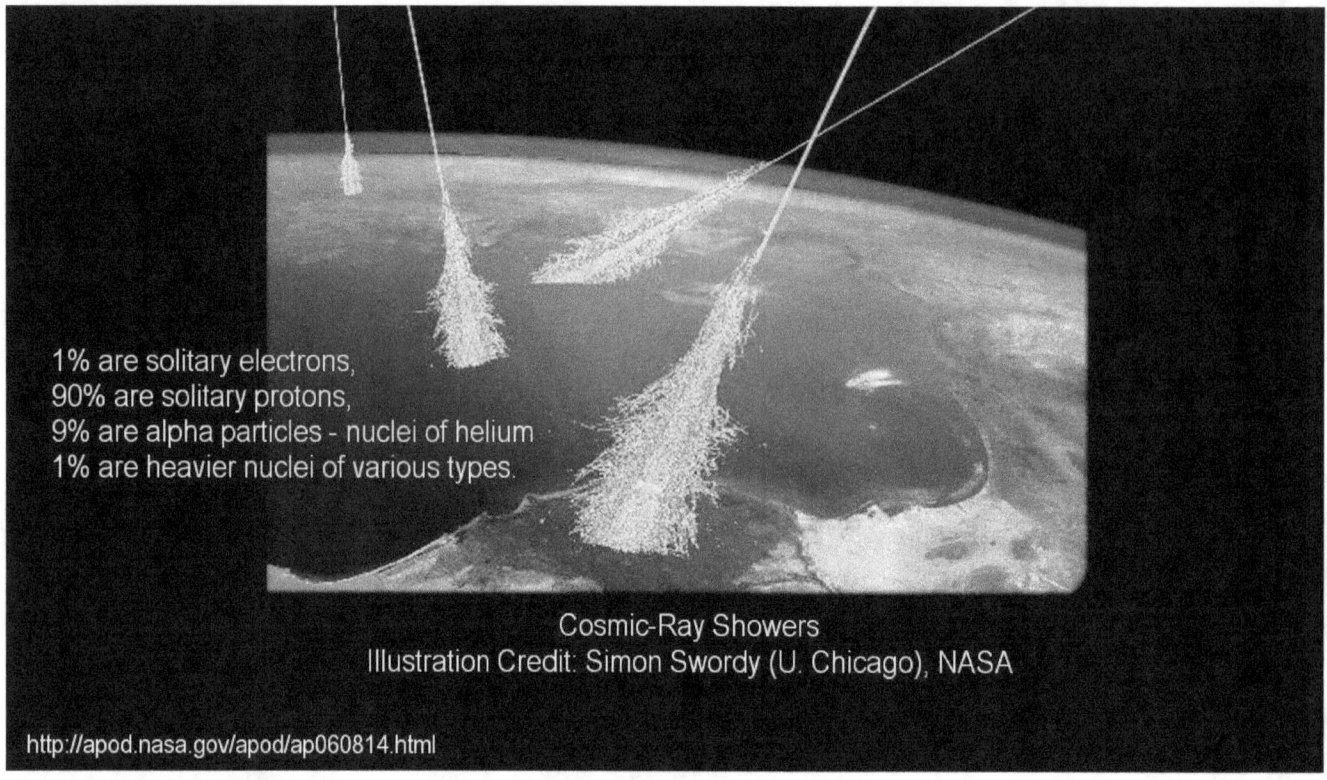

It needs to be noted here that most of the cosmic-ray flux that impacts the Earth flows from the Sun. This fact is rarely recognized. In comparison, the galactic portion is relatively small, though it is deemed to be only source of cosmic-ray flux.

The seeming paradox has been somewhat resolved, experimentally, with measured evidence.

Cosmophysical Factors

Cosmophysical Factors

Cosmophysical Factors

The researcher Simon Shnoll

The researcher Simon Shnoll had conducted a series of identical chemical reaction experiments, continuously repeated for several days.

Each time the results were different

He noted that each time the results were different.

In one case the reactivity spiked tremendously

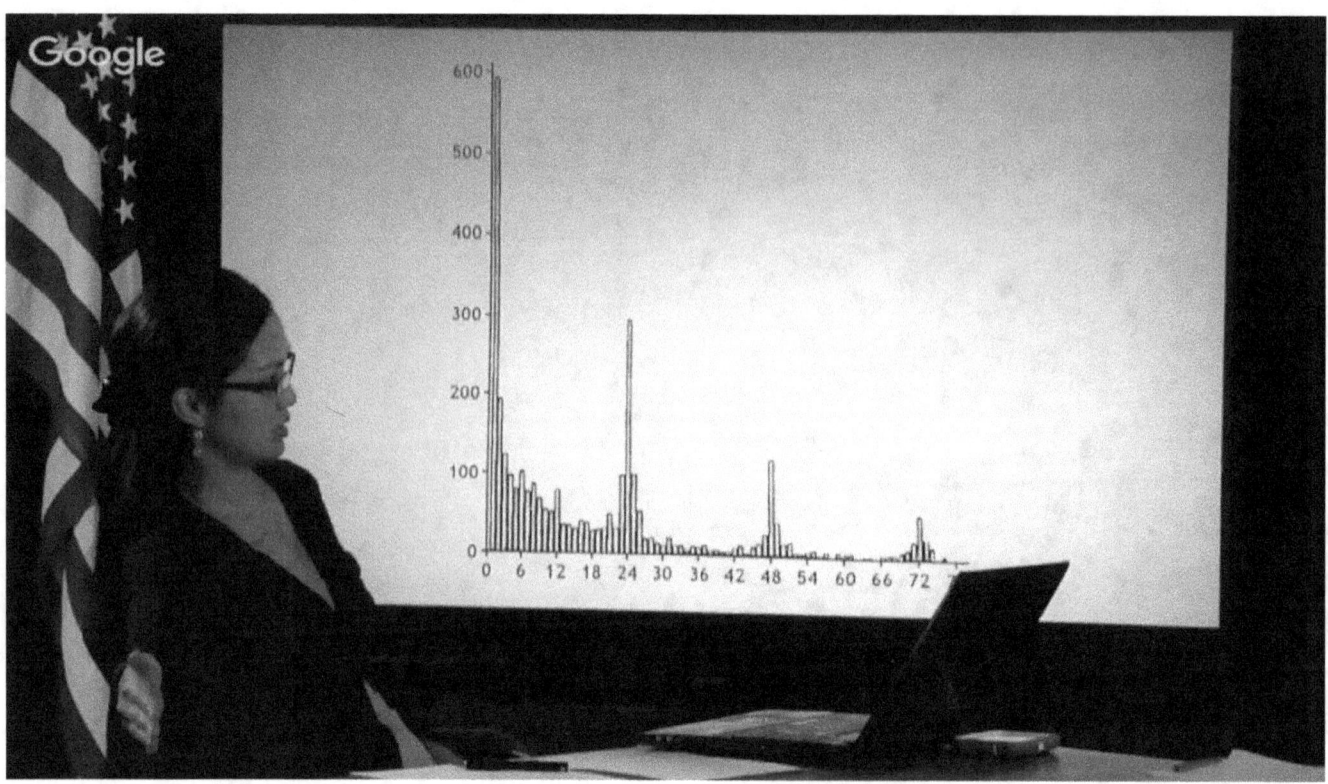

In one case the reactivity spiked tremendously, repeatedly in 24 hour intervals according to the rotation of the Earth, with the repeated-spikes becoming smaller according the rotation of the Sun, as if a coronal hole had emitted strong cosmic-ray flux, which subsequently turned away from its orientation towards the Earth by the rotation of the Sun.

The experiment proves that our Sun is capable of extremely large cosmic-ray-type emissions.

It is generally assumed that all cosmic-ray flux is galactic

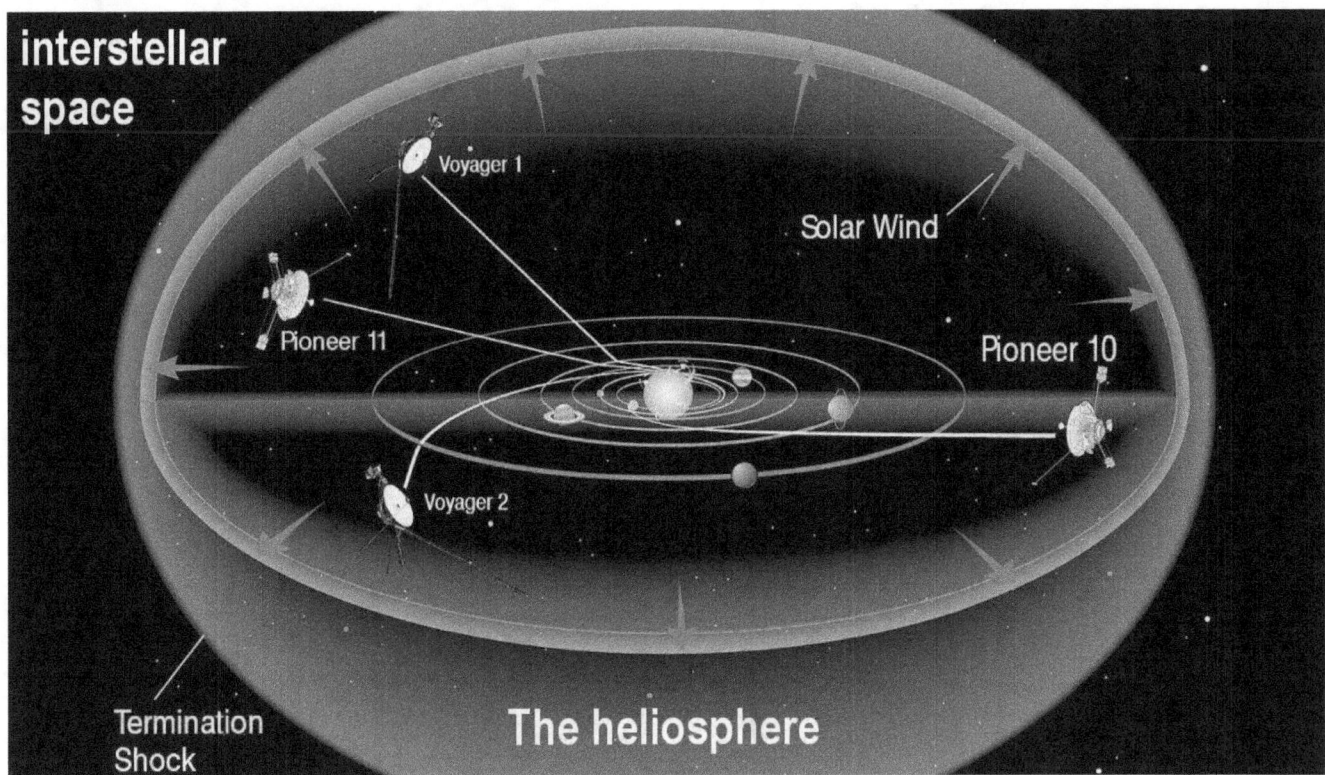

It is generally assumed that all cosmic-ray flux is galactic in origin, so that the fluctuations that we have measured on Earth are assumed to result from the Heliosphere that surrounds the solar system, to be fluctuating in density. This is deemed to alter the volume of the cosmic-ray flux that penetrates it.

Simon Shnoll's experiment disproves this assumption

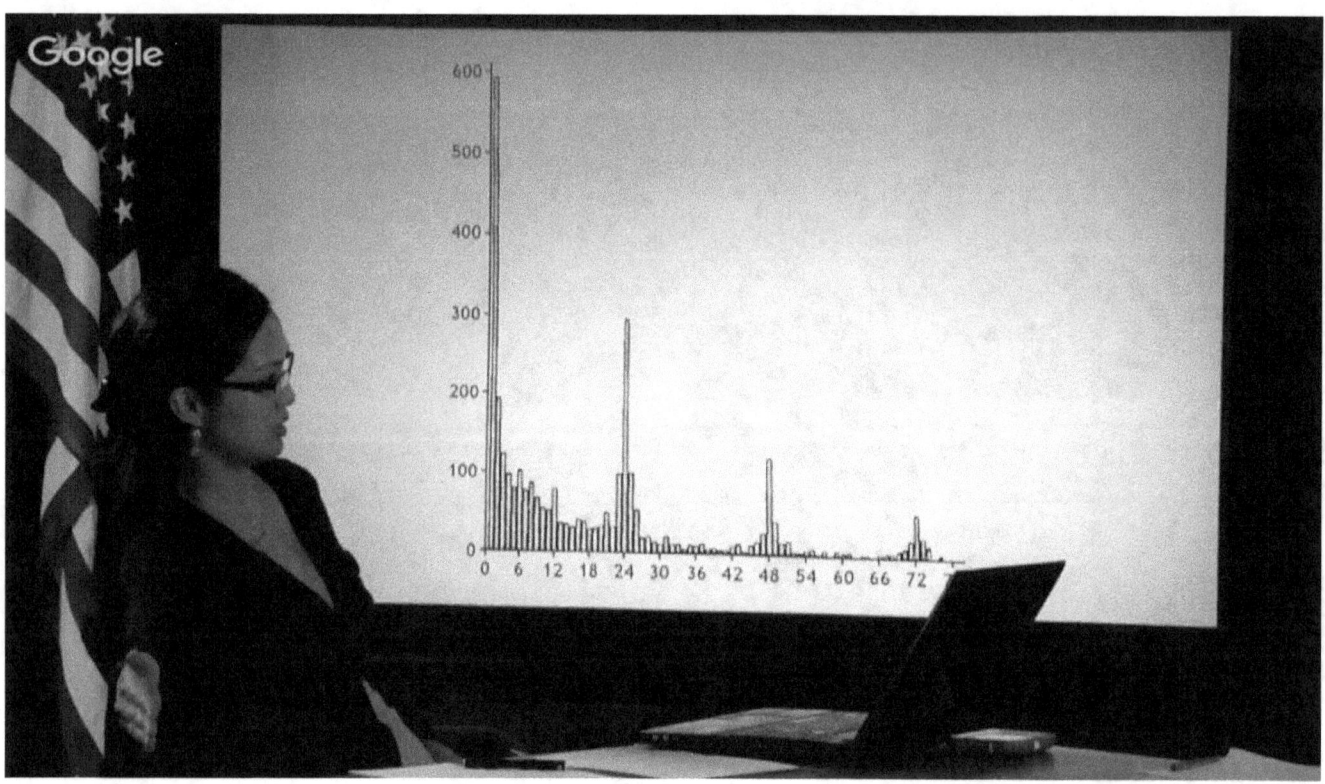

Simon Shnoll's experiment disproves this assumption. The measured fluctuations in his experiment vastly supersede the heliosphere's measured attenuation.

Voyager-1 spacecraft penetrated the heliosphere

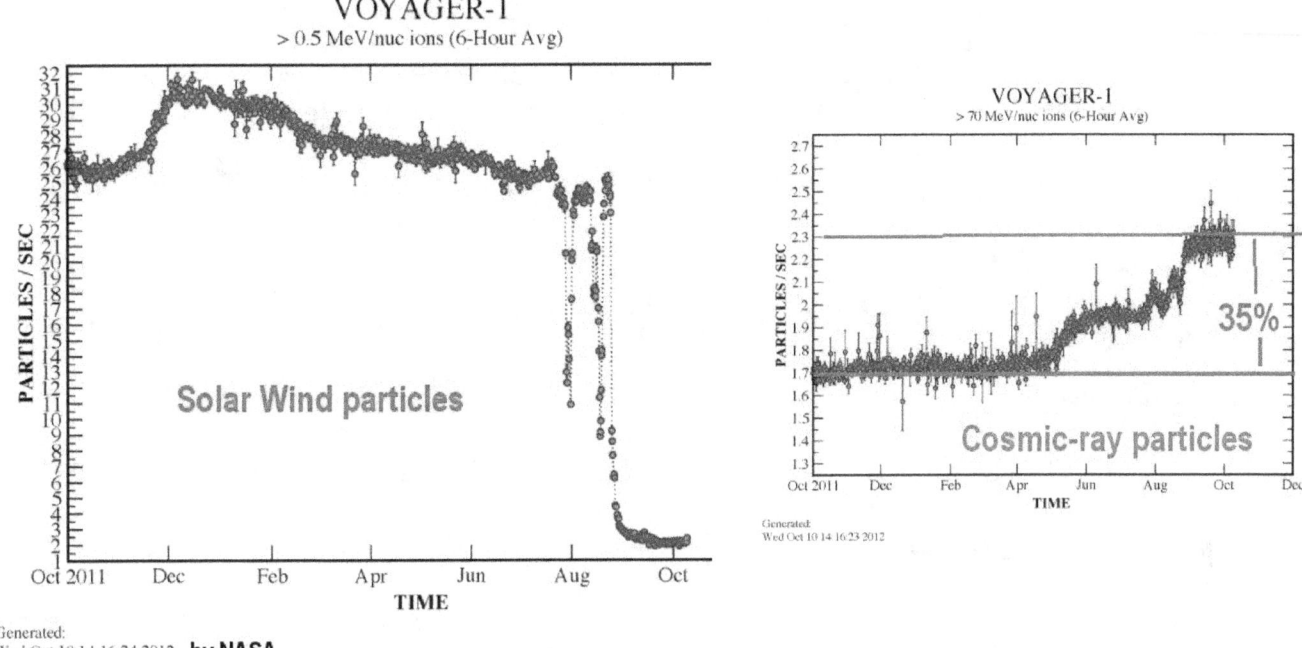

When NASA's Voyager-1 spacecraft penetrated the heliosphere in 2012, it measured a mere 35% increase in the number of cosmic-ray particles encountered. That's the maximum increase that we would encounter if the heliosphere would vanish completely, should this be possible.

In comparison with the attenuation that the heliosphere affords

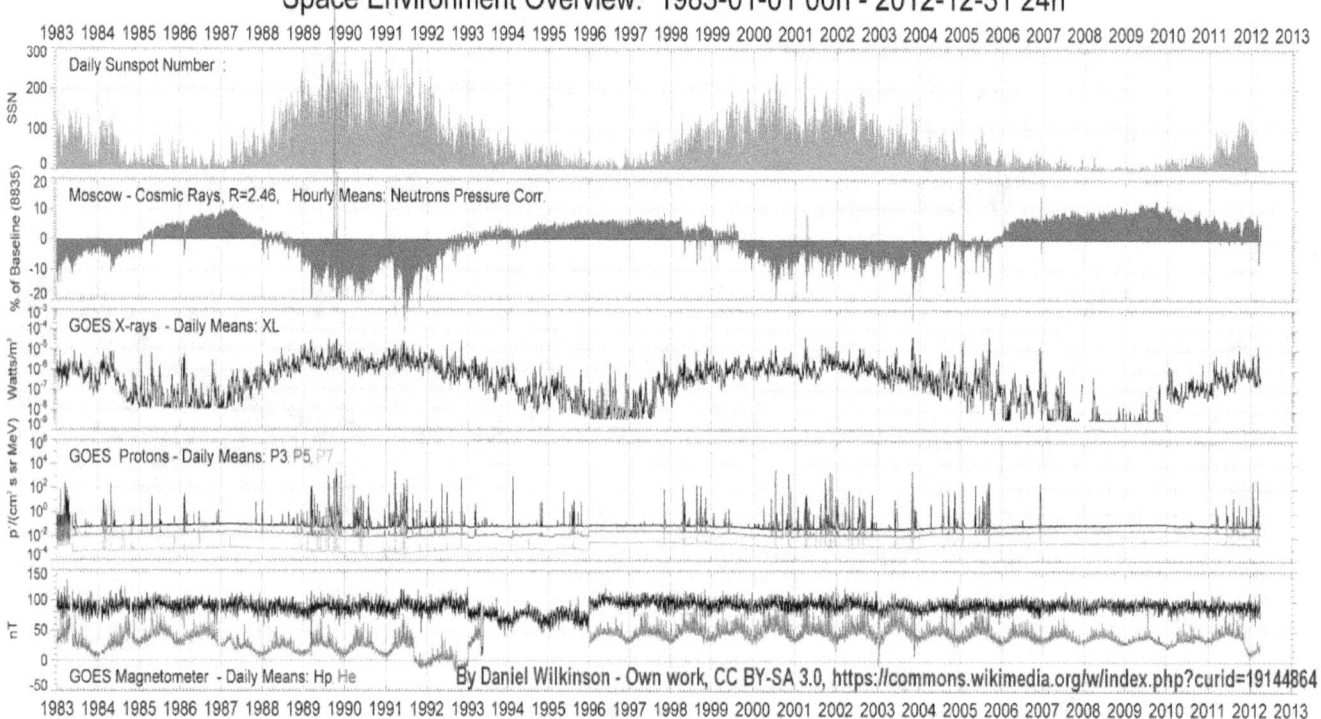

In comparison with the maximum attenuation that the heliosphere affords, large fluctuations of cosmic-ray flux occur at every solar cycle. Fluctuations in the range of up to 30% have been measured.

If we can measure such large fluctuations between the solar cycles, with the heliosphere being unaffected by the soar cycles, we have positive proof on hand that the cosmic-ray flux that affects our climate as dramatically as it does, does originate from the Sun and from nowhere else.

The CERN lab's CLOUD experiment

The CERN lab's CLOUD experiment

measured proof in principle

The CERN lab's CLOUD experiment - measured proof in principle

Demonstrated in laboratory experiments

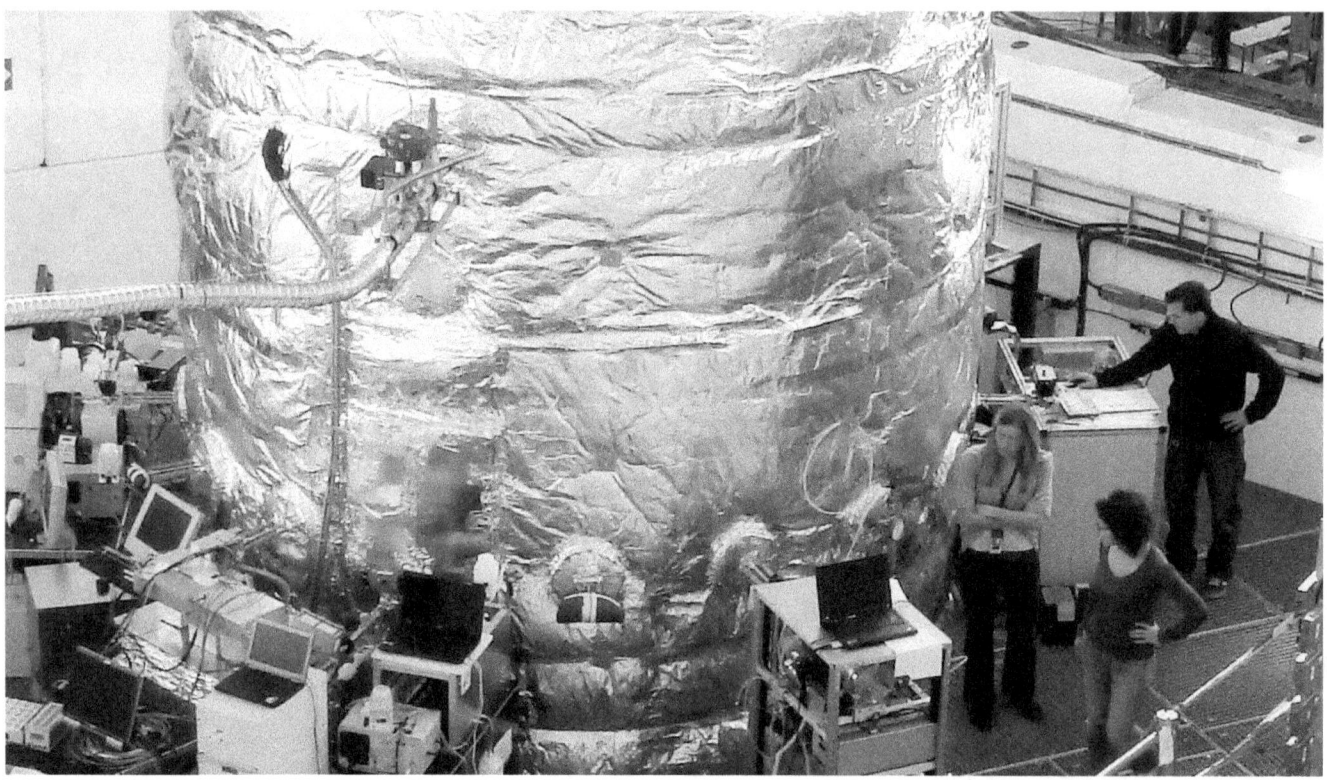

It has been demonstrated in laboratory experiments, with the CLOUD experiment at the CERN lab in Europe, cosmic-ray interaction dramatically increases cloud nucleation.

When artificial cosmic-rays were injected

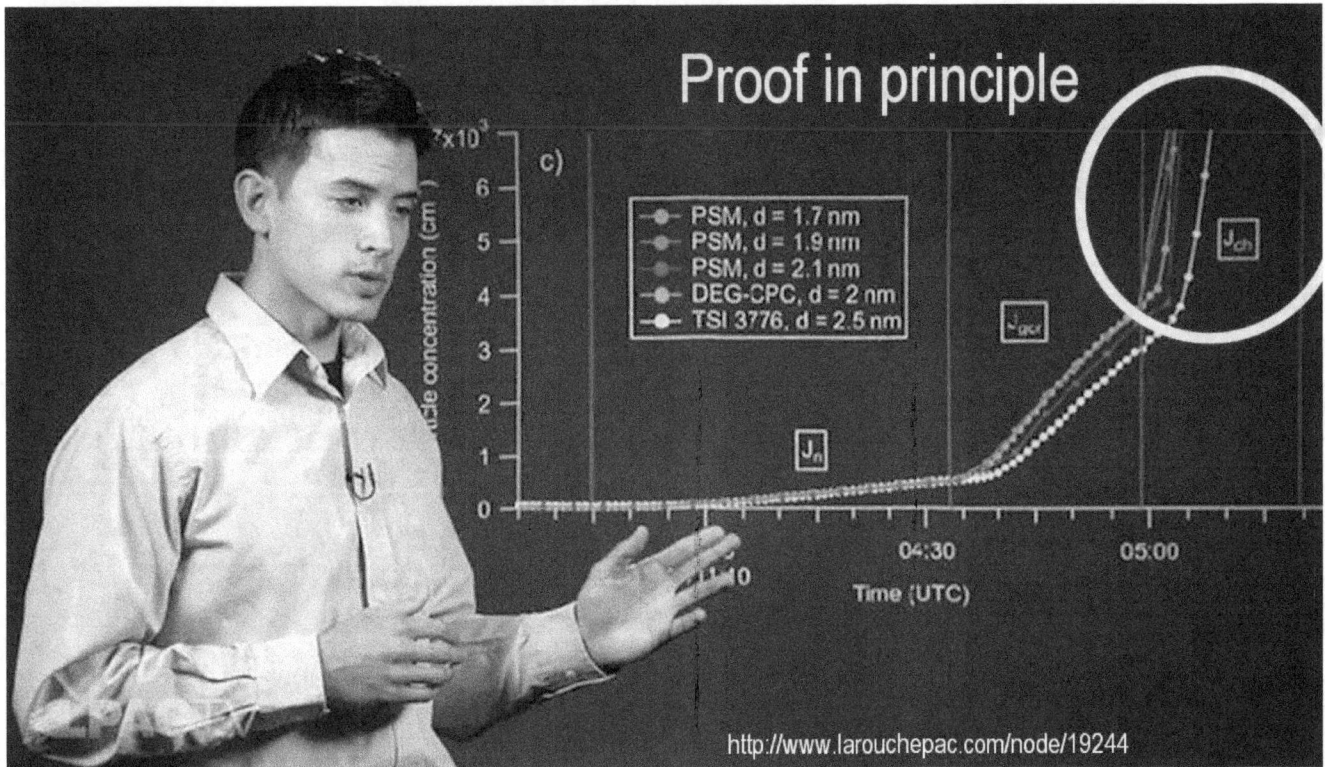

When artificial cosmic-rays were injected into a test chamber at the final step of the experiment, the measured water nucleation went straight up and off the chart.

Thus the experiment proves in principle that the diminishing solar activity, in the boundary zone, is the climate master on Earth via its cosmic-ray effects on cloudiness.

Cloudiness affects everything

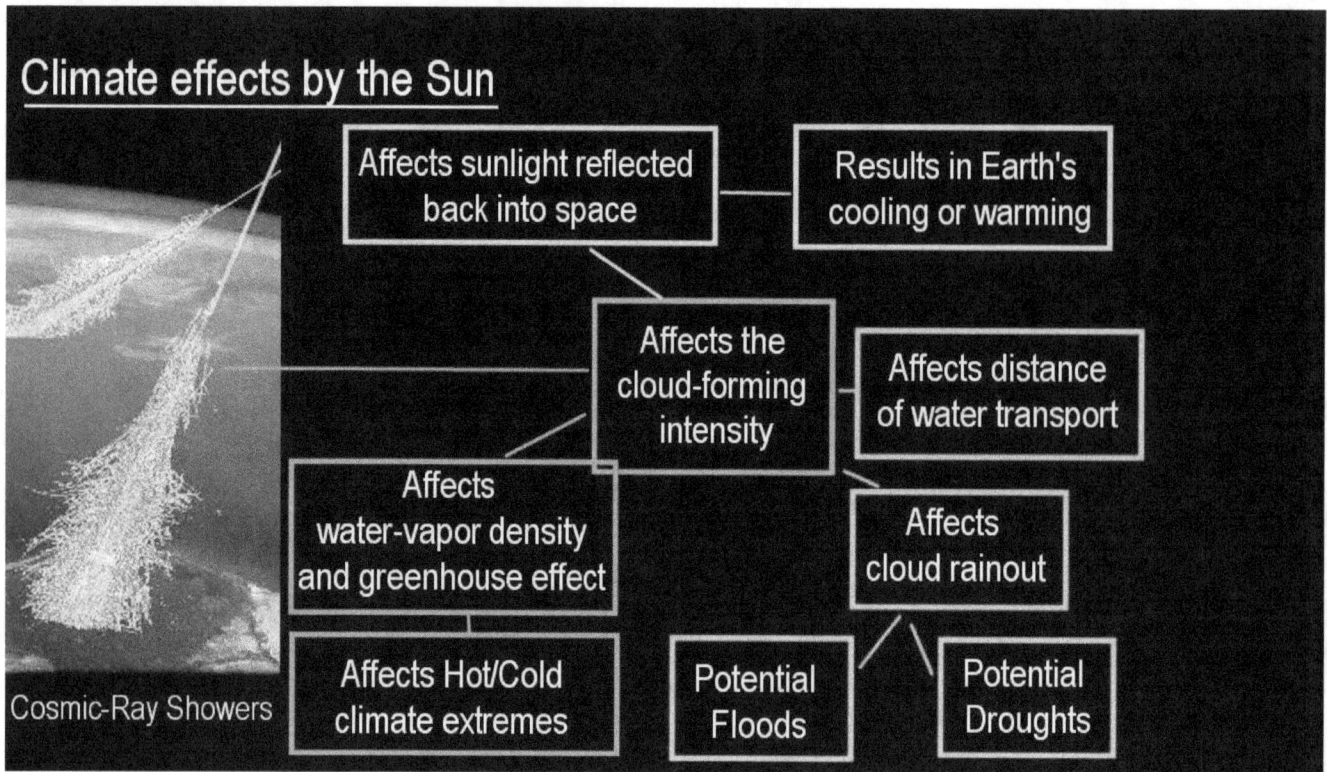

Cloudiness affects everything drought, flooding, global cooling, even temperature extremes. We will see evermore of these effect, resulting from diminishing solar activity and increased cosmic-ray flux, until the soar wind stops and the Sun begins to cool for an additionally challenging effect.

Phase-2 of the solar collapse

Phase-2 of the solar collapse

the diminishing Sun begins to cool down

Phase-2 of the solar collapse - the diminishing Sun begins to cool down

After the solar wind stops

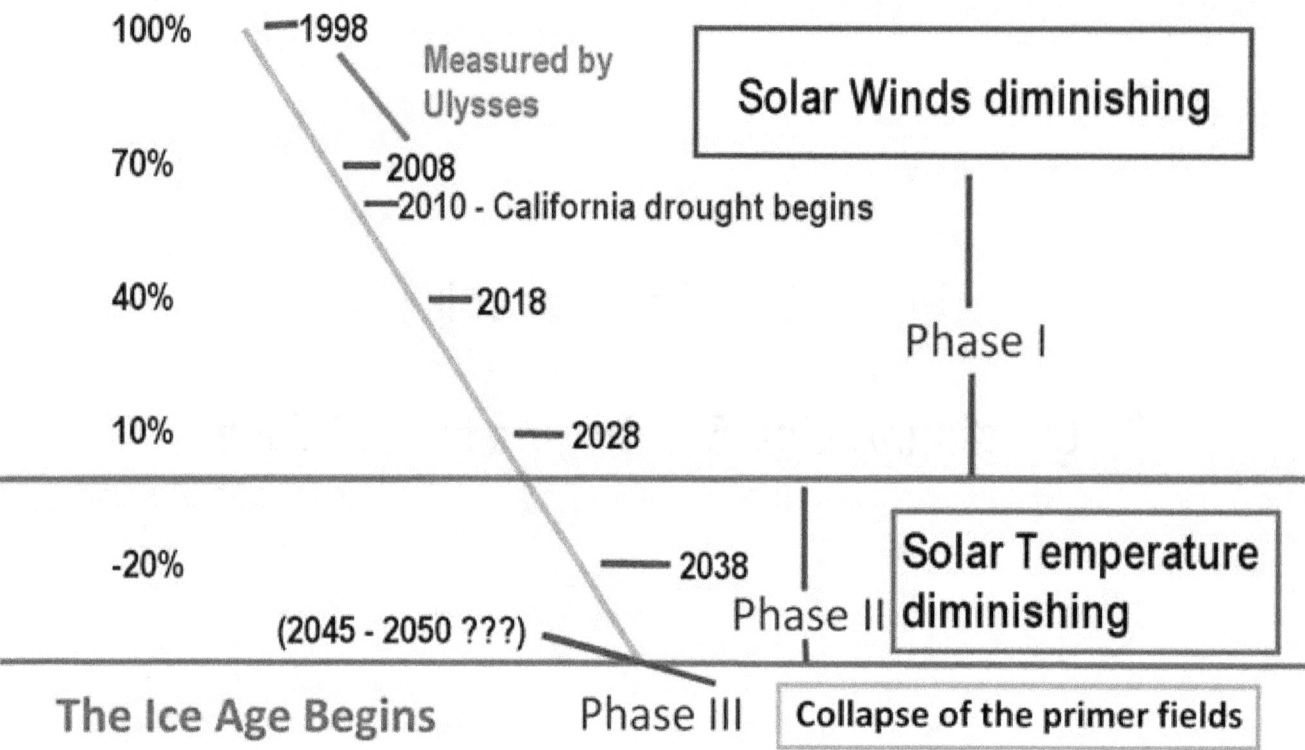

After the solar wind stops, the diminishing Sun begins to cool down to lower surface temperatures. That's the start of Phase-2. Phase-2 ends when the primer fields collapse.

So does the Sun

When the heat is turned down low, after the kettle stops boiling, the water cools down. So does the Sun.

All the way through Phase-2, the Sun will get colder

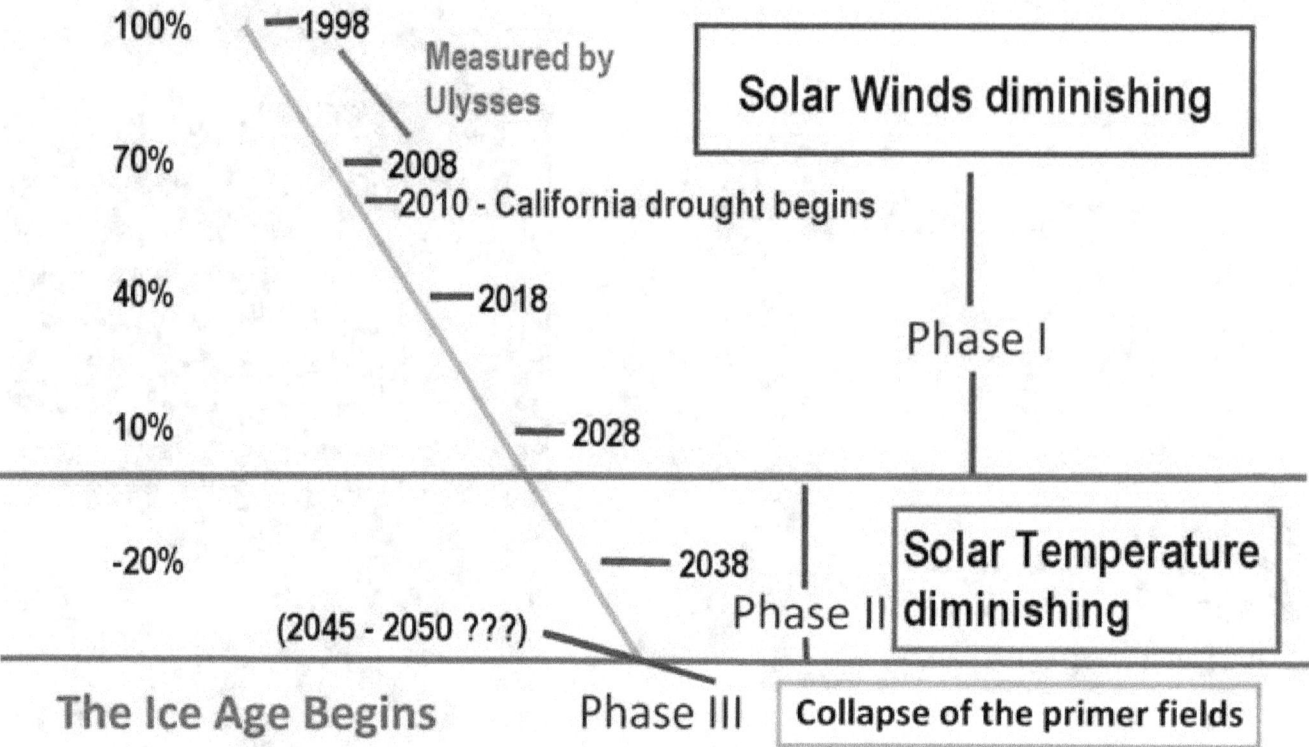

All the way through Phase-2, the Sun will get colder, with the corresponding effects on Earth. The effects can be enormous, such as that the snow no longer melts in the climate volatile regions.

Whatever natural agriculture might still be operating at the start of Phase-2, will likely cease during the course of Phase-2 unfolding, which may be short in duration.

Phase-3 of the solar collapse

Phase-3 of the solar collapse

The Sun's hibernation begins, measured in Be10 ratios.

Phase-3 of the solar collapse - the Sun's hibernation begins, measured in Be10 ratios.

Phase-3 begins when the primer fields can no longer form

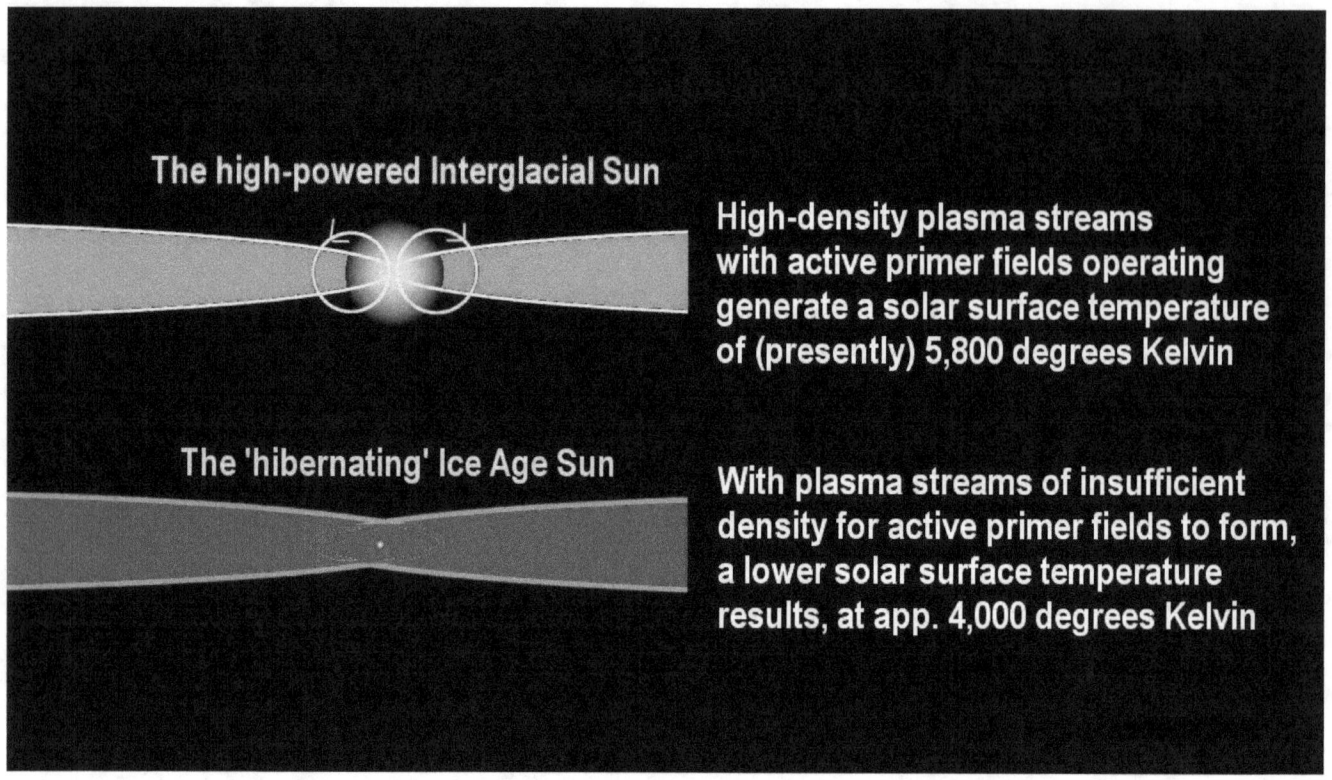

Phase-3 begins when the solar system has diminished to the point that the primer fields can no longer form, and cease to focus plasma unto the Sun.

At this point the Sun begins to hibernate and the next Ice Age erupts on Earth that transforms the world into an ice planet that has never been experienced before in known history, and may not actually be imaginable.

Measurements of Phase-2 to Phase-3 transition

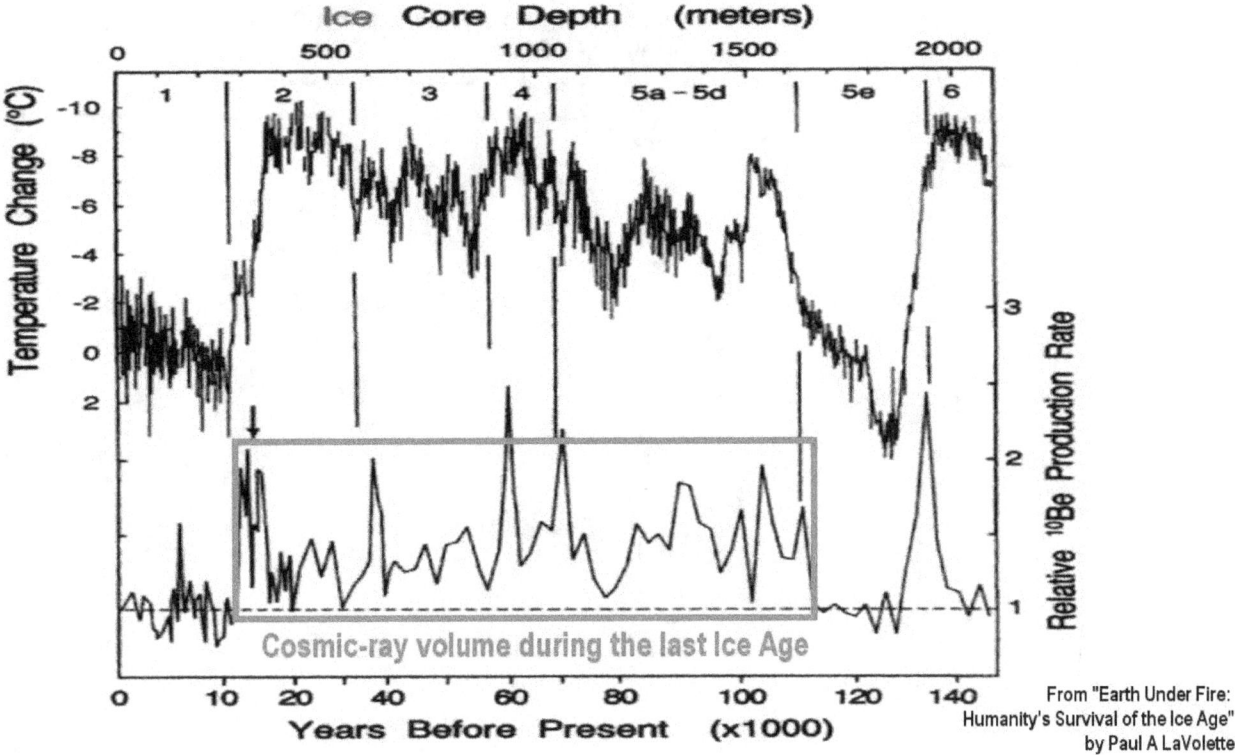

And yes, we do have measurements available of the Phase-2 to Phase-3 transition that puts the Sun into hibernation. The measurements exist in the form of extremely high beryllium ratios that extend all the way through Phase-3, the 90,000-year glaciation phase, the ice planet phase. Only the hibernating Sun that is no longer surrounded by a dense mantle of plasma, can produce the high-volume of solar cosmic-ray flux that generate the high ratios of beryllium-10 that have been measured.

Phase-3 is the phase of the hibernating Sun

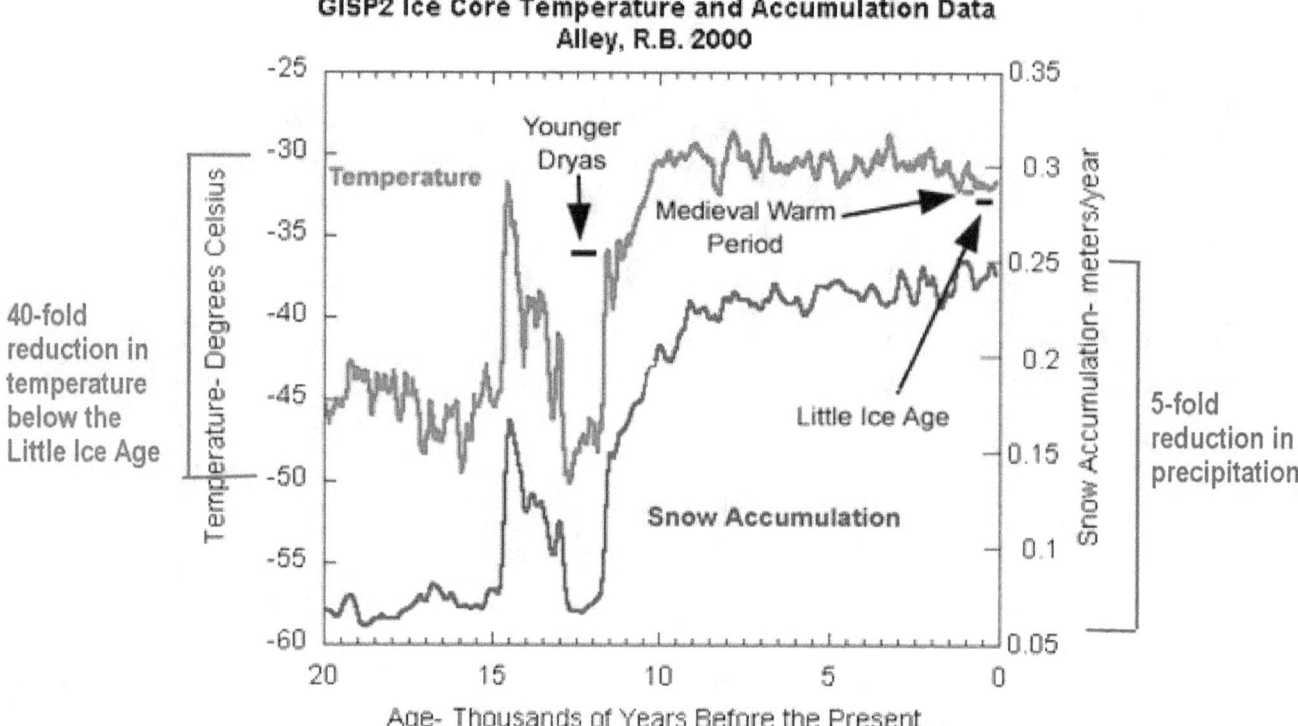

Phase-3 is the phase of the hibernating Sun, a 70% weaker Sun, and a world with 80% less precipitation. While we can prepare us for this type of landscape, by building us a New World with technological infrastructures for human living, and for agriculture, of types that can operate efficiently under Ice Age conditions by placing agriculture and cities afloat on the Equatorial Seas, we won't get anything built if we don't start. In this case it becomes irrelevant to know how soon Phase-3 will start, because then, very few people will be still alive when the Ice Age begins.

We cannot respond in a reactive mode any longer

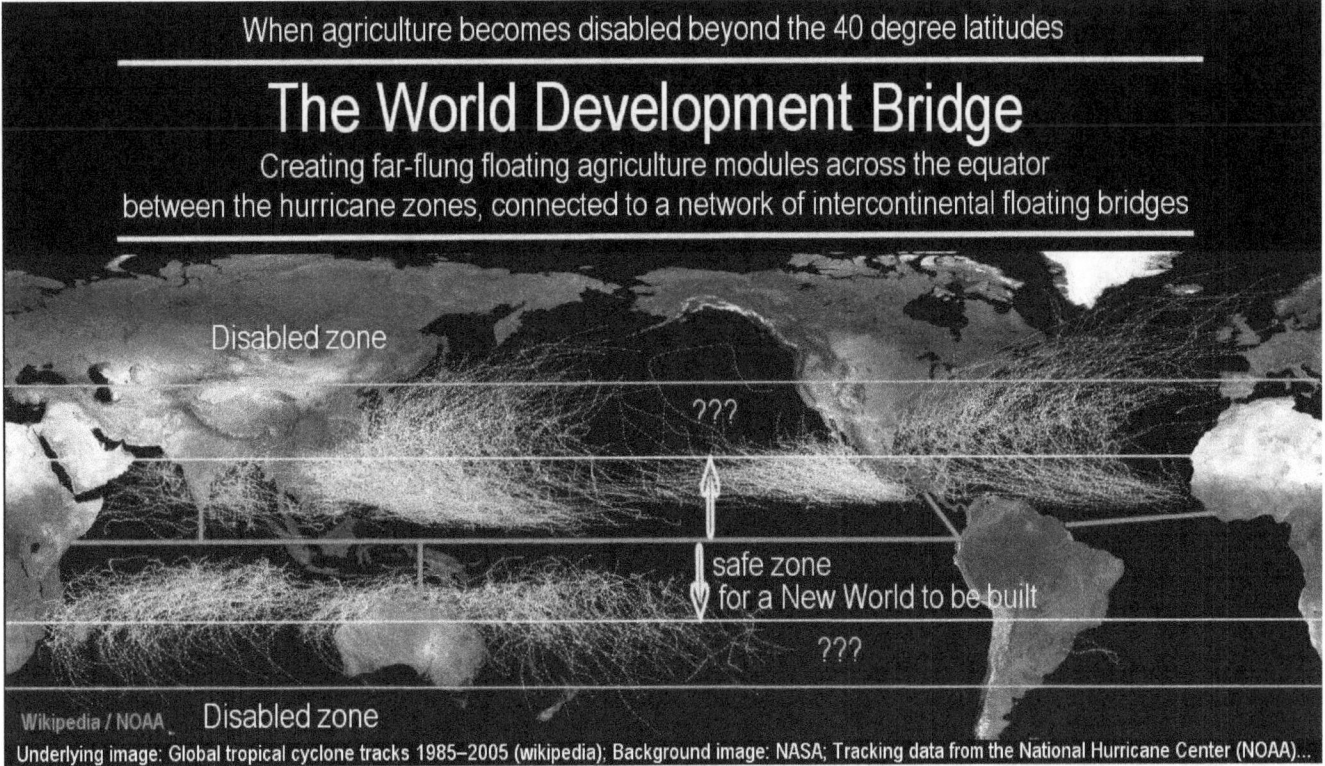

The bottom line is that we need the Plan-B option implemented long before Phase-3 starts. We cannot respond in a reactive mode any longer. because it takes all the time we've got to create the floating agriculture that is able to nourish billions of people, and and build the thousands of new floating cities attached to service the agriculture.

That's our Plan-B, and we need it coming on-line early in the boundary zone, before agriculture is beginning to collapse. If humanity follows the Plan-B path that its science defines as extremely imperative, then it will write itself a ticket to continue to live and prosper - and if not? This option is not worth contemplating. The Plan-A option is to do nothing. Then the people will die.

Phase-1 of the solar collapse is the most critical phase

Phase-1 of the solar collapse

the most critical phase!
Plan-B needs to be implemented in this phase

Phase-1 of the solar collapse is the most critical phase, because Plan-B needs to be implemented in this phase.

Plan-B while we still have a chance

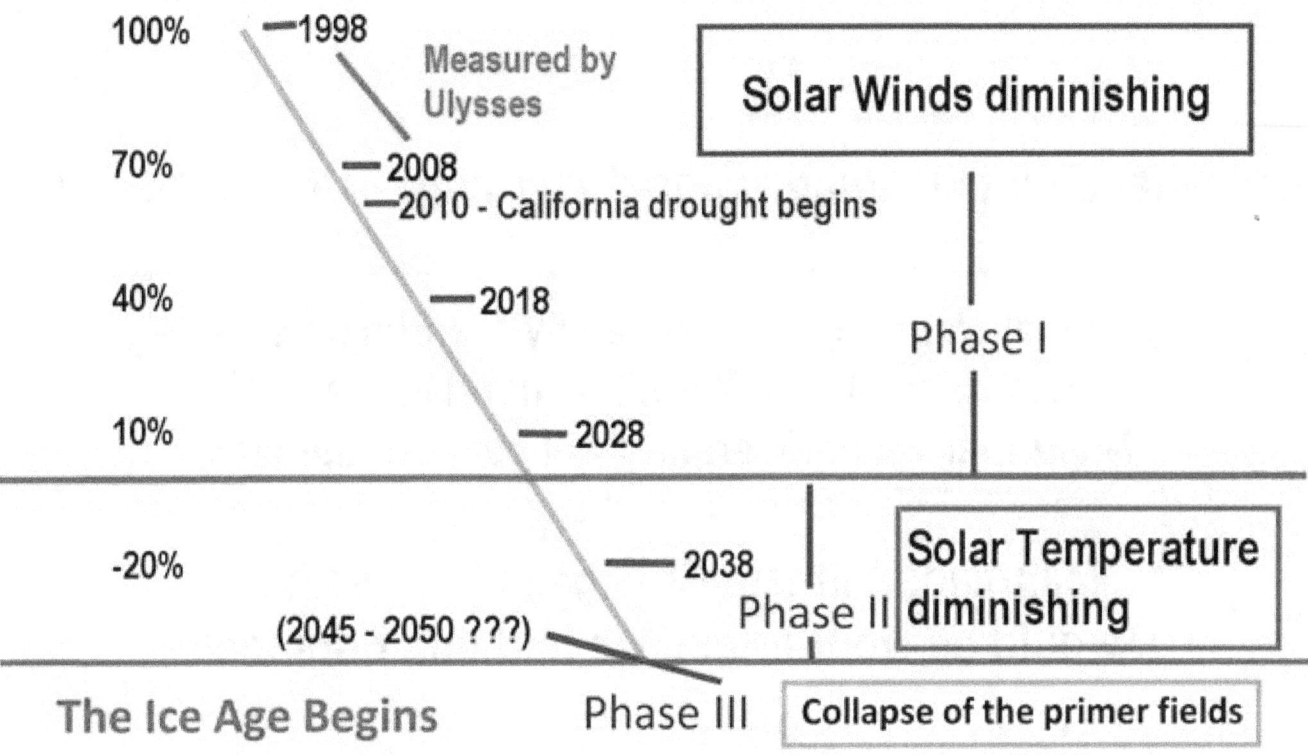

Plan-B needs to be started in the present while we still have a chance to built us out of the already beginning climate-collapse crisis. The time we have remaining is short, and is presently being wasted by society doing nothing and with no intention for stirring its stumps, even while agriculture as we know it is already collapsing to some degree in many parts of the world. Food shortages are on the horizon. The Earth is getting colder year after year. The growing seasons are shrinking.

Ultimately, when agriculture collapses under the climate collapse, entire nations in the volatile regions, such as Canada, Russia, and EU nations, will simply cease to exist, unless the New World is created in which agriculture is technologically protected under the Plan-B created alternative.

But what about Manmade Global Warming - can't it save us?

But what about Manmade Global Warming - can't it save us?

**Every Global Warming was caused by the Sun, exclusively.
The last solar Global Warming ended in the 1990s.
No further Global Warming will come to our aid.**

**Manmade Climate Change is Impossible
It would be wonderful if we had this potential!**

But what about Manmade Global Warming - can't it save us?

Every Global Warming was caused by the Sun, exclusively. The last solar Global Warming ended in the 1990s. No further Global Warming will come to our aid.

Manmade Climate Change is Impossible

It would be wonderful if we had this potential!

The last meaningful historic climate recovery

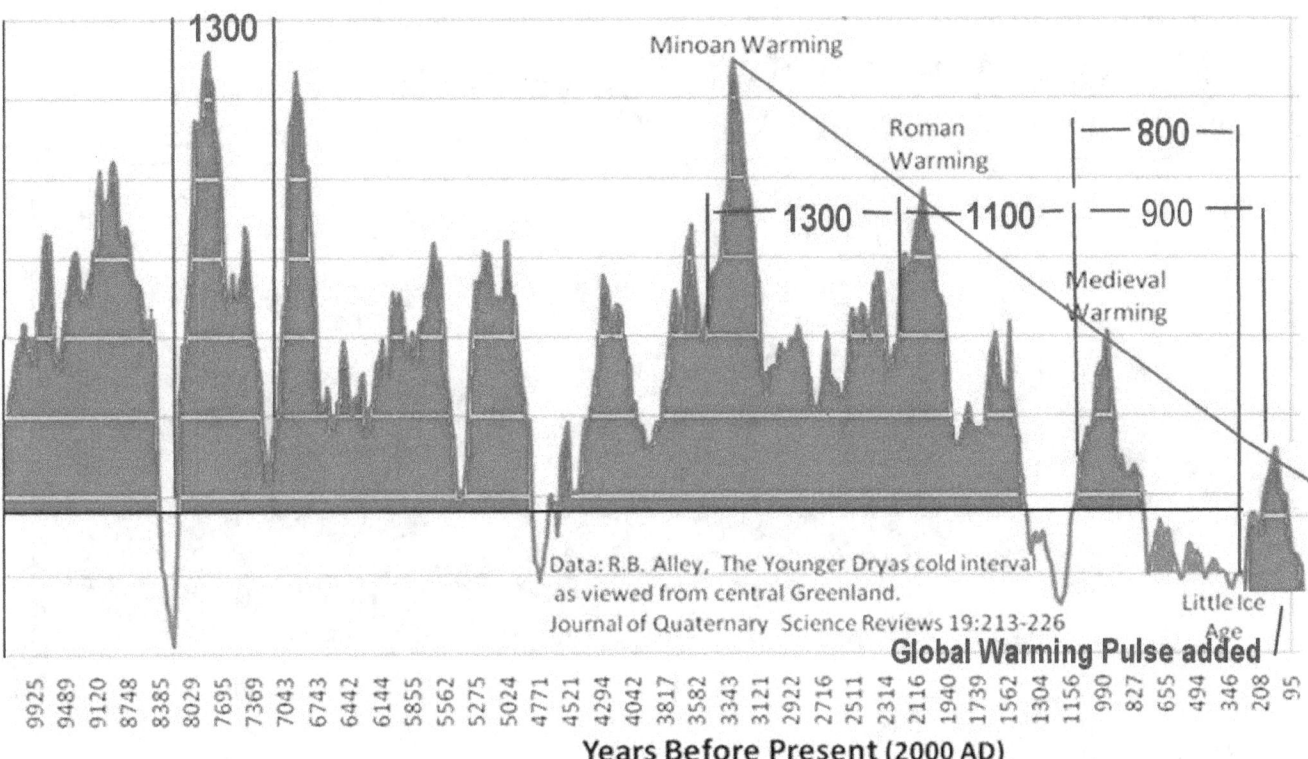

The last meaningful historic climate recovery by real solar Global Warming, had began in the early 1700s. It had peaked in the 1990s, and then began to reverse.

Small as the last warming pulse was

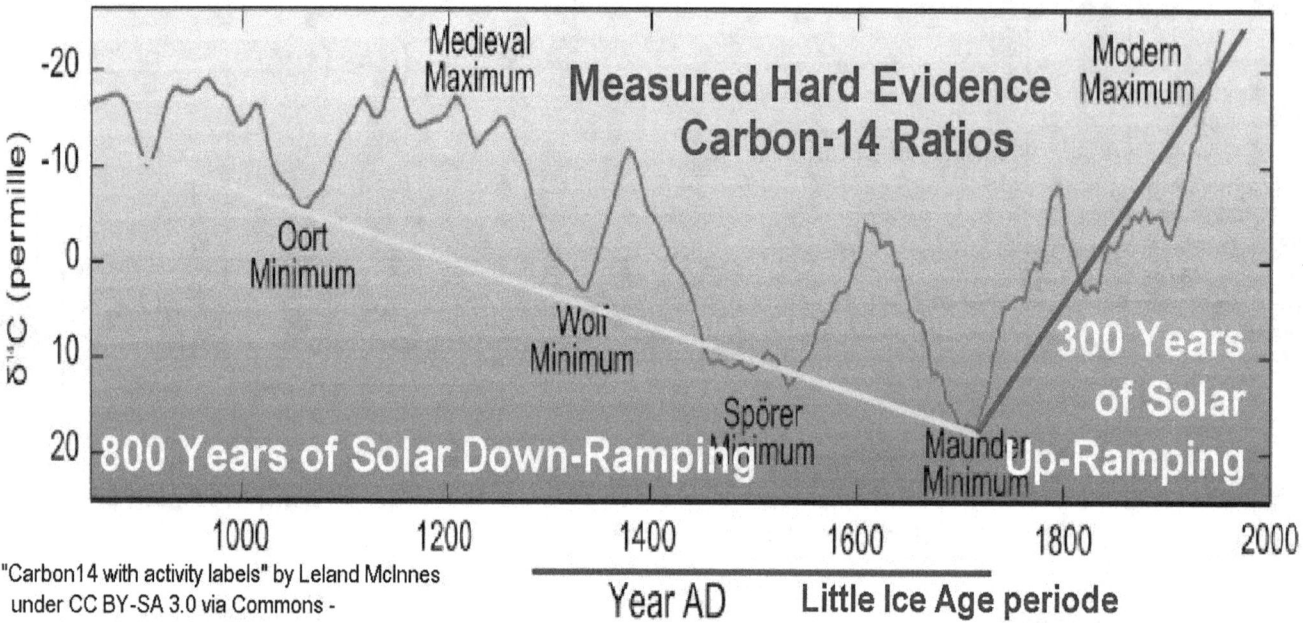

Historic Solar Activity - presented in radio Carbon-14 production ratios

This graph plots solar cosmic-ray flux, NOT earth's temperature fluctuations that follow in step.

Small as the last warming pulse was, it gave us 300 years of amazing global warming that was exclusively caused by the Sun, solidly proved in hard measurements.

Remember this warm period fondly, when the Sun was ramped up. It may be the reason why we all exist. It broke the Little Ice Age, and saved us potentially from the big Ice Age.

This warming is now fading. We are on the fading slope. Nothing will bring it back. Manmade Global Warming is not physically possible, because we cannot affect the Sun. It would be wonderful if we had the capability to master the Sun, which masters the climate on Earth.

Compare the absorption coefficients

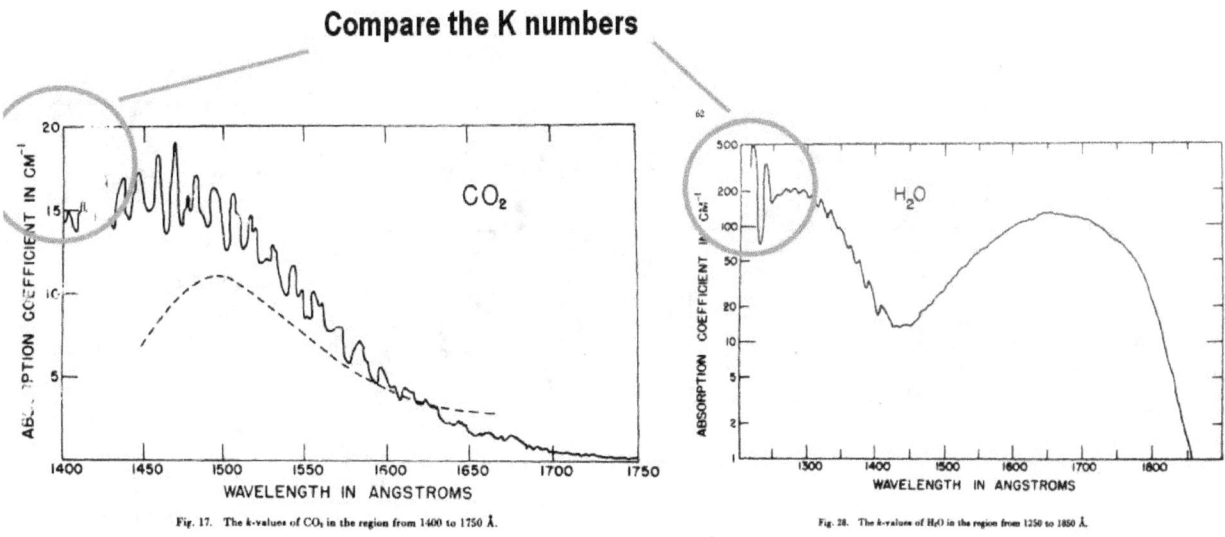

From a 1953 study by the Geophysics Research Directorate of the Air Force Cambridge Research Center Cambridge, Massachusetts - http://www.dtic.mil/cgi-bin/GetTRDoc?AD=AD0019700

Carbon gases in the air are not a significant climate factor, even as greenhouse gases. They pale in comparison with water vapor that causes nearly all of the greenhouse effect. Compare the absorption coefficients. That of water vapor is up to 20 times greater than that of CO2. Then consider further that water vapor is up to 100 times more abundant in the atmosphere.

Water vapor is effective across a vastly wider band

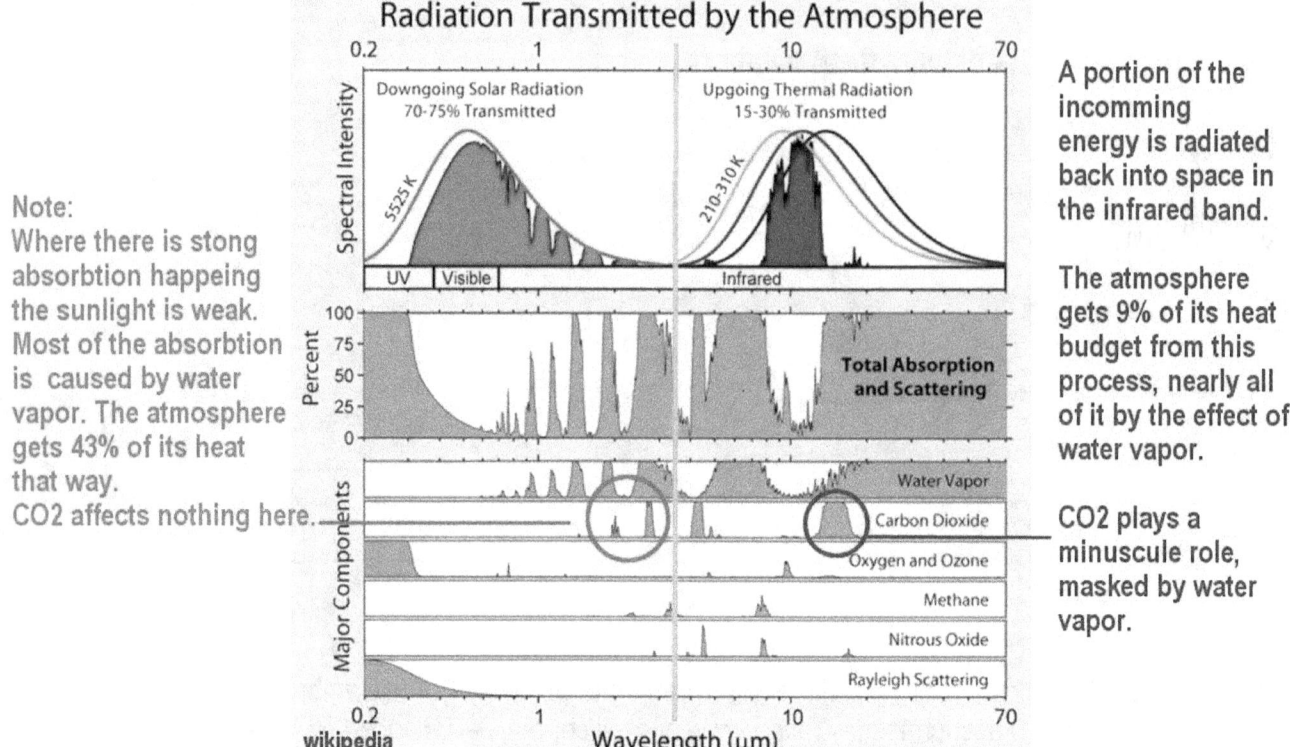

Note:
Where there is stong absorbtion happeing the sunlight is weak. Most of the absorbtion is caused by water vapor. The atmosphere gets 43% of its heat that way.
CO2 affects nothing here.

A portion of the incomming energy is radiated back into space in the infrared band.

The atmosphere gets 9% of its heat budget from this process, nearly all of it by the effect of water vapor.

CO2 plays a minuscule role, masked by water vapor.

In addition, water vapor is effective across a vastly wider band of the energy spectrum. Then add to the comparison the further absorption of solar energy by oxygen and by the Rayleigh scattering effect, which are both effective at the high-energy part of the light spectrum, and it becomes self-evident that CO2 contributes so little to the greenhouse effect of the Earth that it isn't worth the mention.

The CO2 effect is no bigger than a cat

In comparison, the CO2 effect is no bigger than a cat standing beside the World Trade towers in New York, which some people may still remember. Now assume that the cat could be over-nourished to become the size of a horse, it would still be too minuscule to be noticed on the wider scene.

Wonderful if manmade global warming was possible

It would be wonderful if manmade global warming was possible, in order that we would develop this power to block the Ice Age consequences.

While we don't have the capability to stop the Ice Age, we do we have the power to avoid the consequences.

To create a Manmade New World

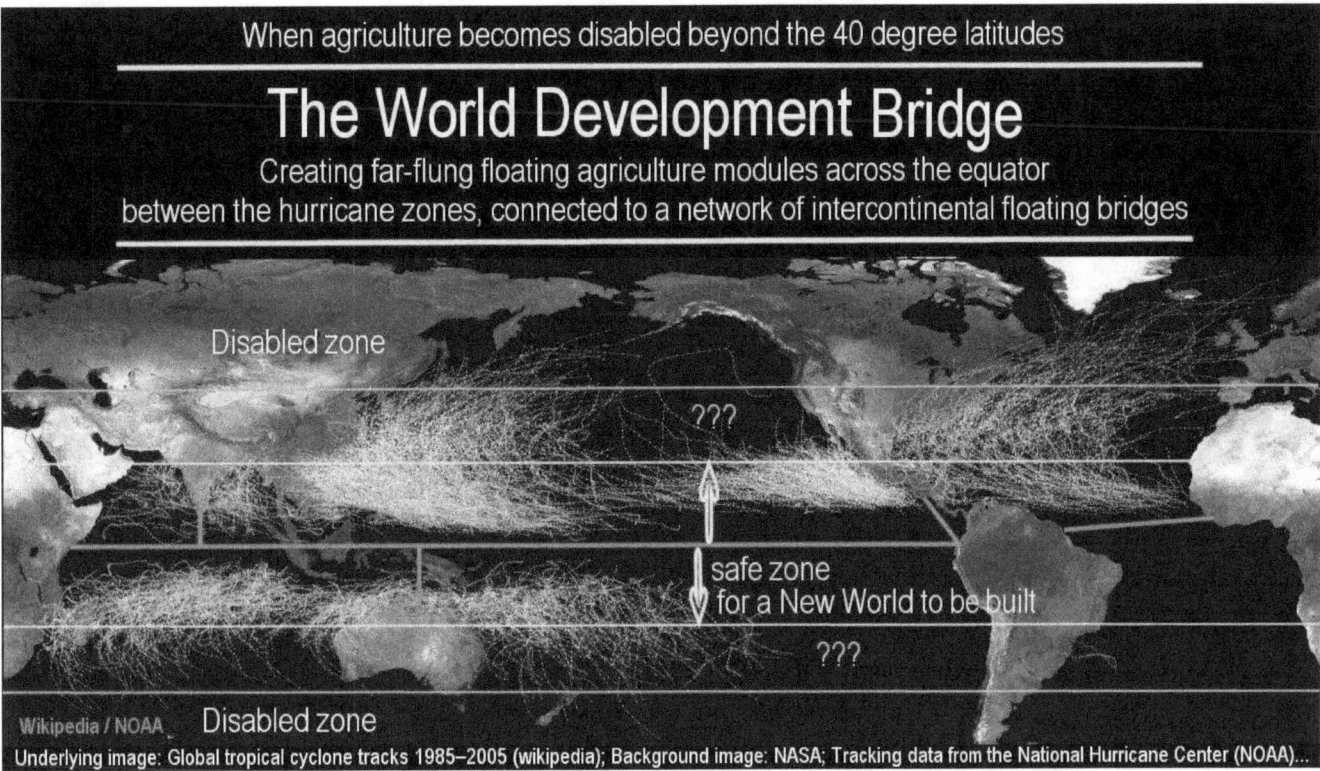

We have the power to create a Manmade New World that the Ice Age consequences cannot affect.

The Strategic Defence of Humanity

The Strategic Defence of Humanity

The Strategic Defence of Humanity

Objective is to block the recognition of the near Ice Age

The manmade global warming doctrine doesn't even pretend to go that far as to block the near Ice Age. Instead, it effectively hides it, and replaces it with apocalyptic scaring. The doctrine's objective is to block the recognition of the near Ice Age, and to likewise block the recognition of the solar cause of the global warming that had saved the world in the 1700s, while it promises doom without recourse.

And still, the treachery goes deeper than merely scaring society into submission.

Renamed to Manmade Climate Change

For this deeper effect, since global warming isn't happening, the doctrine has been renamed to Manmade Climate Change so that it can mean anything that one dares to apply to it.

Similar in effect to Illuminati

Isaiah 14:12 How art thou fallen from heaven, O Lucifer, son of the morning! how art thou cut down to the ground, which didst weaken the nations!

http://www.trickedbythelight.com/tbtl/logospyramid.shtml

The result is similar in effect to that of the term, Illuminati, the light of the star that fell from heaven and became Satan, which became attached to a range of secret underground organization, mostly without an identifiable face, but of reputed fierce power and wide influence, so that resistance is deemed futile.

The Manmade Global Warming effect without a possible solution

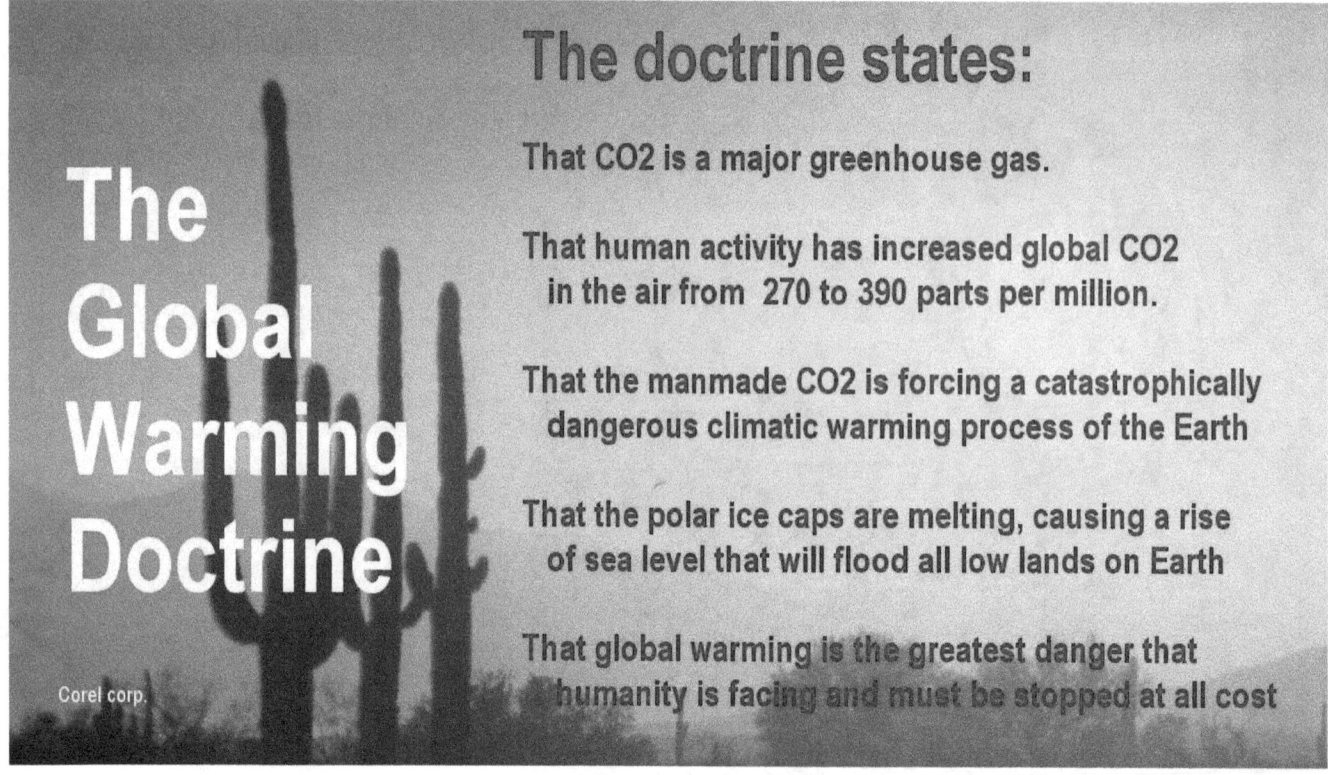

The Manmade Global Warming effect is similarly said to be dangerous, and ultimately without a possible solution.

An existential challenge without an identified face

When the human mind is confronted with an existential challenge without an identified face, like that of Manmade Climate Change, which is a challenge without hope, - a challenge for which no solution is deemed possible, except the self-termination of humanity - then the mind reverts backwards to a lower-level of thinking, to a more infantile thinking, which becomes a form of self-defence.

This mental regression the intention of nearly all psychological warfare games.

Against the imposed, reduced-mentality background

A shocking statement was made by a United Nations official Christiana Figueres at a news conference in Brussels. Figueres admitted that the Global Warming conspiracy set by the U.N.'s Framework Convention on Climate Change, of which she is the executive secretary, has a goal not of environmental activists to save the world from ecological calamity, but to destroy capitalism. She said very casually:

> "This is the first time in the history of mankind that we are setting ourselves the task of intentionally, within a defined period of time, to change the economic development model that has been reigning for at least 150 years, since the Industrial Revolution."

She even restated that goal ensuring it was not a mistake: *"This is probably the most difficult task we have ever given ourselves, which is to intentionally transform the economic development model for the first time in human history."*

posted in a blog on Climate, Feb 3, 2017 by the renowned American economist Martin Armstrong.

"Global Warming is about Destroying Capitalism"

https://www.armstrongeconomics.com/world-news/climate/global-warming-is-about-destroying-capitalism/

Against the imposed, reduced-mentality background, the most devastating impositions can be forced on society.

The impositions are forced

Mr. Armstrong quotes Christiana Figueres
Executive Secretary of the United Nations Framework Convention on Climate Change

"This is the first time in the history of mankind that we are setting ourselves the task of intentionally, within a defined period of time, to change the economic development model that has been reigning for at least 150 years, since the Industrial Revolution."

The impositions are forced, politically with regulations, economically with carbon taxes, physically with deindustrialization, and socially with the mass-burning of food under the biofuels project that society willingly participates in at the gas pump.

Killing upwards to 100 million people a year with starvation

In a world that has a billion people living in chronic starvation, the mass-burning of food at a rate that would normally nourish 400 million people, is likely killing upwards to 100 million people a year with starvation.

The effect is - and this is intended

The effect is - and this is intended - that the Manmade Climate Change doctrine effectively robs society of its humanity by staging it to fight against itself, unwittingly, as in the ancient wars by empire against humanity that are still-ongoing in some form.

The Greek playwright Aeschylus

Prometheus Bound - Dirck van Baburen

The Greek playwright Aeschylus dealt with the Illuminati type issues, or Manmade Climate Change type issues, already as far back as 2,400 years ago. He addressed the concepts that are deliberate mind-killers in his trilogy, Prometheus Bound.

To raise his audience to a higher-level self-perception

Heinrich Füger (1751–1818)
Prometheus Brings Fire to Mankind
- wikipedia

His solution to the issues was to raise his audience to a higher-level self-perception with the perception of the truth.

Manmade Climate Change is physically impossible

For us, the equivalent scientific recognition is, that Manmade Climate Change is physically impossible. This obvious fact becomes apparent when the 'Illuminati' devils' voices are banished from the mind, whereby the simple truth becomes apparent without fail.

The Strategic Defence of Humanity requires

Heinrich Füger (1751–1818)
Prometheus Brings Fire to Mankind
- wikipedia

The Strategic Defence of Humanity requires that humanity raise itself above the false doctrines to higher levels of scientific perception, even it self-perception and a recognition of its power.

Under the freedom of the truth

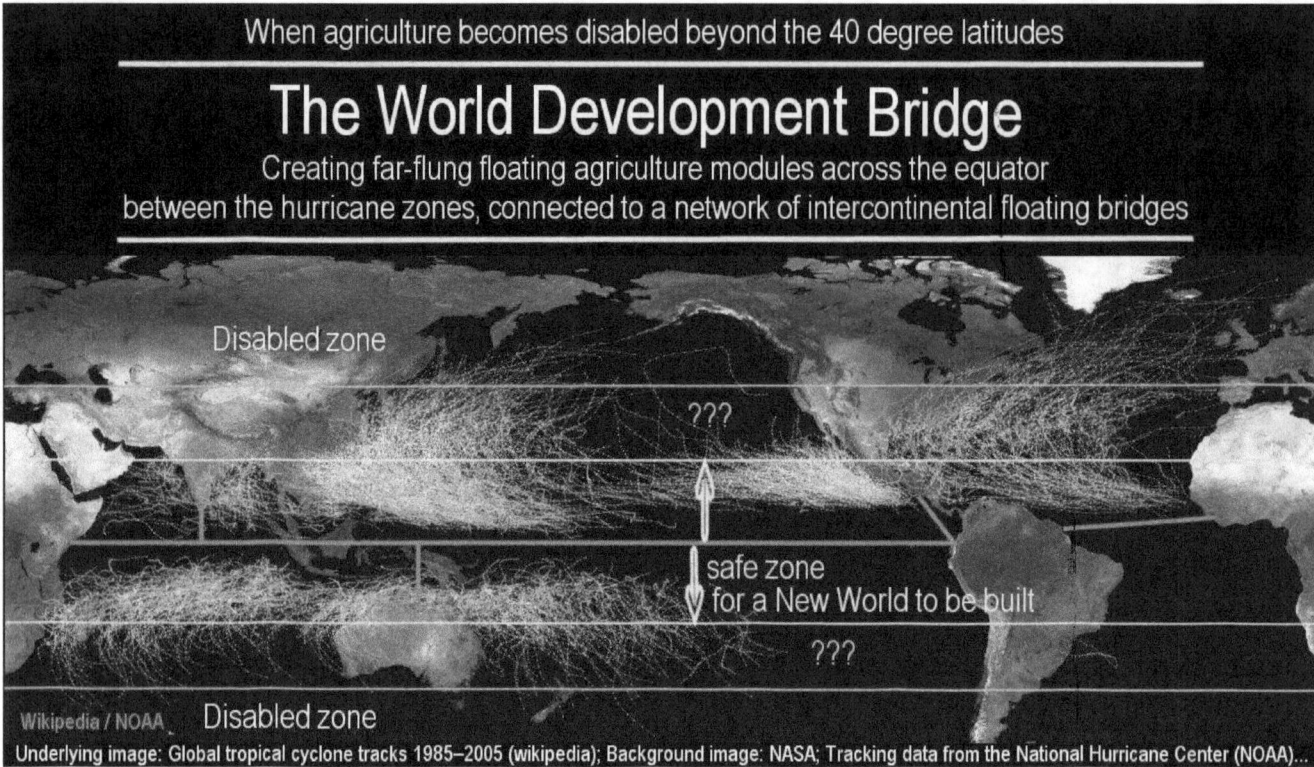

Then, under the freedom of the truth, the Plan-B option will be pursued, and humanity will have a future.

However, the truth poses a big challenge for scientific recognition

However, the truth poses a big challenge for scientific recognition, because the diminishing greenhouse effect, and its reduced climate moderation, enable hotter temperature extremes in the summers that are mistakenly accepted as proof for manmade global warming climate change, while the opposite is the case.

When the greenhouse-effect becomes reduced

When the greenhouse-effect becomes reduced by reduced water-vapor density, resulting from cosmic-ray increased cloudiness, larger climate extremes do occur, experienced as hotter sunshine.

The defence of humanity against the false notions that are mistakenly accepted as real, is possible with a more expanded scientific understanding of the astrophysical principles, and of the truth of their effects.

Tragically, this understanding is still lacking, even in the science community that almost universally remains trapped by the mind-choking dictates of the science-Illuminati,

Consequently nothing gets build.

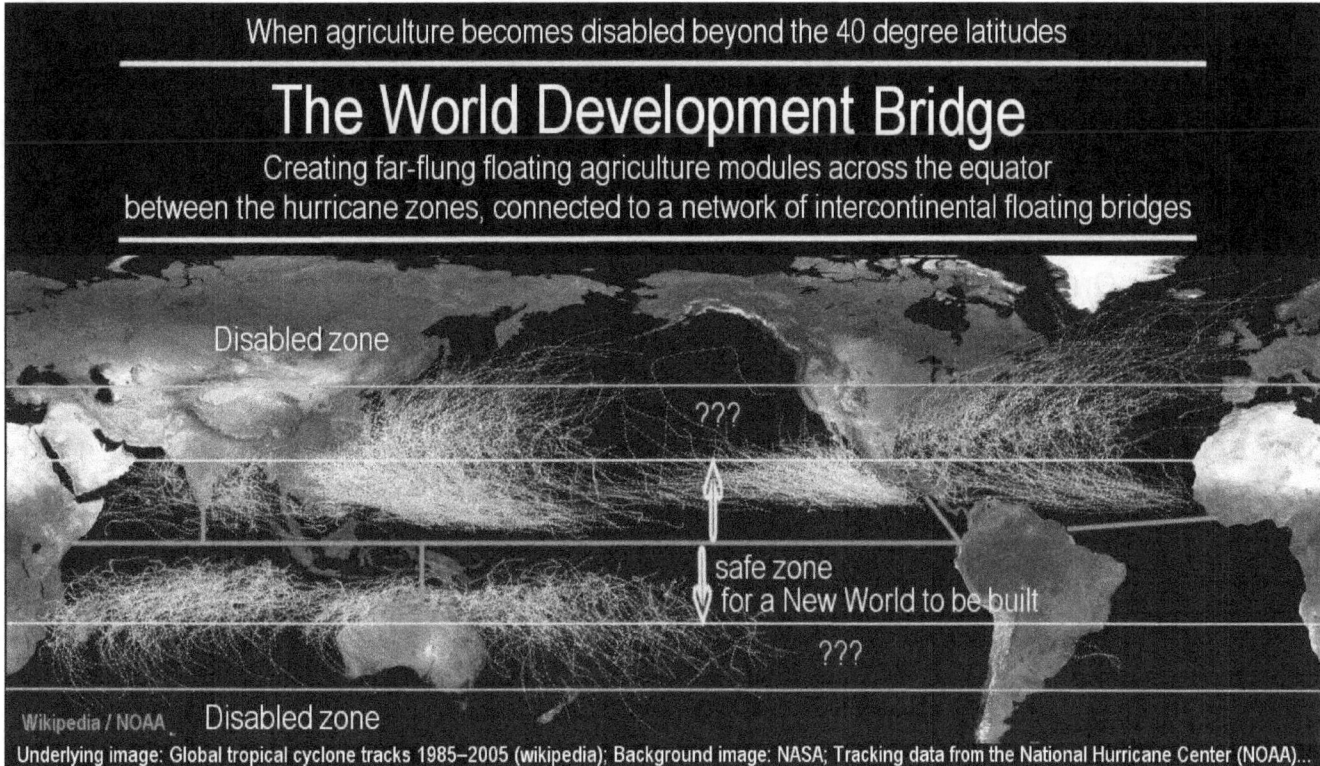

Consequently nothing gets build.

More than 50,000 protest 'signatures' had been uttered

Over 50,000 signatures from the science community opposing the Manmade Global Warming doctrine

1992 The Heidelberg Appeal
- signed by 4000 scientists from 69 countries, including 63 Nobel Laureates

1997 The Leipzig Declaration
- signed by 110 climate specialists

1998 The Oregon Petition Project
- signed by 17,000 scientists (organized against the Kyoto Protocol)
The petition was organized and circulated by Arthur B. Robinson, president of the Oregon Institute of Science and Medicine

The Kyoto Protocol met with an 85% rejection across the world by 2004

2006 Statements Opposing the Doctrine of Manmade Global Warming
- put on record by - U.S. Senate Committee on Environment & Public Works

2007 The U.S. Senate Report:
- Over 400 Prominent Scientists Dispute Man-Made Global Warming Claims - listed by name in detail

2008 New Oregon Petition Project
- online, and still ongoing - signed by over 31,000

Details at: www.ice-age-ahead-iaa.ca/alternate_healing/lovescapenovels/climate_change_opposition.html

For decades in the past, more than 50,000 protest 'signatures' had been uttered by the science community worldwide in opposition of the Manmade Global Warming doctrine, while the detailed arguments remained limited to within the mystic framework established by the science-Illuminati.

Just imagine the boundless freedom that this simple science recognition

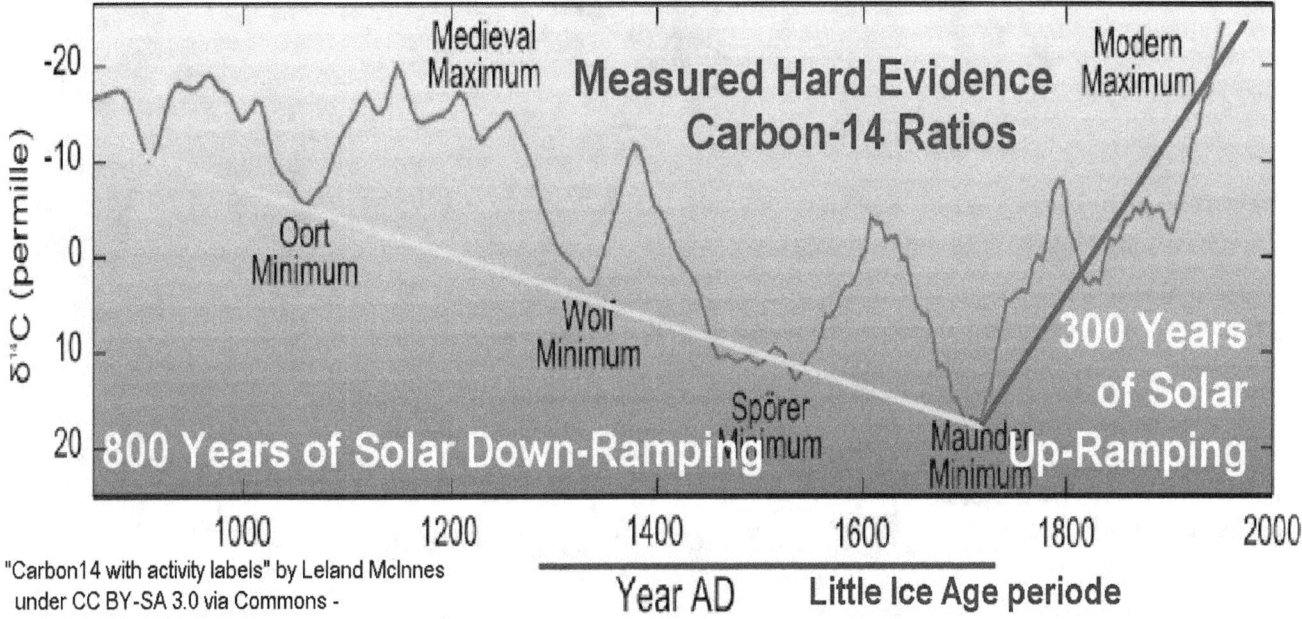

The protest movement should have build on the measured fact that the ever-changing Sun is the climate driver on Earth.

Just imagine the boundless freedom that this simple science recognition, built on measured fact would have brought to humanity, and could bring to it even now. It would end carbon taxes, reverse the vilification of humanity, cancel the biofuels murdering by food-burning, and it would open the mind to the recognition of the near Ice Age on the horizon and inspire the building of a new world for it, for our future survival on an Earth that is becoming largely an ice-planet from the 2050s onward.

Earth's greenhouse effect is diminished by cosmic-ray increase

Earth's greenhouse effect is diminished by cosmic-ray increase, cloud nucleation reduces water vapor, and causes deeper droughts and a lesser greenhouse

With the mind becoming free to accept the truth, the recognition would dawn that the greenhouse effect of our atmosphere is diminishing instead of increasing.

The most important service that the greenhouse effect of the Earth's atmosphere provides, is to moderate climate fluctuations.

Without the greenhouse effect of the atmosphere, we would encounter the same temperature fluctuations as we do on the moon, which has no atmosphere. The nights would drop to -170 decrees Celsius, and the days be hotter than boiling water, at +117 degrees. The moderating effect of the greenhouse narrows these extremes to a nicely liveable climate. However, since up to 90% of the greenhouse effect is contributed by water vapor in the atmosphere, a small reduction of the water vapor density, resulting in reduced greenhouse effects, can have large consequences. And that's what we are are beginning to experience evermore now, as the result of the increasing cosmic-ray flux that flows from our weakening Sun.

The currently increasing solar cosmic-ray flux, which increases the cloud forming intensity, and the rainout of the clouds primarily over the oceans, is diminishing the greenhouse effect. It is destructive of it. The increased water nucleation into clouds, takes more water vapor out of the atmosphere than evaporation can replenish. The thereby reduced water-vapor density reduces the greenhouse effect.

As the result of the reduced greenhouse-moderation

As the result of the reduced water-vapor density and reduced greenhouse-moderation, we experience increasingly larger climate fluctuations. We experience more sweltering hot summer days, and with them larger droughts and ever-more forest fires, which are both amplified by the sweltering heat, while winters become colder and snowier in many parts of the world.

In Australia wheat harvests fell 40%

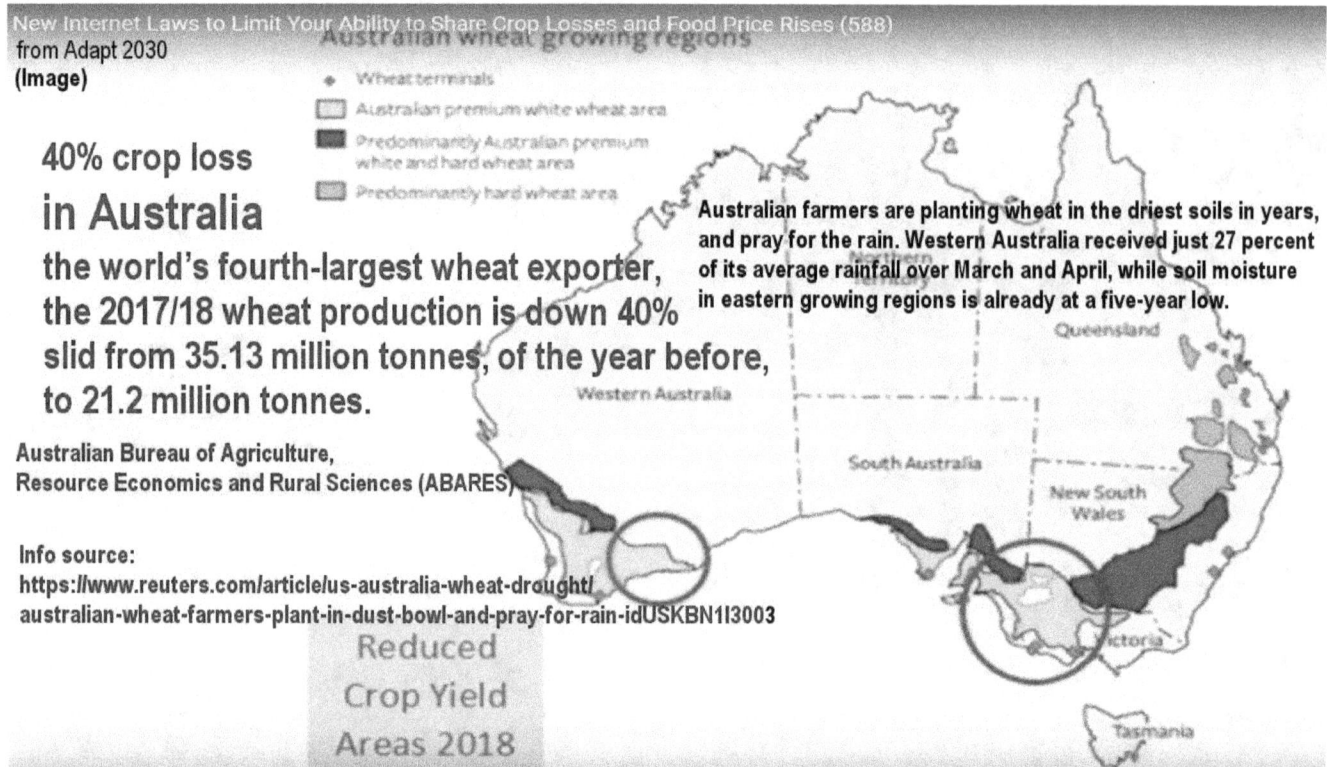

These effects are already being felt. They are real, and they are increasing. In Australia for example, which is the world's fourth-largest wheat exporter, wheat harvests fell 40% in the 2017/18 season, because of drought and too much heat as the result of the weakening Sun and its increasing cosmic-ray flux.

This year Australian farmers are planting their wheat in the driest soils in years, and pray for the rain. So far Western Australia had received just 27 percent of its average rainfall over March and April, with soil moisture in the eastern growing regions being at a five-year low.

We should see these types of effects as warning bells ringing, as a warning to the world of how the boundary zone to the near Ice Age is unfolding.

The removal of a dozen floors from the towers

 In the previous comparison the total greenhouse effect is represented by the world-trade towers, compared to a cat representing the CO2 contribution. While the overfeeding of the cat, to become a horse, if this was possible, would have no significant effect on the total scene, the removal of a dozen floors from the towers would have a major effect. The shrinking of the buildings would represent the removal of water vapor from the atmosphere by increased solar cosmic-ray flux ionization. That's, in principle, how the Sun affects our greenhouse as a big variable factor.

The result of the shrinking greenhouse effect

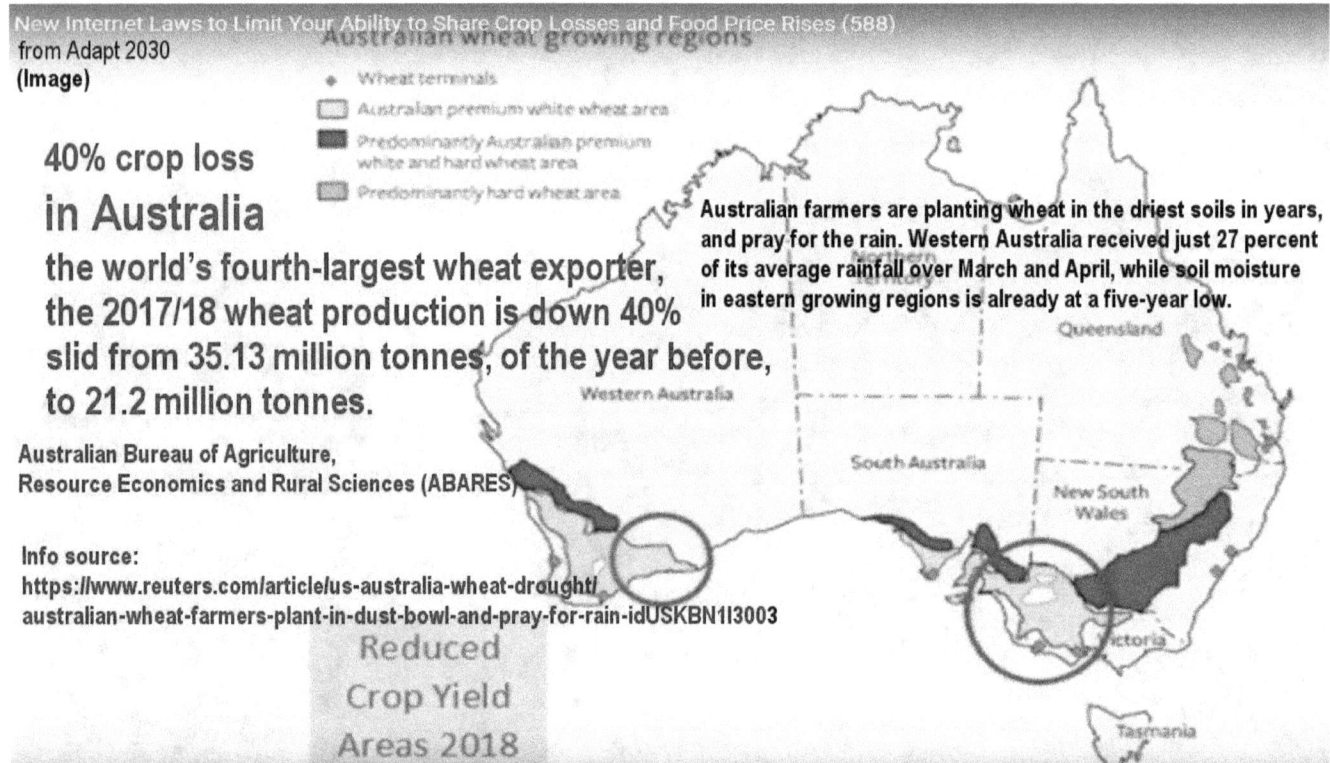

The result of the shrinking greenhouse effect, as it was felt in Australia, will be felt worldwide in ever-larger measures over the next 15 years, because of the ever-increasing solar cosmic-ray flux that affects our atmosphere, resulting from the weakening Sun.

Now consider precipitation diminishes by 80%

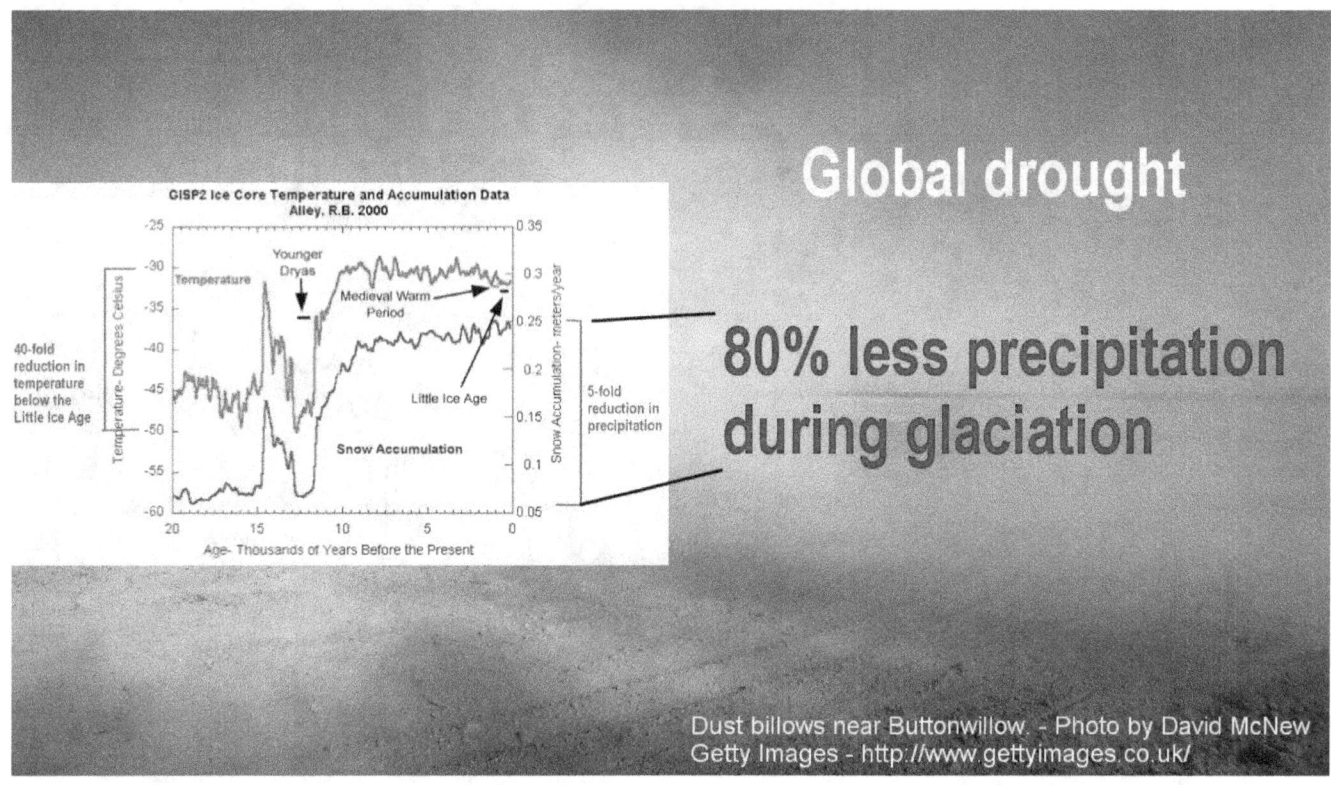

Now consider that during glaciation conditions, precipitation diminishes by 80%, because a colder Sun evaporates less water.

That's like taking 80 stories off the buildings

New York City Skyline - World Trade Center - wikipedia
A work of the United States Federal Government

This means that our greenhouse effect might be up to 80% weaker. That's like taking 80 stories off the buildings, which leaves us only a very small greenhouse remaining. The reduced greenhouse would compensate somewhat for the colder Sun, but it would make the nights tremendously colder.

These effects are sure happen. While no historic measurements for them exist, they will affect our living and our agriculture in large measures if these are not technologically protected.

The diminishing greenhouse effect may come to our aid

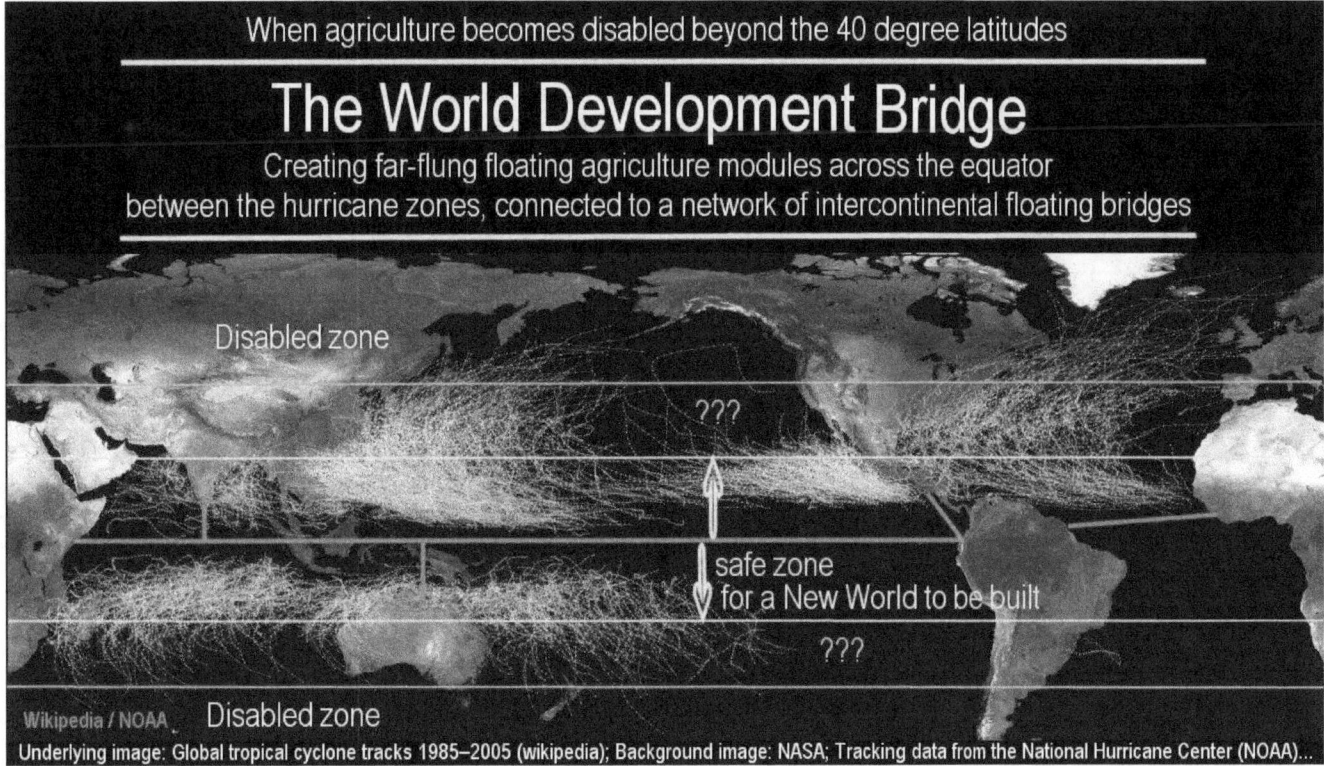

That's where Plan-B begins for building the protecting technological infrastructures.

Ironically the diminishing greenhouse effect may come to our aid, when the increasing consequences might yet inspire society's willingness to face the Ice Age challenge and respond to it with building Plan-B infrastructures.

None of the climate transformations are manmade

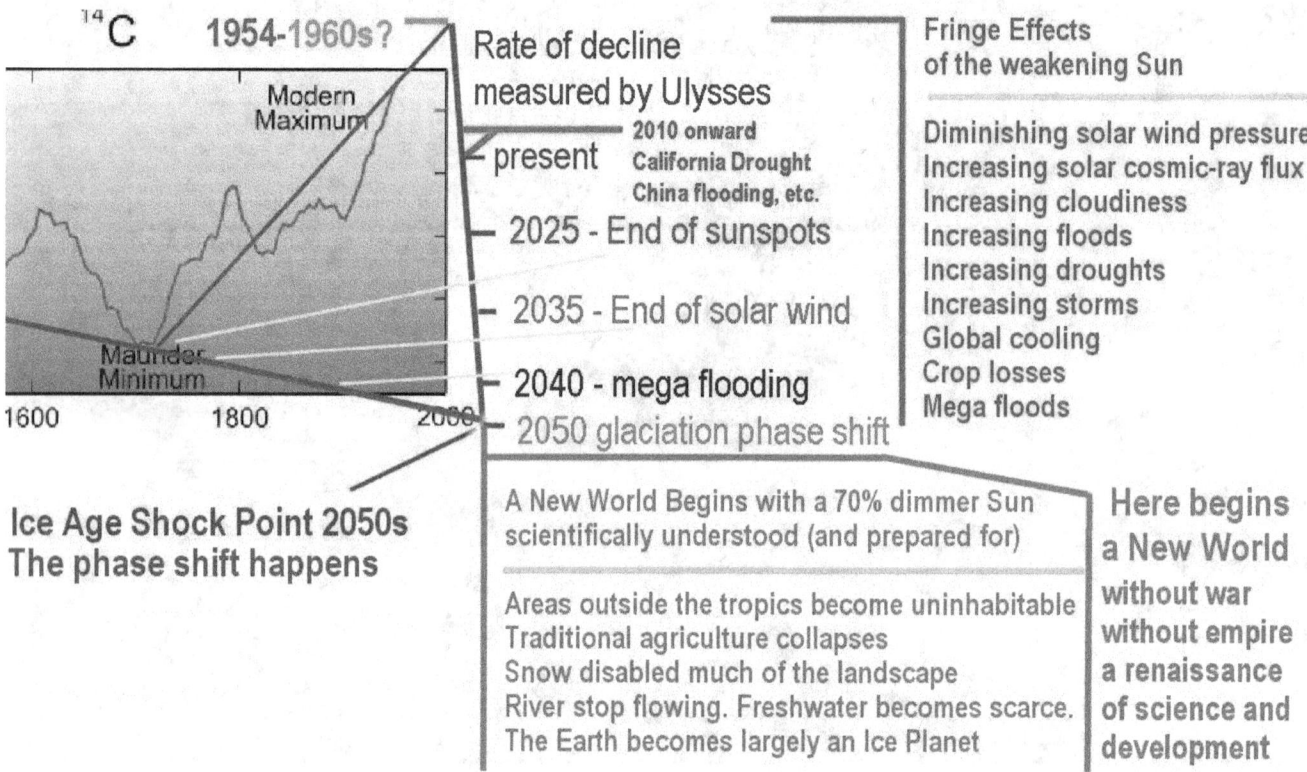

We live in the boundary time-zone leading up the next Ice Age glaciation cycle that potentially begins in the 2050s. None of the climate transformations that we encounter in the boundary zone are manmade. Nor can they be prevented. But we can bring our living into line with the unfolding dynamics by building us a new world. We have the power to do this technologically.

However, where are the movements in this direction? Who stands in the hustings fighting for a response? Where is the march of the dollars? The hustings are empty. The dollars are horded out of reach. The entire scene of society protecting its future, is presently a dead scene.

The entire climate scene is wrapped up in dreams

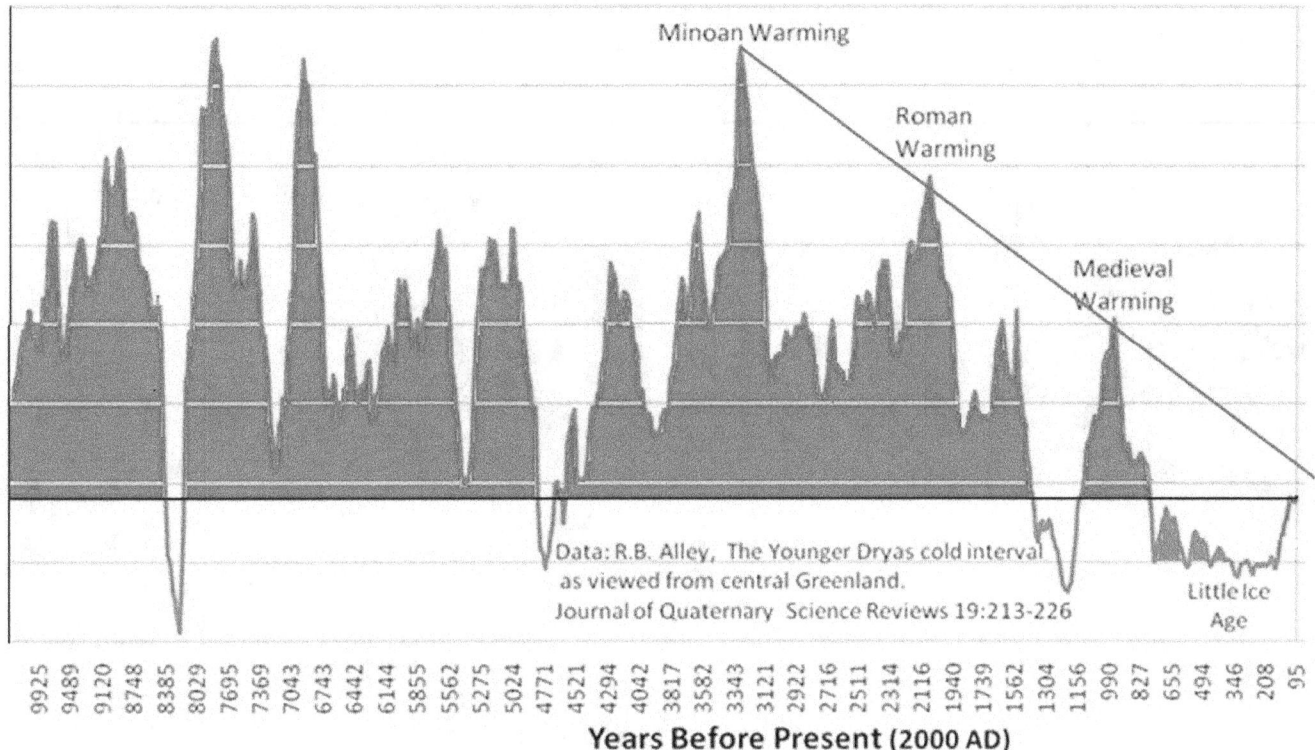

The entire climate scene that determines our collective future is wrapped up in dreams, which are mostly dreams of denial of the scientifically obvious.

The last Global Warming pulse that occurred in the 1700s

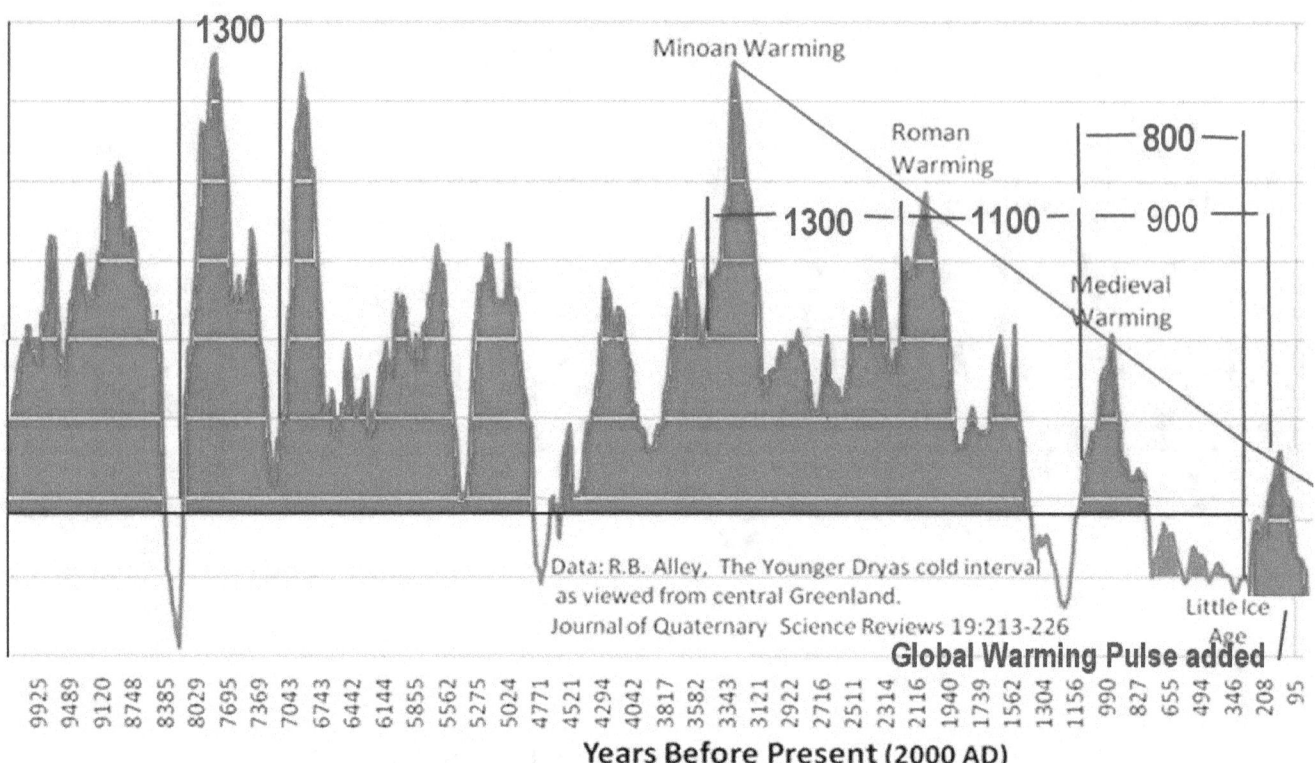

For example, the last Global Warming pulse that occurred in the 1700s, that had saved the world from the Little Ice Age, and possibly from the big one too at the time, was the most recent of the big warming pulses of the last 3,500 years. These pulses are the result of a diminishing internal resonance feature within the solar system.

The climate warming pulses are of short duration

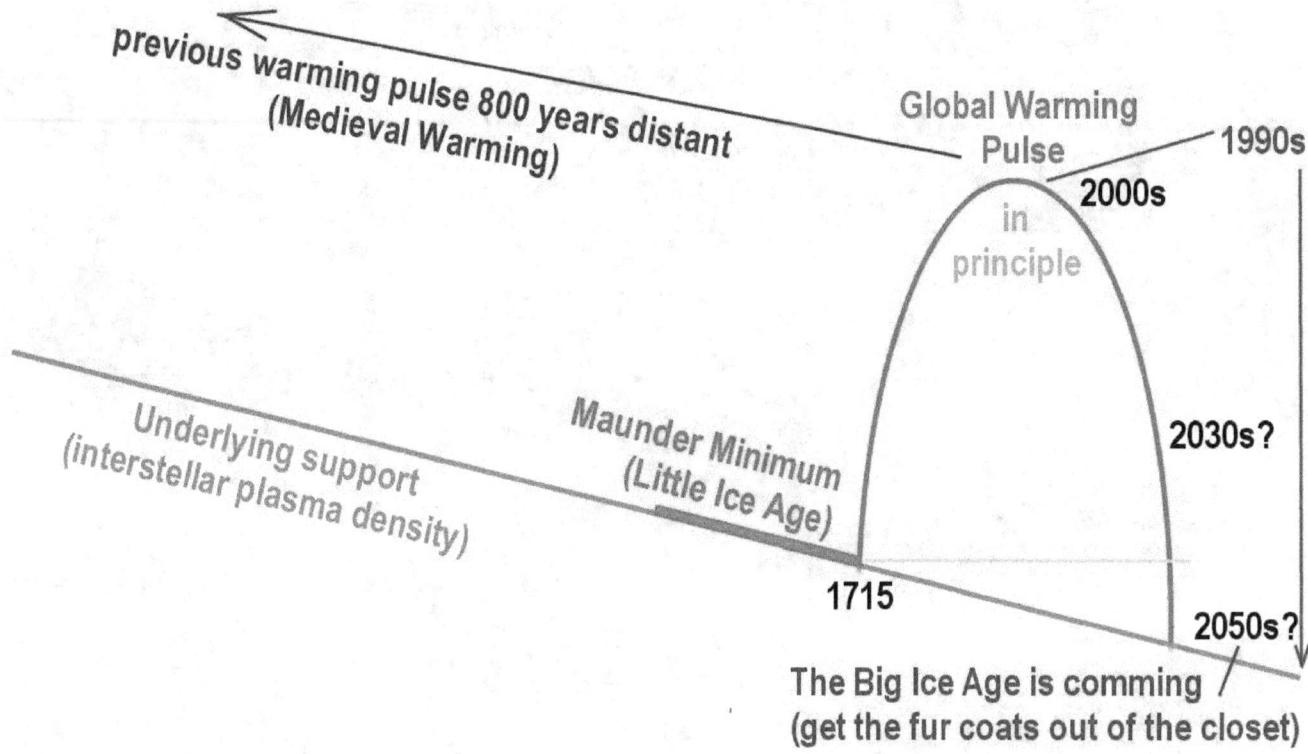

By being resonance pulses, the climate warming pulses are of short duration, and since we are presently near the end of the current one, its lingering effect is fast vanishing.

This means that we are nearing the conjunction of two major plasma structures, which have been both simultaneously diminishing. The warming pulse is ending, which takes us all the way down to the underlying background, which has been diminishing likewise.

The combination of the two dynamics adds up to a free-fall collapse now in progress towards the inevitable glacial climate in the 2050s. In the unfolding boundary-time zone, we will experience evermore fringe effects along the way, of which increased drought and reduced greenhouse effects, are a part of, that are already being experienced. And we will experience these fringe effects with ever-increasing severity.

With the last global warming pulse now fast ending, everything is happening correspondingly, increasingly fast, with changes erupting in ever-smaller timeframes.

The fast progression puts the onus on us for fast-action responses

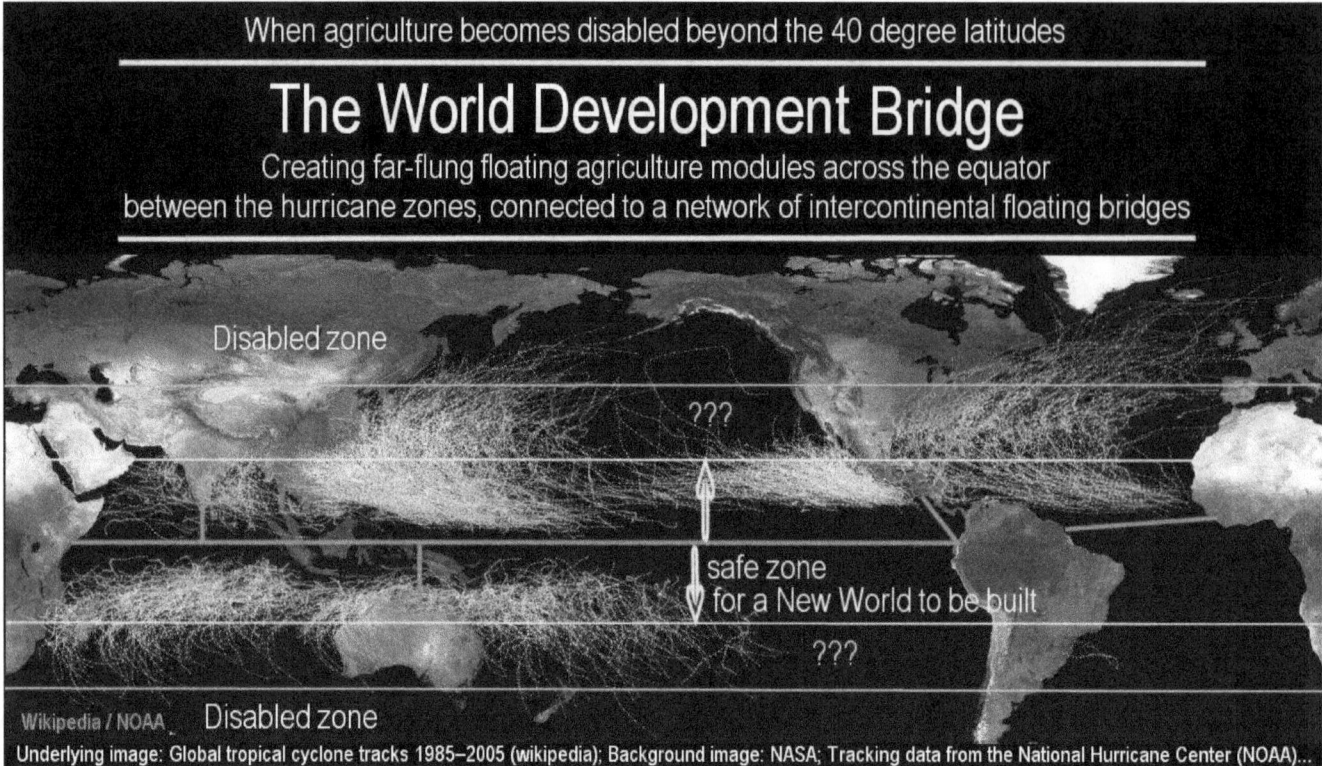

The fast progression puts the onus on us for fast-action responses in implementing a Plan-B option for the Strategic Defence of Humanity, which should be on the global agenda already and be driven by the scientific imperative to protect our food supply.

When the global climate is collapsing under the diminishing Sun, our agriculture that depends on specific climate conditions will be collapsing likewise. Our response must be to build us a new world with technological infrastructures that the collapsing climate cannot touch. If this requires that the new world be placed afloat across the equatorial seas then this must be our development plan to assure our future existence. We are in the type of boundary zone where nothing less will do, where the current Plan-A, to do nothing, adds up to suicide.

How soon humanity will get to the point where it should be already, depends on the individual efforts in society to get there. In this respect too, we live in a type of boundary zone. This one is of a different type that is not determined by time and by cosmic dynamics, but is determined by the dynamics of our humanity where time is not an absolute factor, where instead 'time' simply means "space for repentance."

Most people consider the wonderful warm climate to be normal

Most people consider the wonderful warm climate that we had enjoyed in the last hundred years, going into the year 2000, to be the normal climate on Earth that will never change. In reality, this idyllic climate is an anomaly that is fast fading. Science makes this rather plain.

A fortunate anomaly had rescued us from the Little Ice Age

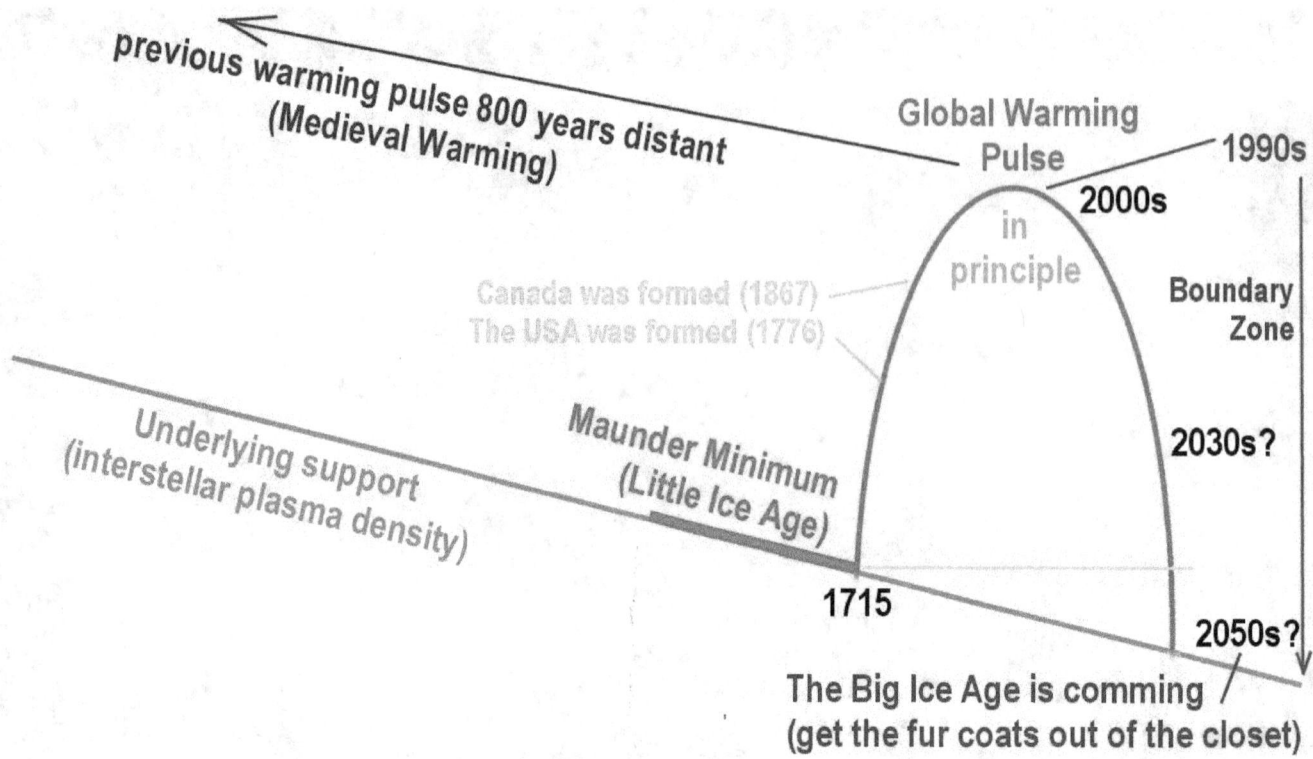

It was a fortunate anomaly that had rescued us from the Little Ice Age in the 1600s, and from the drought conditions that are known to have occurred in this time frame under the extremely weak Sun. The growing conditions were so poor at the time that starvation had become the norm, and cannibalism is said to have occurred.

Now, we are on the fast track of slipping back to the low-density plasma background that had existed during the Little Ice Age, which has continued its diminishing.

The current plasma-density background is way below what it had been during the Little Ice Age. This means that we are on the way towards experiencing, and dropping below, the type of cold climate that the solar-global-warming anomaly had saved us from. Those harsh historic climate conditions are already beginning to shine through. We simply won't survive in this unfolding climate collapse that the ever-declining 'normal' base-line has in store for us, during the boundary zone.

Plan-B is critical for our 7-billion world population

Phase-1 of the solar collapse

the most critical phase!
Plan-B needs to be implemented in this phase

The Plan-B is critical for our 7-billion world population. But it isn't optional - it is the human thing to pursue. Plan-B is the human thing to pursue, because to deny us this option, is a form of self-denial. That's why to fail, promises catastrophic consequences.

Fortunately we won't go this route of deadly self-denial. Self-denial is not the norm for humanity. We can step away from self-denial.

What had enabled the phenomenal world population growth

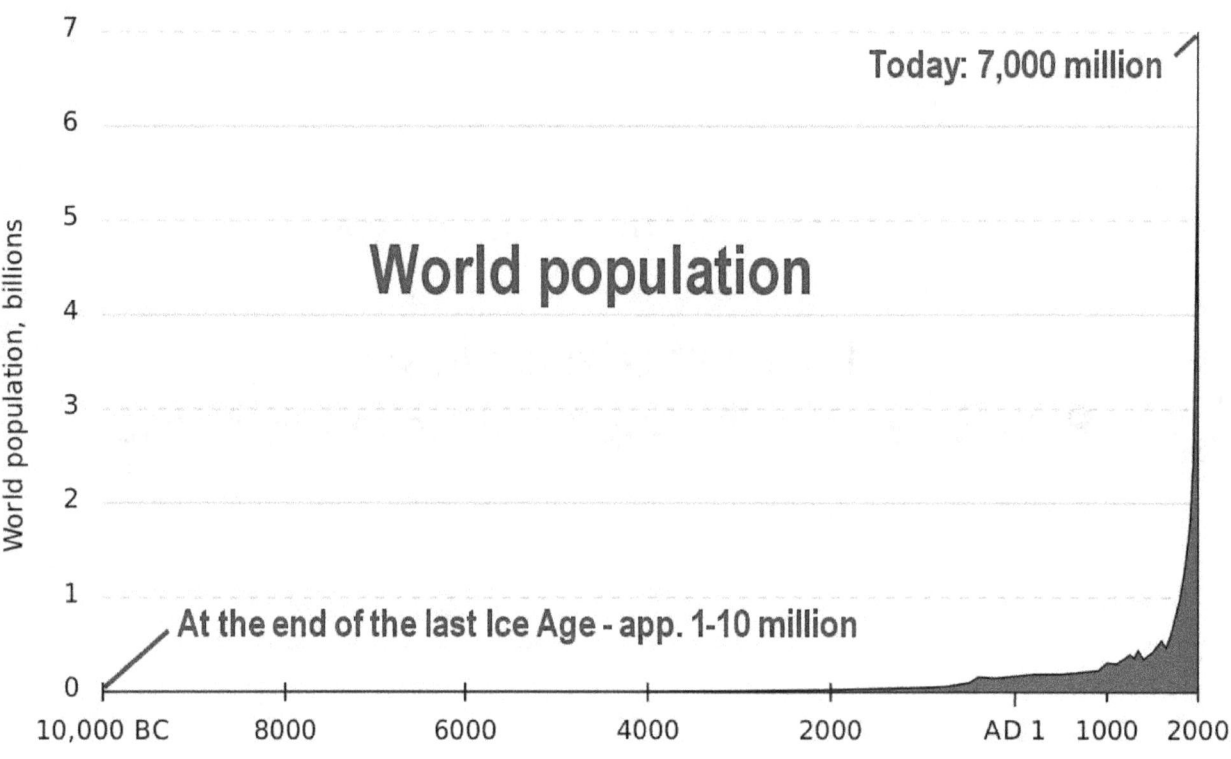

Just consider what had enabled the phenomenal world population growth in the last 300 years, from very low population levels. This population gain didn't just happen. Nor was it the result of improved breeding habits in society.

The product of science and technologie

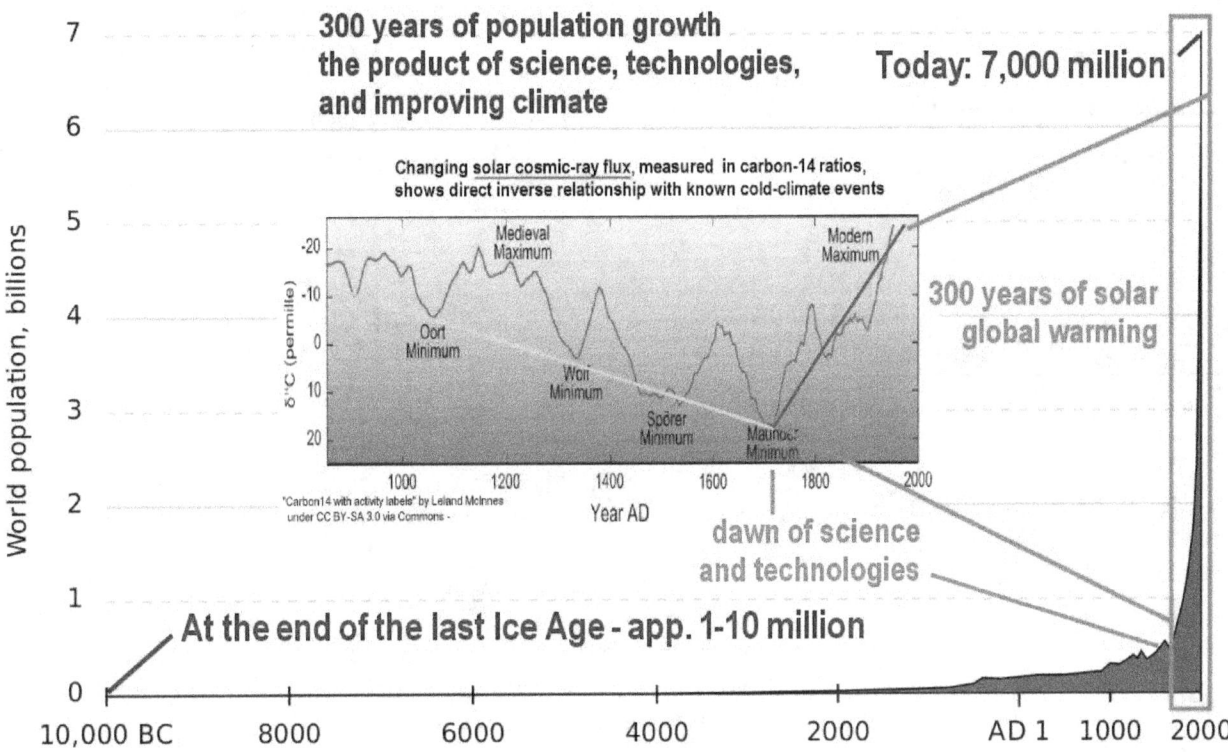

The increase was created as the product of humanity's science and technologie, in conjunction with the rapidly improving climate. The big population increase began with the general dawn of science and technologies, and later with industrialization. And it also coincided with the 300 years of the Sun-caused global warming that broke us out of the little Ice Age. The improved climate created ideal growing conditions that the technological agriculture had been able to build on. Without these two factors, the population level would have remained as flat as it has been before.

With the climate collapsing like a falling stone

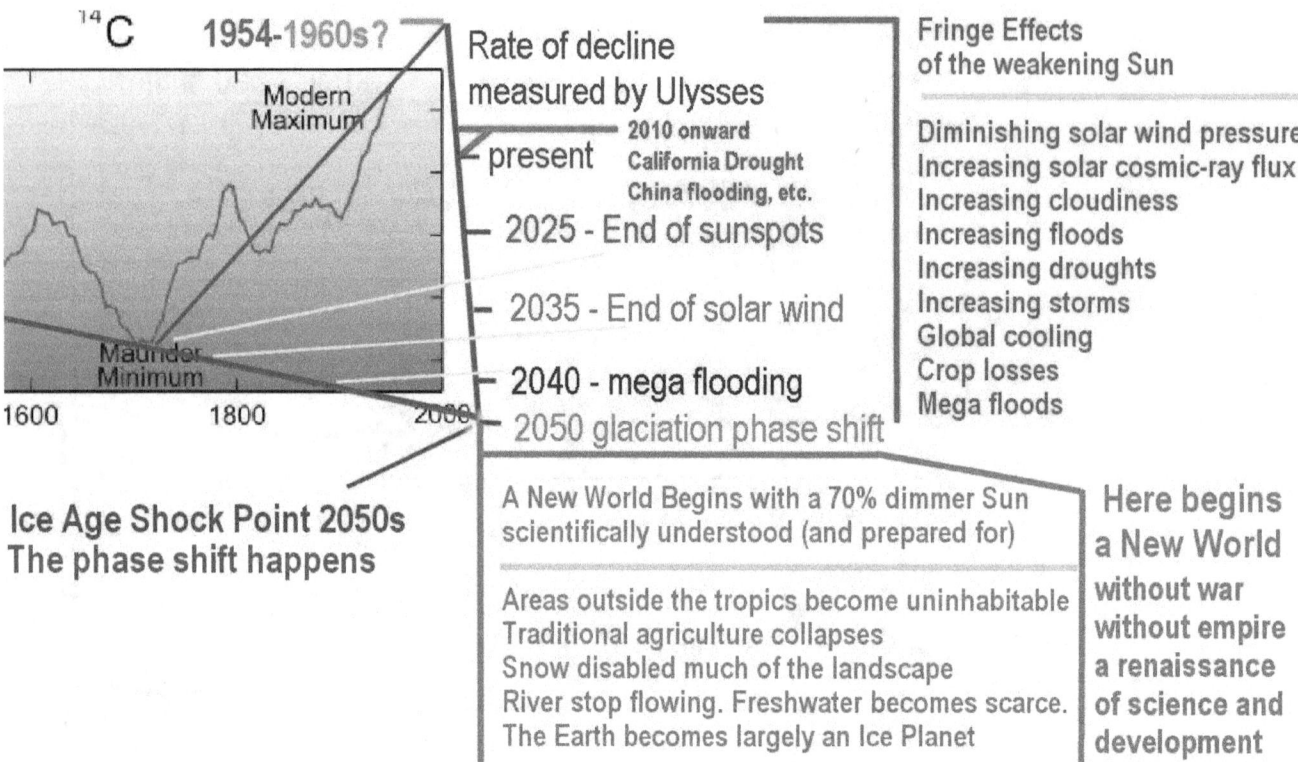

Now, with the climate collapsing like a falling stone, the only factor that we have in our hands, which we can control, is our science and technology. The general recognition of them needs to be mobilized if we hope to win the 'war' to compensate for the collapsing climate, especially in agriculture. And this can be accomplished.

Our power to create climate independent agriculture

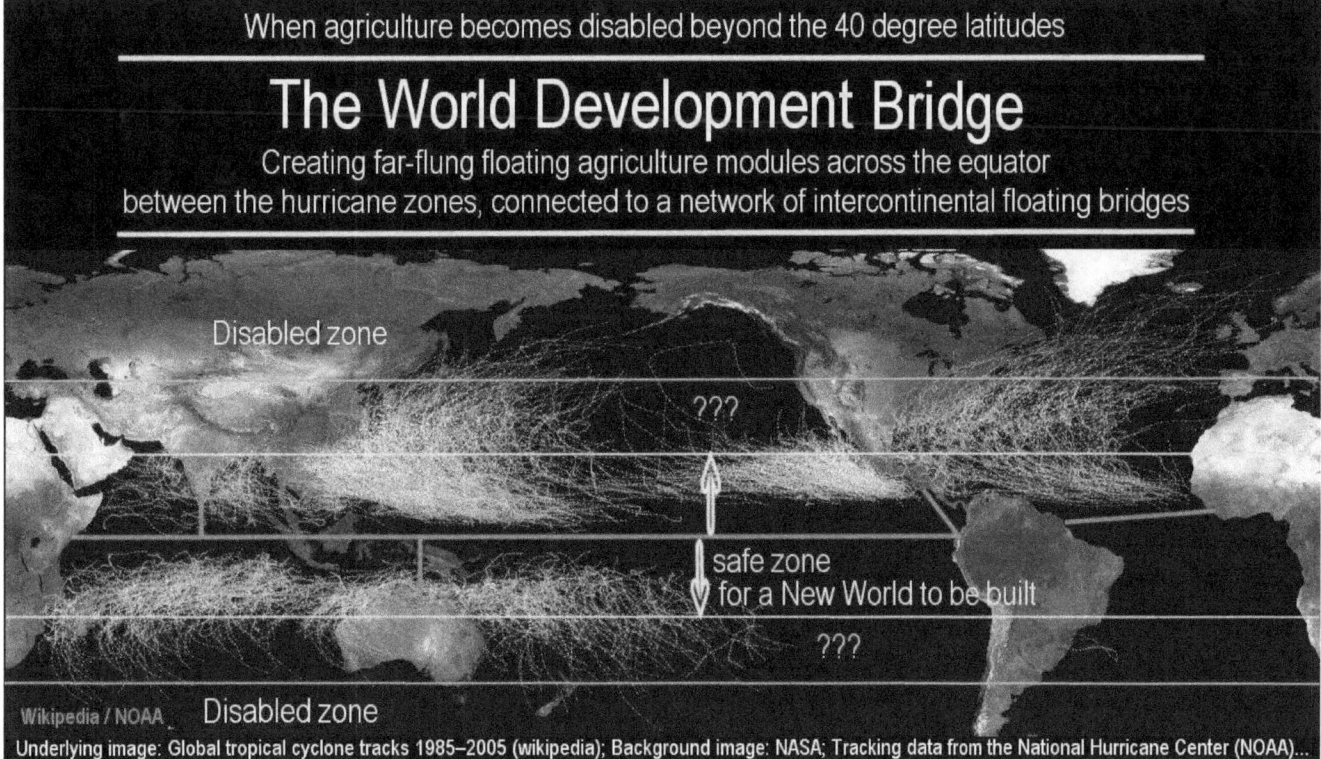

We have it in our power to create climate independent agriculture. It is not a human thing to exploit the natural world. It is the human thing to step above the natural world and raise it up to higher forms of power. Building protected floating agriculture, placed afloat across the Equatorial Sea, with new cities attached to them with new industries and universities to optimize growing conditions, would be an example of humanity steeping up to greater forms of productive power. A 50-fold increase in produced yield throughout the agricultural system may be within reach. It would be an example of advanced humanist development potential.

Of course, the task on this scale is so vast in scope that no nation can carry the torch alone. All nations must carry the torch together, and benefit together.

The total eradication of nuclear weapons will happen

On the Plan-B productive path, we will find that the world's nuclear weapons arsenals will simply be recycled and be burned up as nuclear energy fuels. The ice age challenge can accomplish this step with ease, which nothing else has so far been able to accomplish.

I think it is safe to say that the total eradication of nuclear weapons will happen in the not-so-distant future. Science will assure this. It has the power to impel politics.

When a country like China looses a portion of its food crops

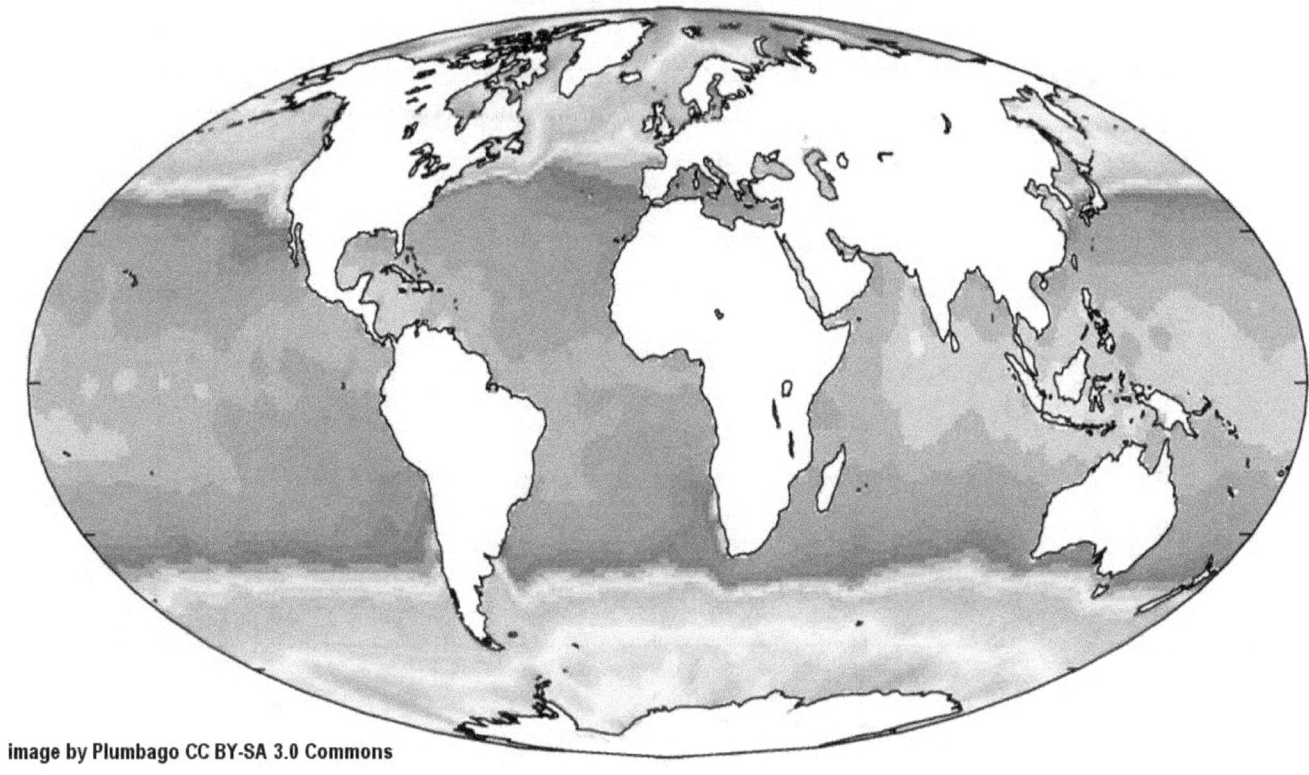

image by Plumbago CC BY-SA 3.0 Commons

When a country like China looses a portion of its food crops to drought or colder climates, the politics will become adjusted to enable Plan-B options, that total nuclear disarmament is an inherent part of.

Part 3 - 50-Years Ice Age Boundary Zone

Canada, Europe, and Russia will likely take the lead

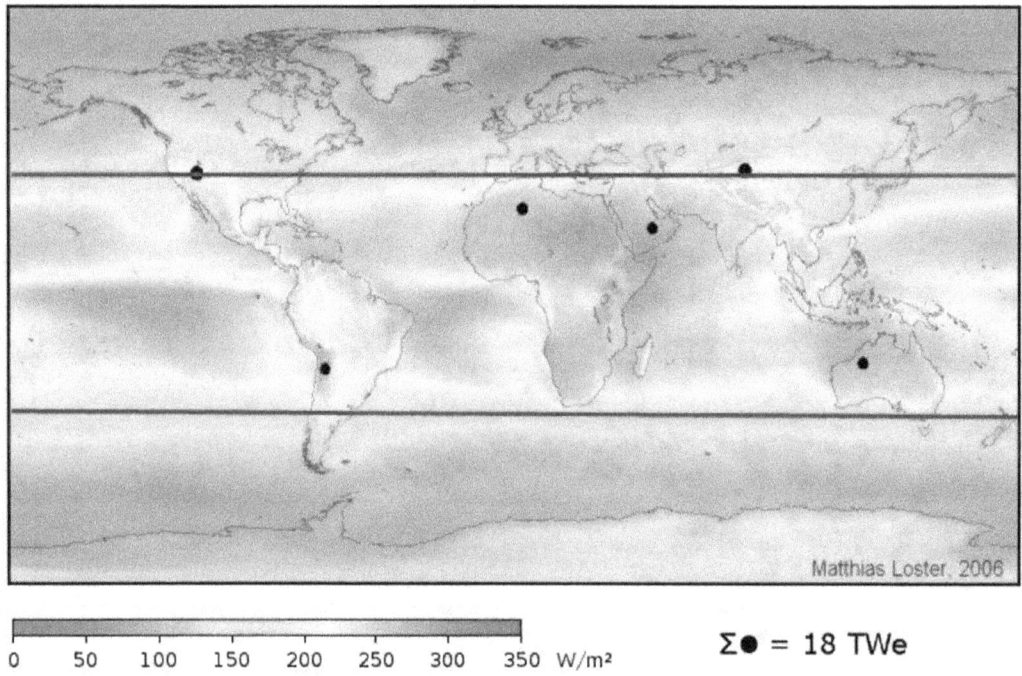

Canada, Europe, and Russia will likely be in the critical climate-position sooner than China, and take the lead on this. Once the process is moving, the lid is blown off false paradigms that keep humanity divided.

When critical physical imperatives begin to force the issues

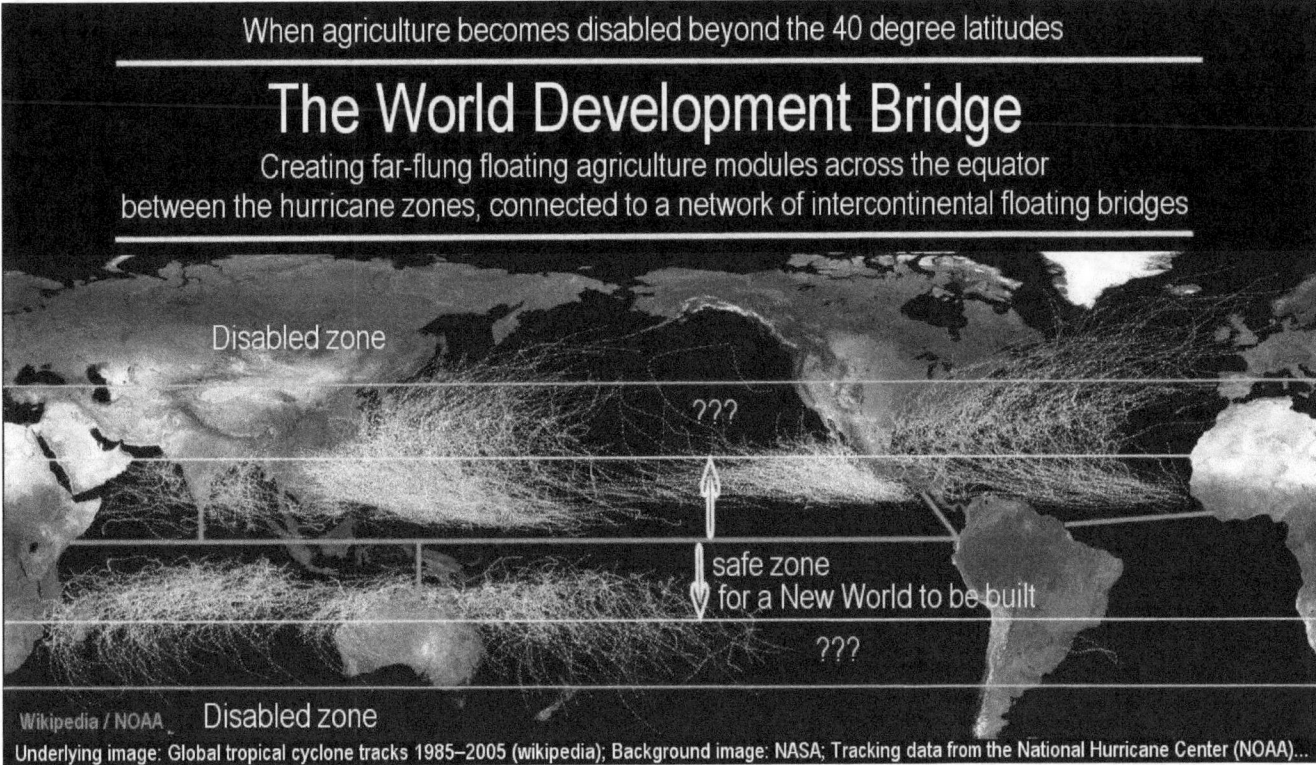

When critical physical imperatives begin to force the issues, things will begin to move, and move evermore rapidly.

Part 4 - Climate-Change, 5 to 15 years into the future

Part 4 - Climate-Change History in 'nano time' - five to fifteen years into the future.
Things are changing ever-faster now on the solar level, even while we are only half-way through Phase-1.

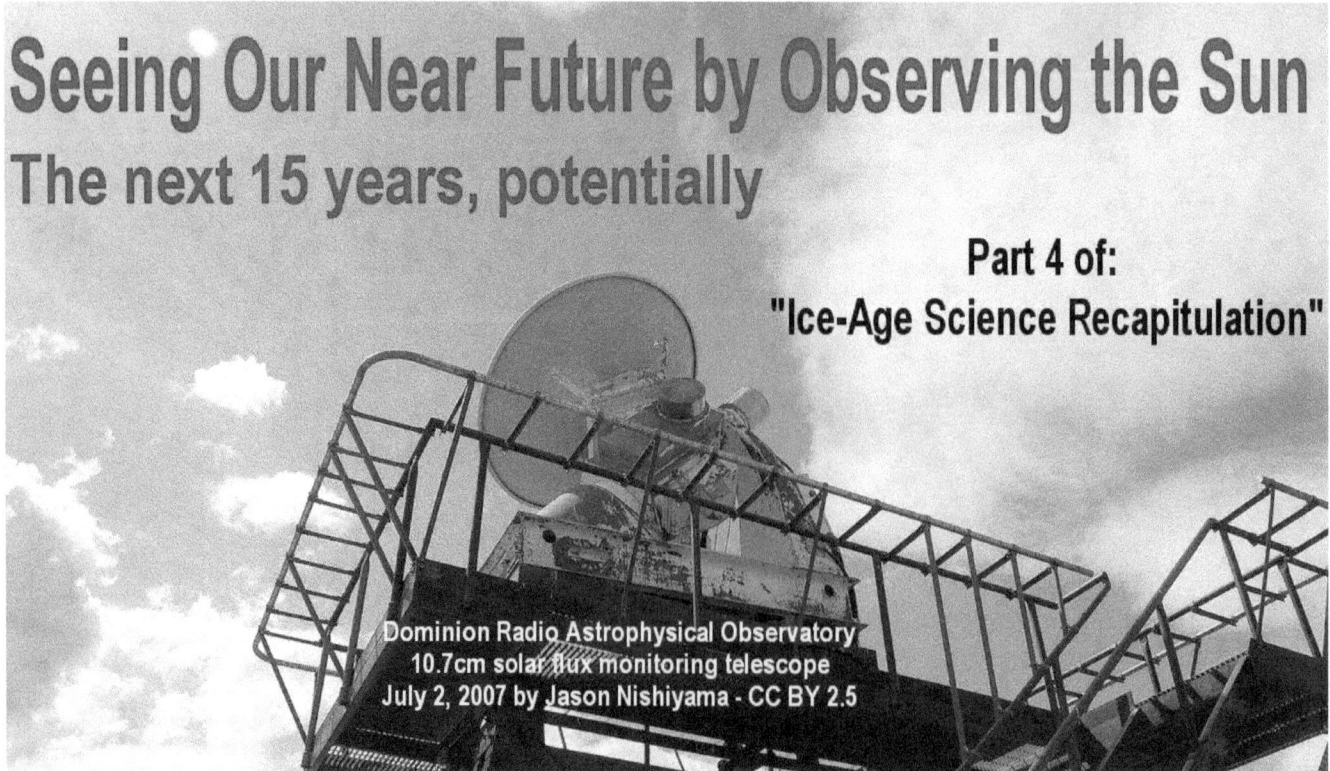

Part 4 - Our Near Future - The next 15 years, potentially

Huge winter blizzards in late spring

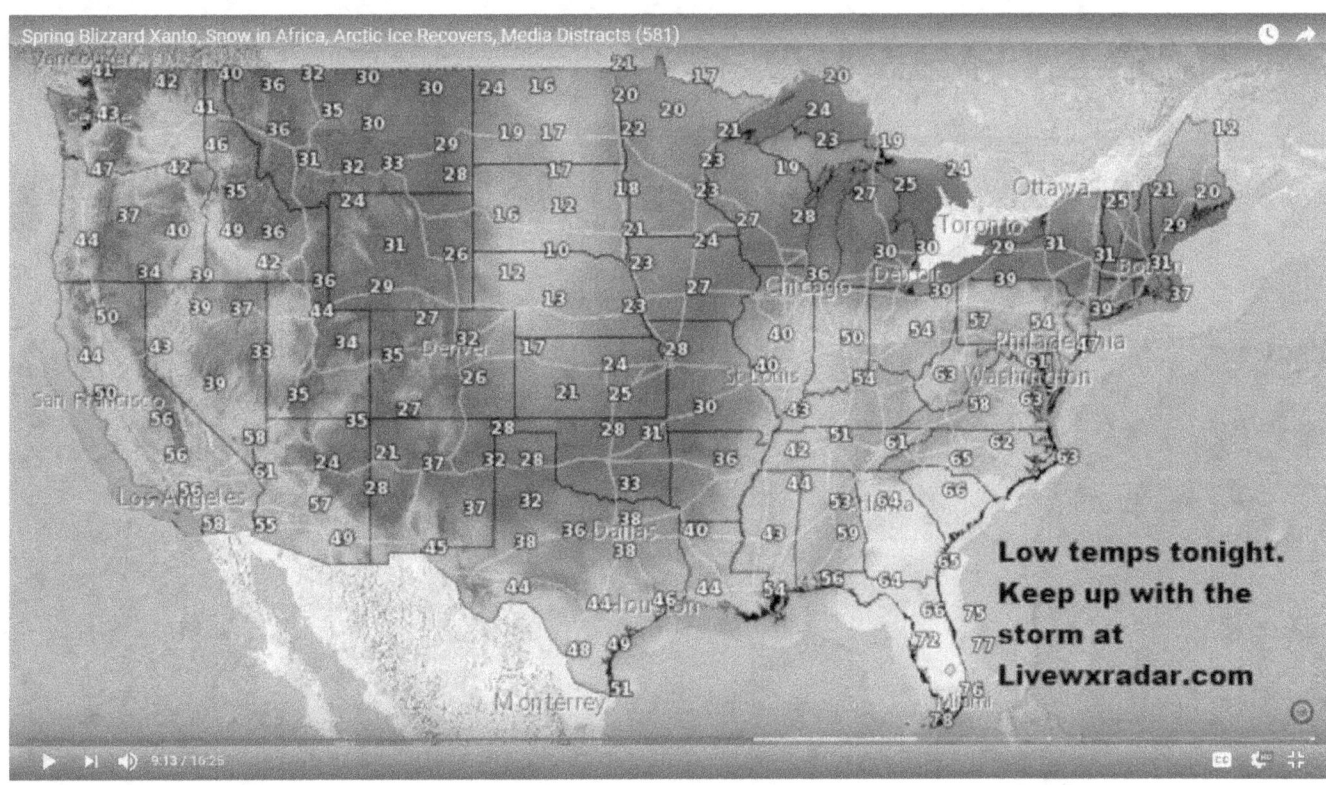

We have had already huge winter blizzards in late spring, sweeping across the grain belt of North America in April in 2018.

Massive snowfall when spring planting should be in progress

Massive snowfall swept through the region when spring planting should be in progress.

At the same time droughts ravished the American Southwest

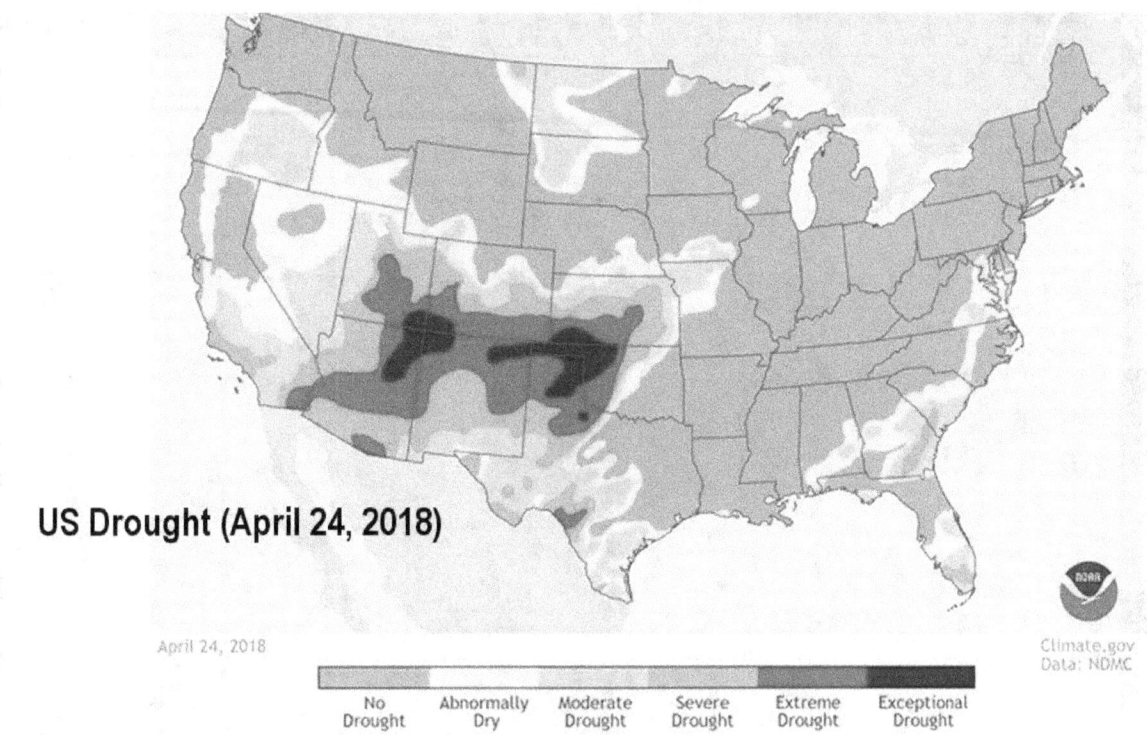

https://www.climate.gov/maps-data/data-snapshots/data-source-drought-monitor

At the same time droughts ravished the American Southwest.

Snow in North Africa in January 2017, in the Sahara

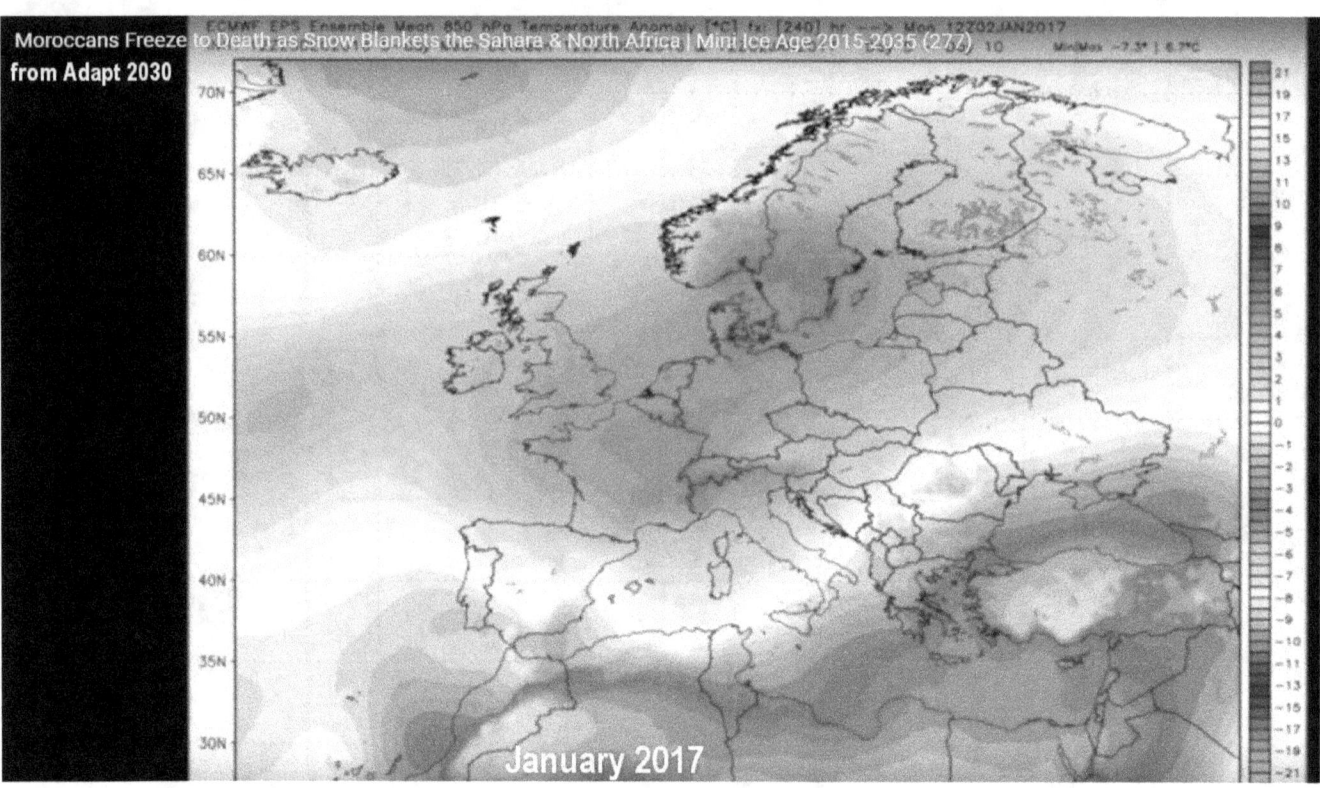

We have also had snow in North Africa in January 2017, in the Sahara, in which pilgrims were trapped and froze to death.

Wheat crop losses in Australia in the same season

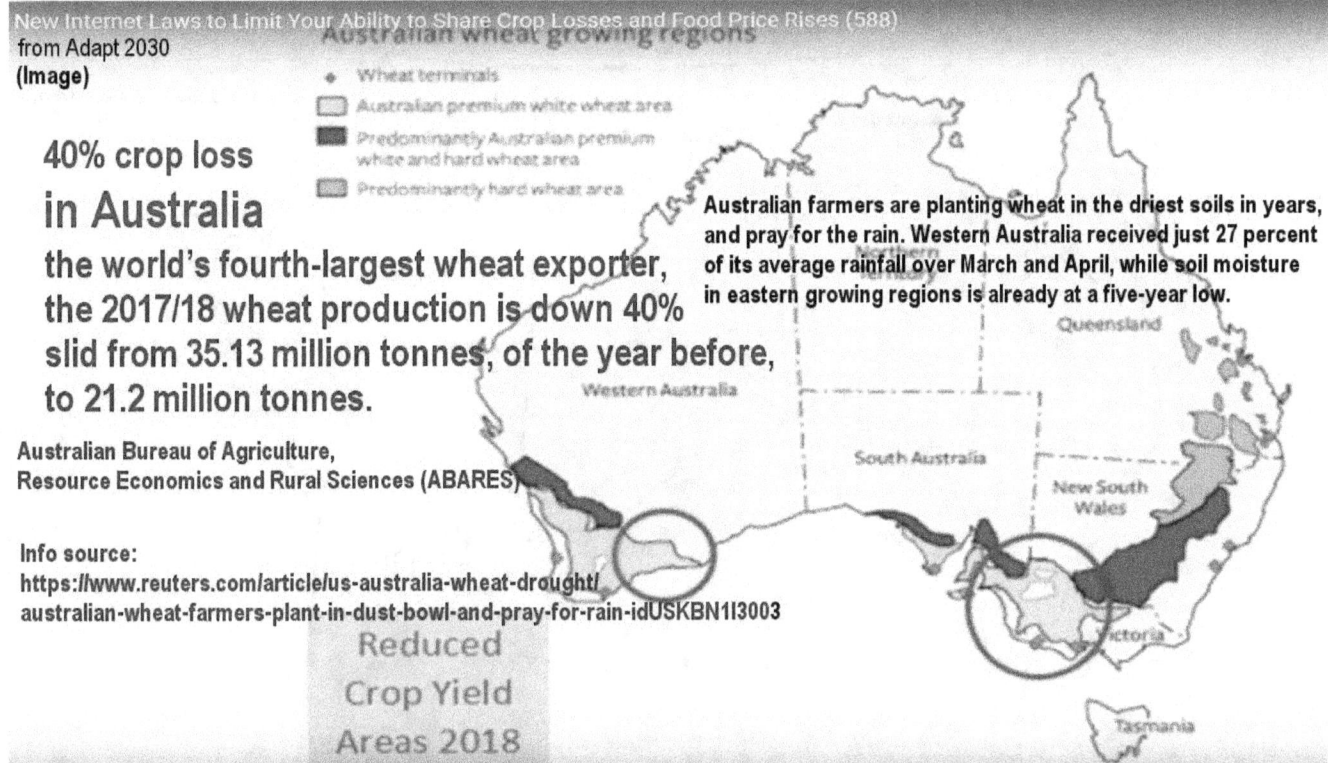

We have seen wheat crop losses in Australia in the same season, in the range of 40% due to drought and too much heat.

Part 4 - Our Near Future - The next 15 years, potentially

When climate collapse narrows the growing season

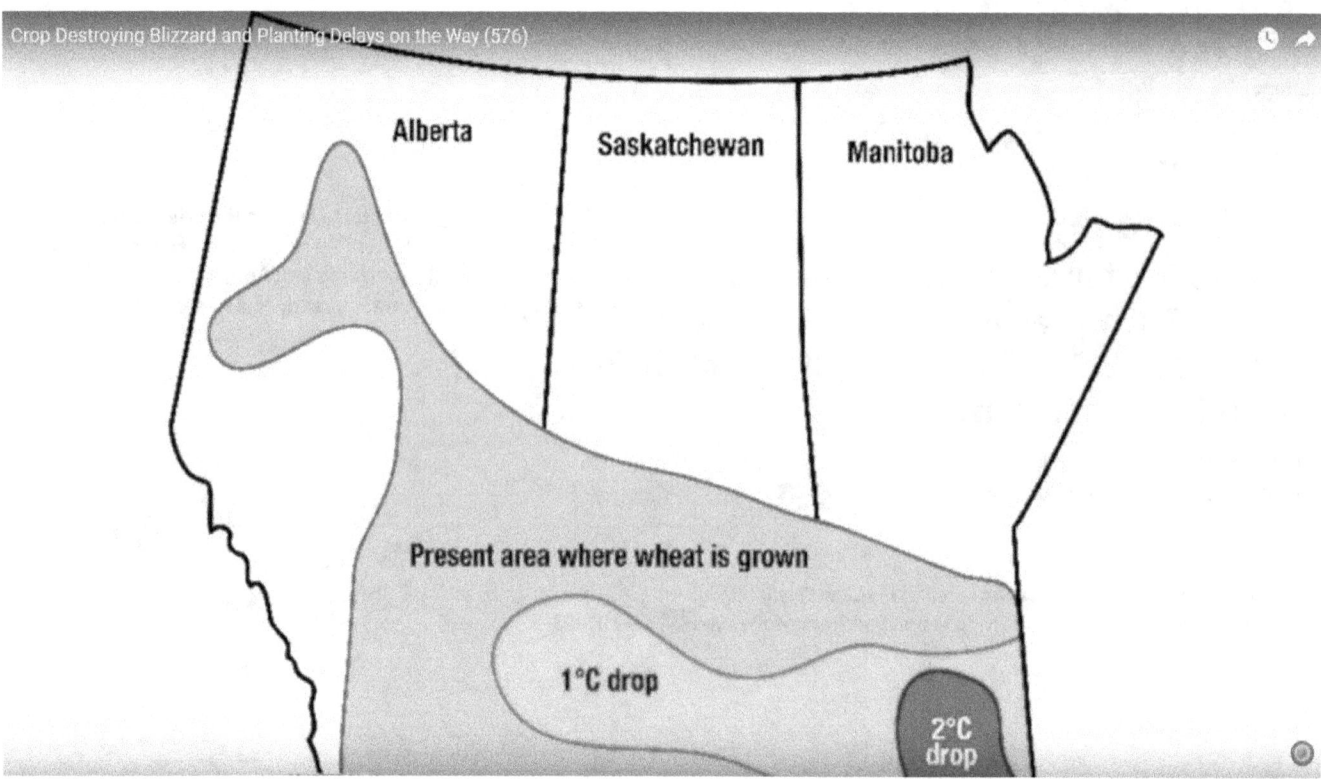

Agriculture is such a fragile thing, that when climate collapse narrows the growing season, enormous crop losses can be incurred, and eventually will take entire regions out of food production altogether. With these types of uncertainties on the horizon, how can one measure the future?

Part 4 - Our Near Future - The next 15 years, potentially

Measuring the solar dynamics in real-time

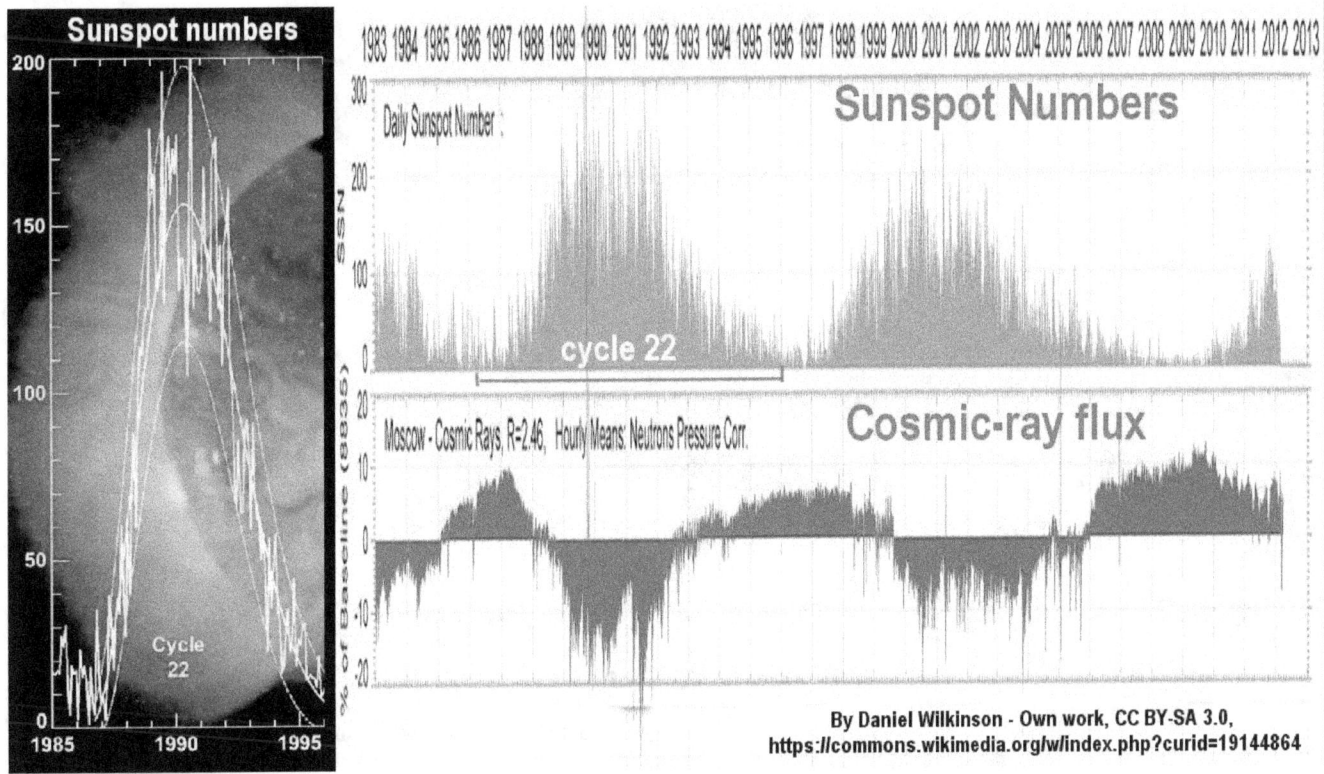

Some forms of measurements, to measure the future, became possible by measuring the solar dynamics in real-time, in the present, and thereby measure their direction of change. This became possible by measuring the changing ratios of solar cosmic-ray flux in terms of measuring the density of the neutron flux that cosmic rays generate in the atmosphere. Certain types of cosmic-ray interaction with the atmosphere generate free neutrons. Thus, the changing cosmic-ray flux can be measured in real time by measuring the neutron density fluctuation.

High-flux periods between the solar cycles

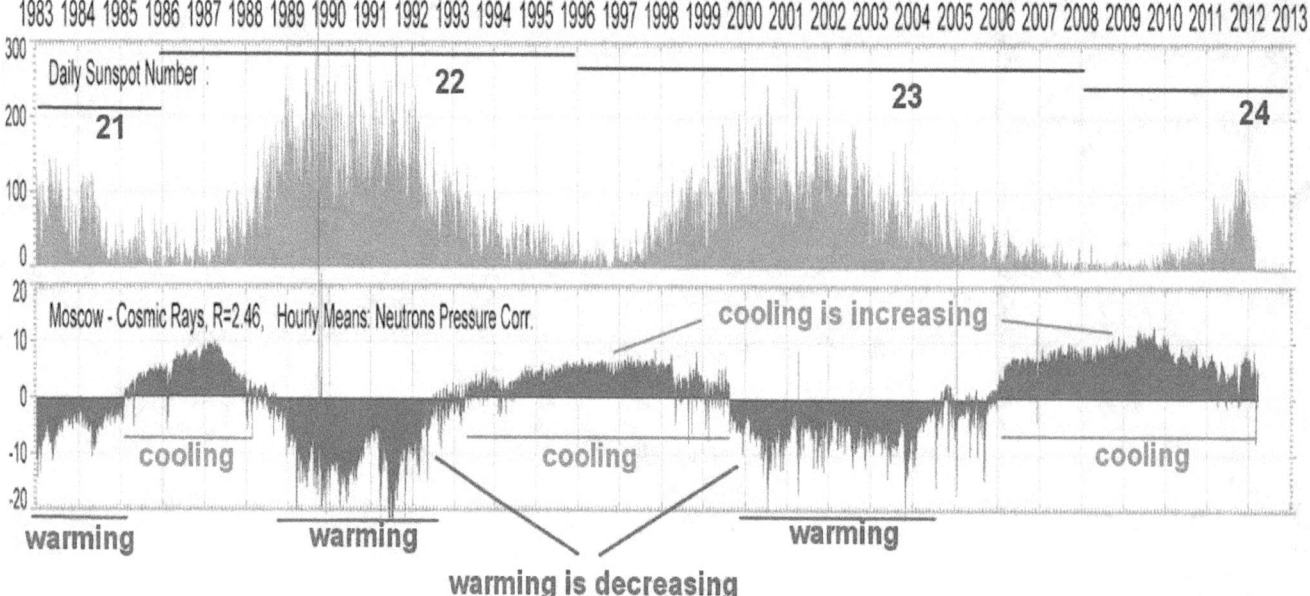

The measurements are telling us that the high-flux periods between the solar cycles, which produce colder climates, are getting larger, while the low-flux times, during the peaks of the solar cycles, which generate warmer climates that offset the cooling, are getting smaller, as the solar cycles themselves are getting weaker..

It is worth noting that for cycle 24, the cosmic-ray flux didn't diminish significantly during the peak period, but remained high all the way through the peak times. Which means that no offsetting warming had occurred at all during the peak time of this solar cycle.

If we look at the full cycle 24, to-date

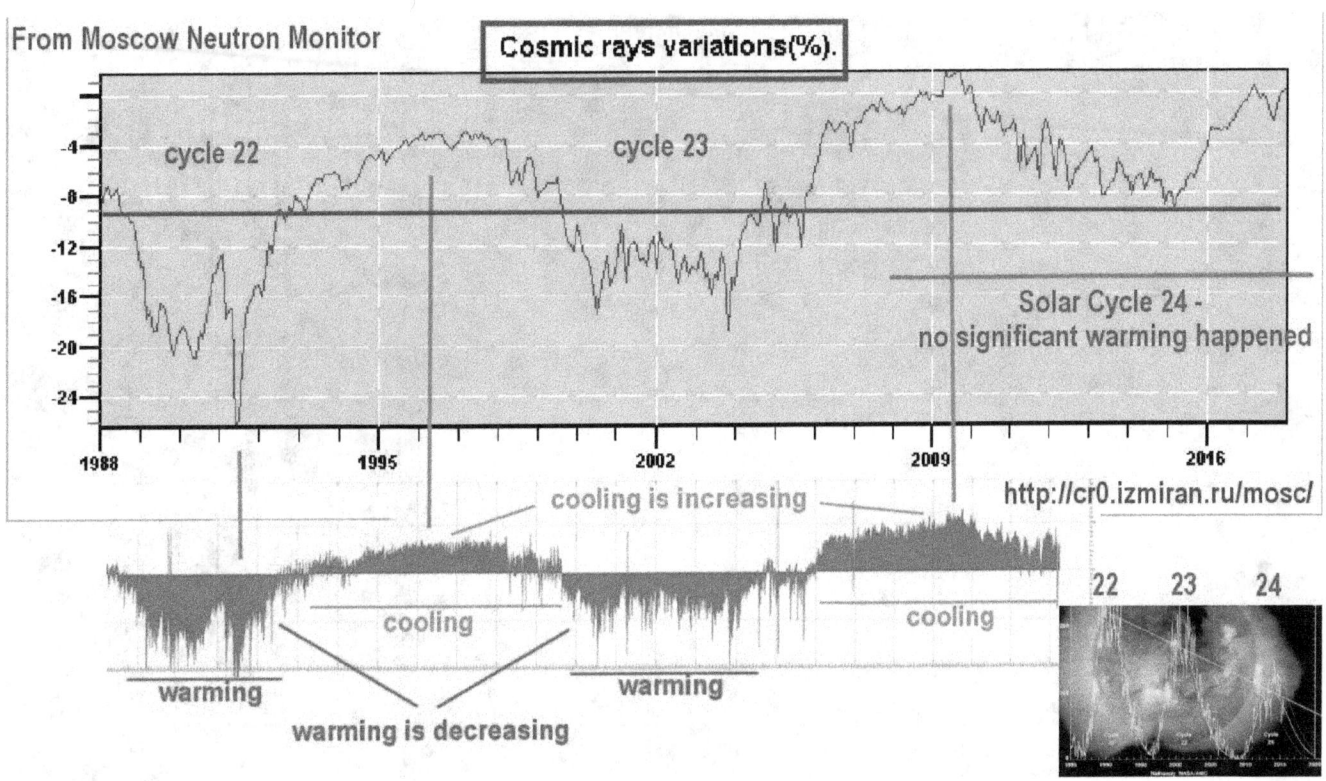

If we look at the full cycle 24, to-date, the cosmic-flux took a a little dip where warming should have occurred, and then it shot up again to extremely high levels. This means that cycle 24 was too weak to cause any re-warming during its peak period, and weaker still, will be the cycles thereafter.

In other words, the progression of the measurements tells us that the climate on Earth will be getting progressively colder, year after year, with no recovery in sight that would stop the climate collapse that has begun, of which we have seen just a few fringe effects so far, with evermore to come.

The very heartbeat of the solar system itself, is slowing down

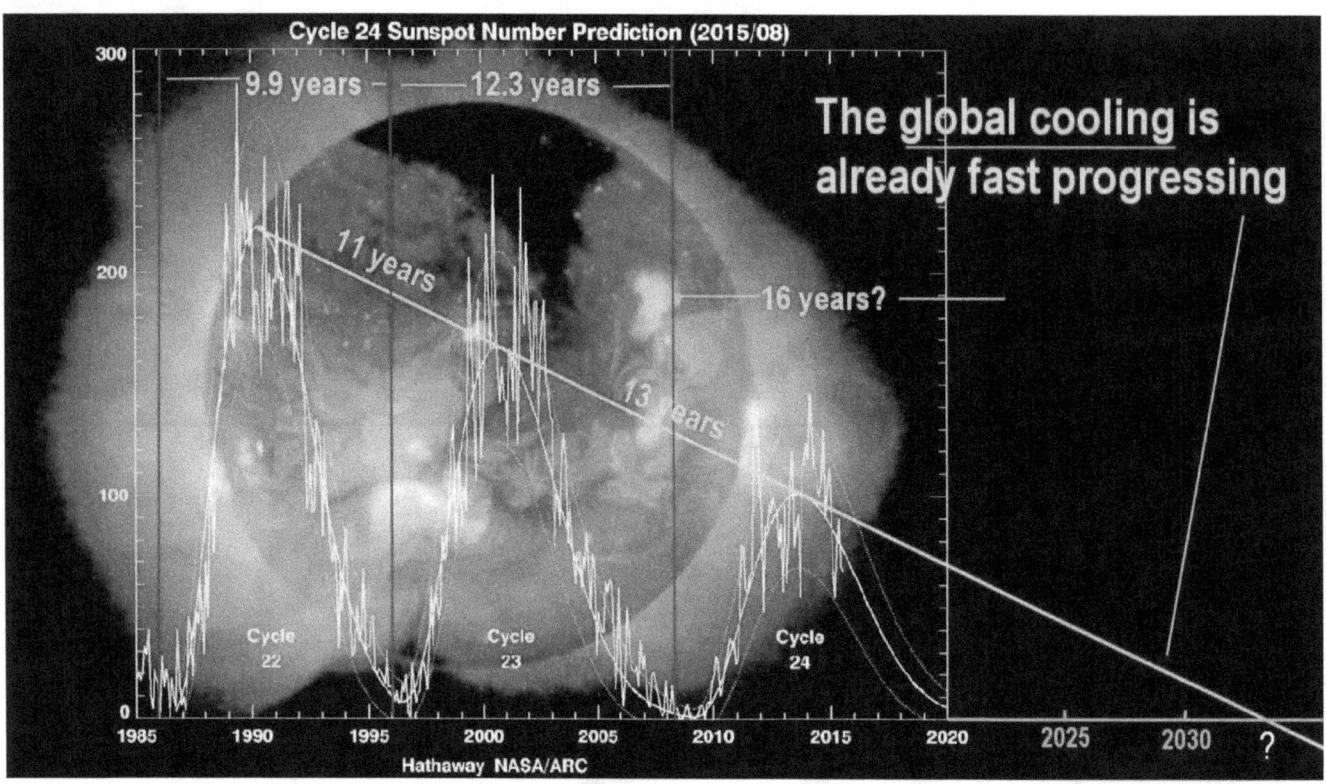

The weakening of the solar system is not as superficial as the weaker solar cycles and their increase in solar-cosmic-ray flux indicate. The very heartbeat of the solar system itself, is slowing down. The interval between the solar peak times has been increasing from the traditional 11 years to 13 years. At this rate of increase we can expect the current solar cycle to be potentially 16 years in duration, extending to 2024, and the next solar cycle peak to occur in 2030, if it occurs at all.

We used to measure the duration of solar cycles as the time between the minimums in sunspot numbers. But now, in modern times when for 817 days no sunspots had occurred in cycle 23, the old convention to look for the minimum in sunspot numbers is no longer practical. It becomes more practical therefore, to measure the time between peak sunspot numbers, or between dips in the neutron counts that occur during peak times, in these cases when sunspots no longer occur.

We may also locate the solar-cycle peaks with radio telescopes, when sunspots no longer occur.

Cycle 25 may be a candidate for this. Cycle 25 will likely be so weak that it doesn't produce any sunspots anymore, so that technology must be applied to measure the solar cycles that have become invisible.

Part 4 - Our Near Future - The next 15 years, potentially

Radio-flux measurements have collapsed by half for cycle 24

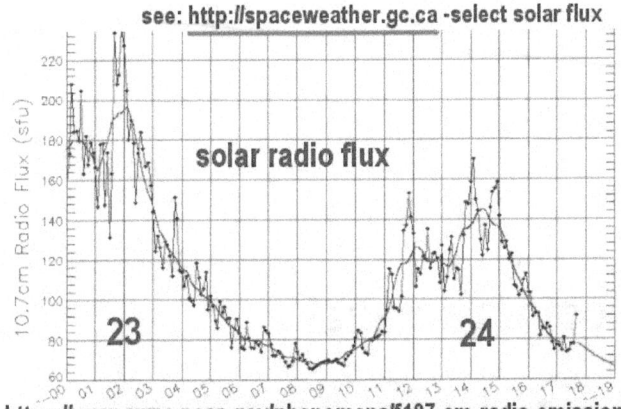

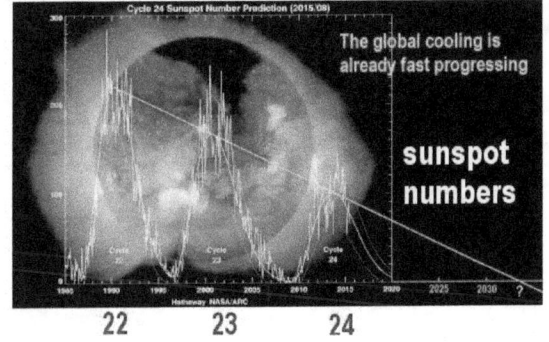

Radio telescopes are useful for that. When the radio signal from the Sun is measured in the 10.7 cm band, the measured signal strength matches the sunspot numbers closely, and follows the solar cycle into the sub-visible level. The radio-flux measurements have collapsed by half for cycle 24, in comparison with previous solar cycles. One researcher suggests that by the time we get to the end of cycle 24, potentially in 2024, the Sun will be 'asleep,' whatever this may mean.

When nothing is predictable anymore, it can be said with certainty that we "haven't seen anything yet", as the saying goes, especially in terms of climate collapse and corresponding crop failures, or even agriculture collapse altogether, in the volatile regions. Plan-B anyone? Yes, Plan-B becomes critical in the near future, in the shadow of the weakening Sun.

Now, the Sun has lost its 'top hat'

Now, the Sun has lost its 'top hat'
where will it all end?

Now, the Sun has lost its 'top hat' - where will it all end?

The Sun's polar magnetic fields have diminished

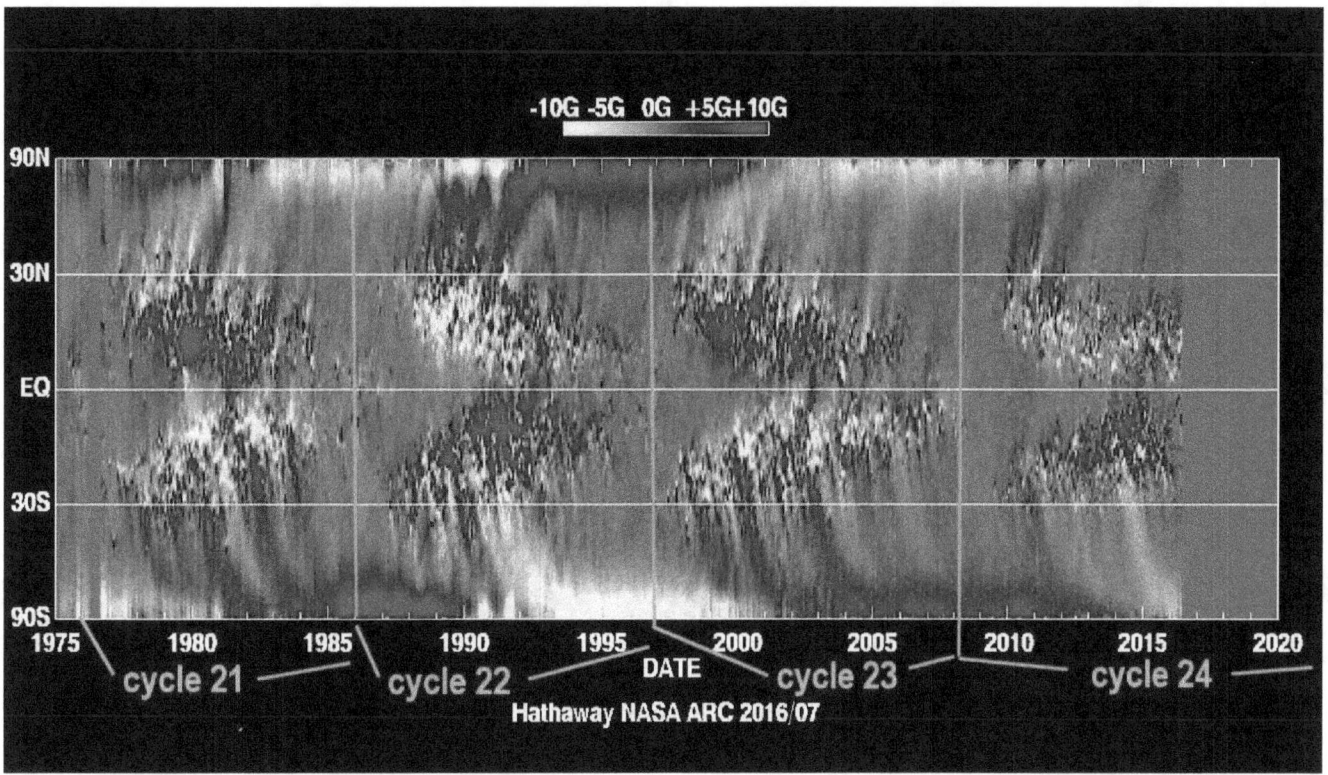

Even the magnetic fields of the Sun are getting weaker, especially the Sun's polar magnetic fields. The polar magnetic fields near the 90 degrees latitude, flip their magnetic orientation at the peak of the solar cycles, which is evident in the reverse of the colors in this magnetograph, for every solar cycle. It is interesting to note here, that the intensity of the Sun's polar magnetic fields have diminished. The fields for cycle 23 were weaker than those fore cycle 22. And for cycle 24, the southern field was weaker again after it flipped in 2014, while the northern field, when it should have flipped, simply vanished.

Part 4 - Our Near Future - The next 15 years, potentially

A dynamic feature of the Sun has suddenly failed

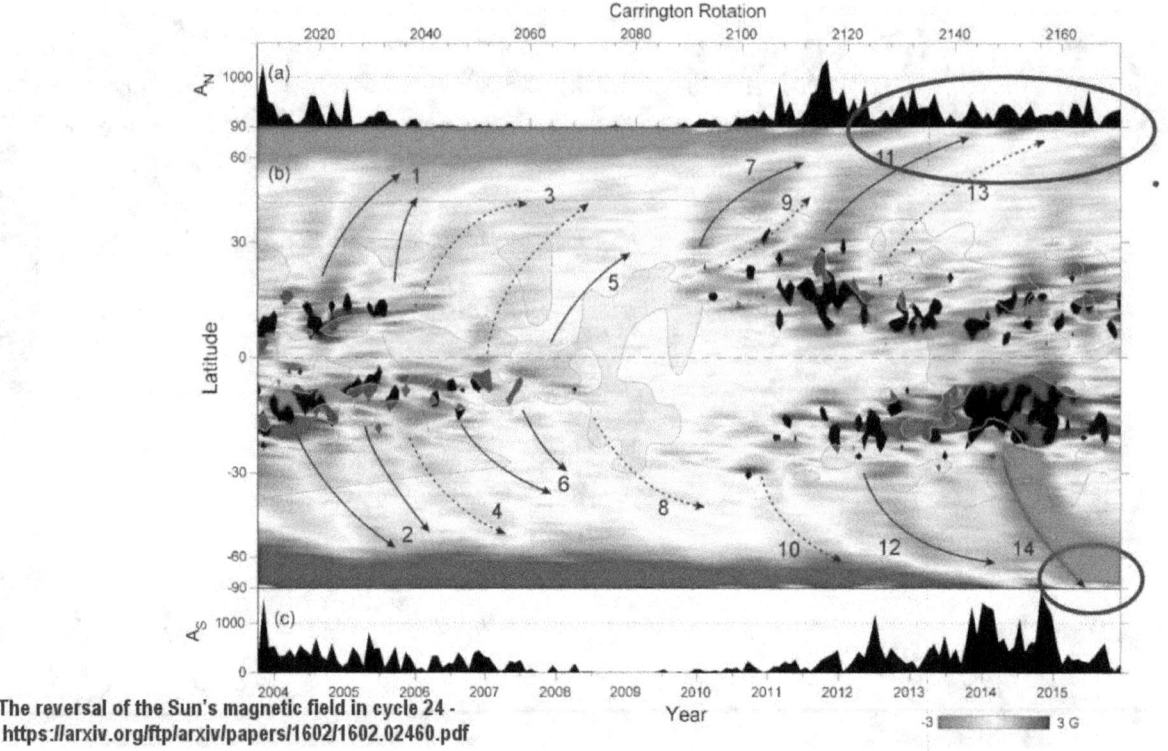

The reversal of the Sun's magnetic field in cycle 24 -
https://arxiv.org/ftp/arxiv/papers/1602/1602.02460.pdf

The entire system that powers the Sun and its magnetic fields, has become so weak that a dynamic feature of the Sun that has never failed. has suddenly failed. It vanished as if it never happened before.

How close to the End are we?

How close to the End are we?

When will the boundary time-zone end and glaciation begin?

How close to the End are we? When will the boundary time-zone end and glaciation begin?

The start-up of Ice Age glaciation caused by the Sun changing states

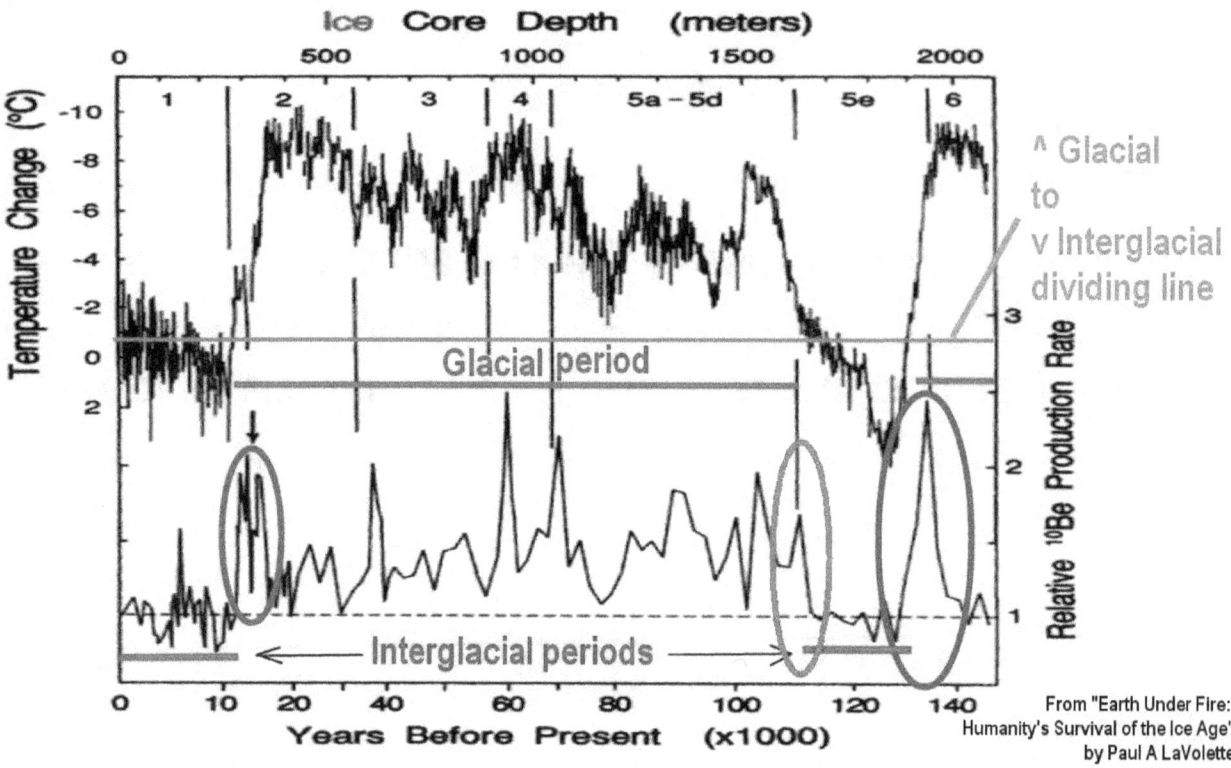

From "Earth Under Fire: Humanity's Survival of the Ice Age" by Paul A LaVolette

How close the ongoing weakening of the solar system that powers the Sun has brought us to the inevitable collapse of the primer fields by which the next Ice Age begins, cannot be precisely forecast for the lack of historic references. We only know with certainty that the start-up of Ice Age glaciation (encircled green), and also the start-up of interglacial periods (encircled blue), were caused by the Sun changing states. This is what the measured beryllium-10 isotope ratios are telling us, because beryllium-10 is produced exclusively by cosmic-ray interaction with the atmosphere.

When the last interglacial period ended 110,000 years ago, a dramatic increase in the Beryllium-10 ratio occurred, as one would expect for the Sun going into hibernation by it loosing its plasma mantle with the collapse of the primer fields, so that the full solar cosmic-ray flux was then reaching the Earth. Coincident with these effects happening, the glacial period began. The evidence indicates that all Ice Age theories that are not founded on the Sun changing states, are necessarily false.

In this context it becomes interesting to note that when the Sun is re-activated out of its hibernation mode, with increasing plasma density, large cosmic-ray events occur until the primer fields are formed anew, and the Sun's plasma mantle is re-established. Once this is done, the cosmic-ray flux drops back to interglacial levels.

In the case of the start-up of the current interglacial period, around 12,000 years, we see two beryllium spikes preceding it.

The Younger Dryas re-glaciation

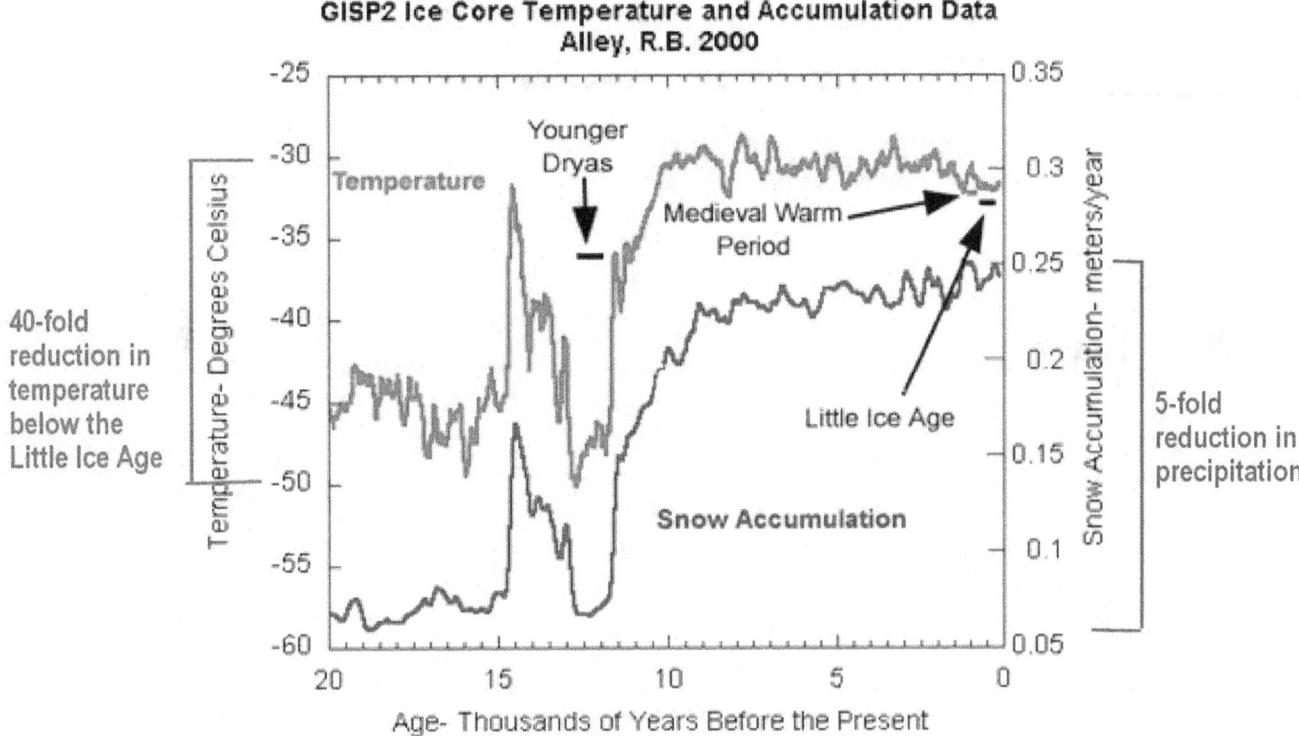

In this case the initial start-up had evidently failed, as there probably wasn't enough plasma density building up at the time, for the primer fields to form. The Younger Dryas re-glaciation was the result of that failure to form the primer fields. When the plasma-streams recovered, 3,500 years later, the start-up succeeded.

It is widely believed in mainstream science that the Younger Dryas re-glaciation was the result of melt-water effects and ocean current fluctuations.

Measured Beryllium ratios obsolete numerous theories

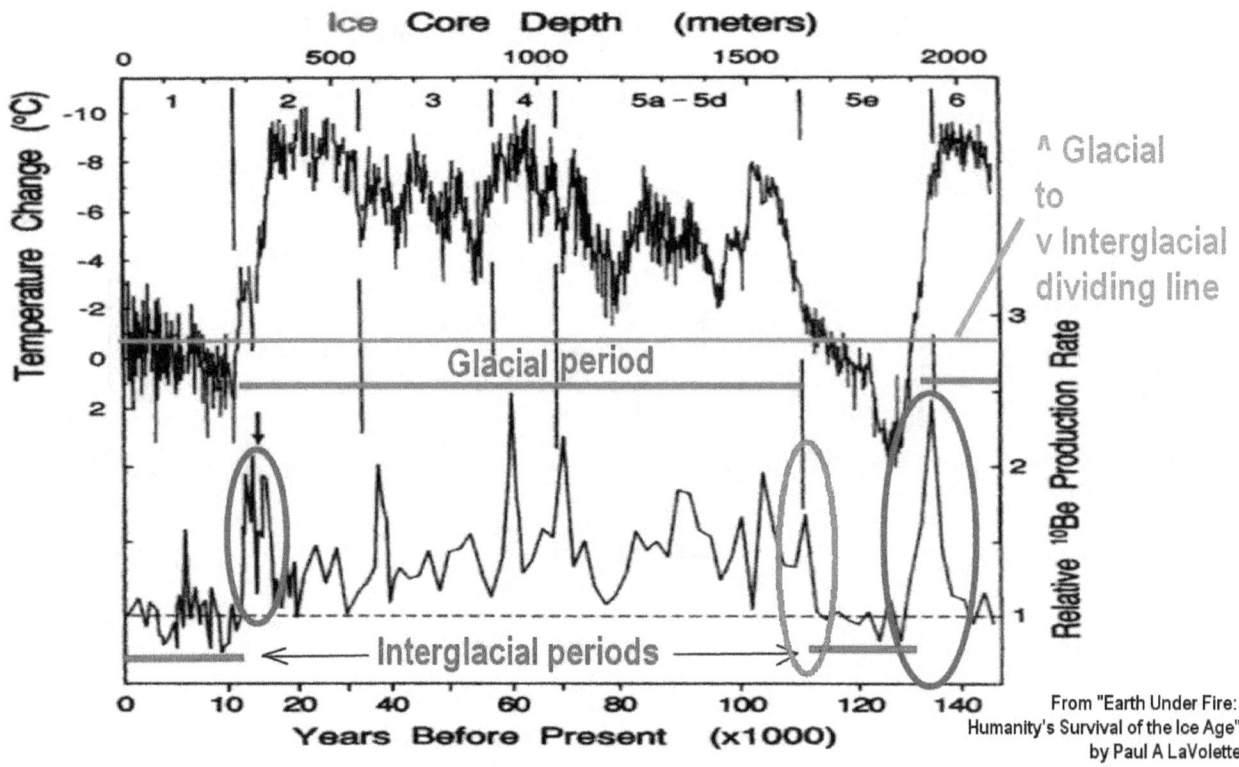

The physically measured Beryllium spikes obsolete these theories.

The measured Beryllium ratios also obsolete the numerous theories of gradual and long-term Ice Age start-up transitions. The isotope ratios speak of a rapid, digital-type, on-off phase shift in solar dynamics, as starters for the Ice Ages, rather than long drawn-out transitions spanning thousands of years.

Dr. Zbigniew Jaworowski suggested in his 2003 paper, "The Ice Age is Coming," that the phase shift to glaciation will likely be swift, as swift as a single year, and will likely begin without a special warning.

The swarm of the increasing fringe effects that we encounter

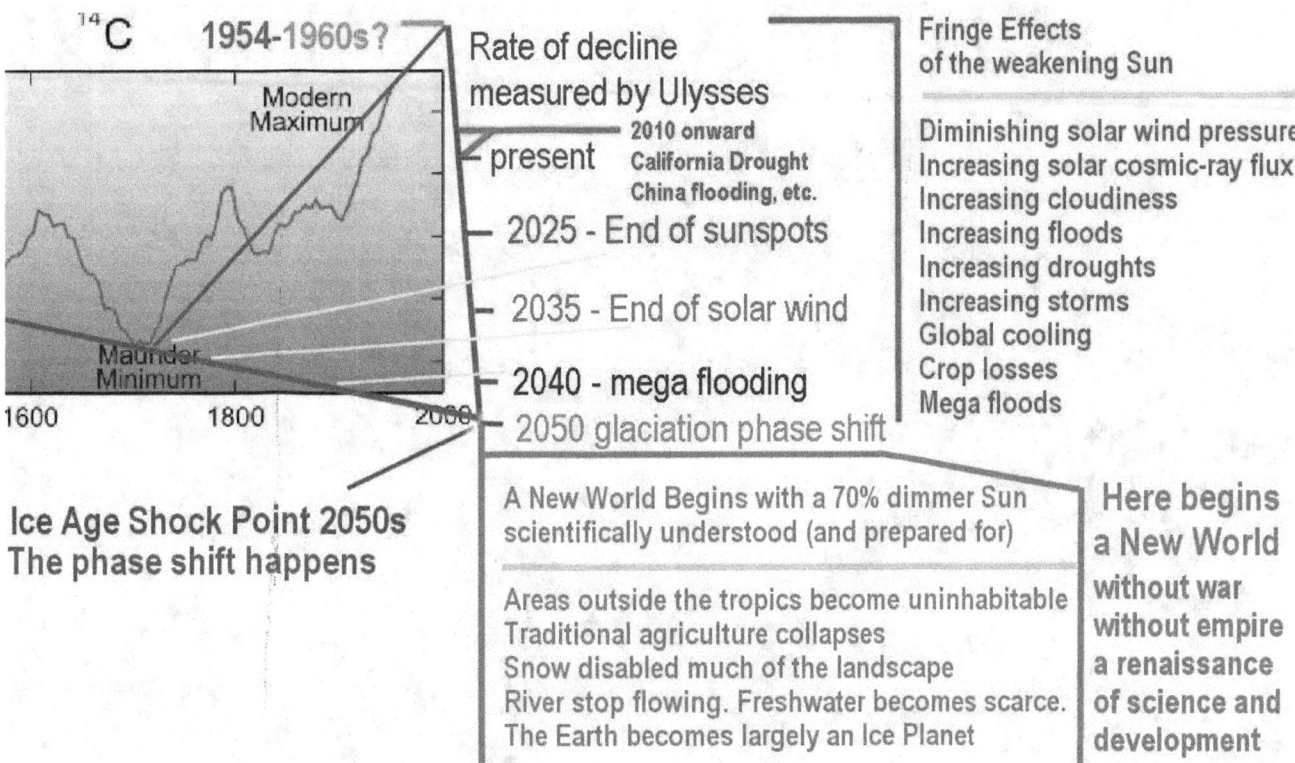

The closest that we can come to a warning written on the wall, is the swarm of the increasing fringe effects that we encounter in our world the deeper we move into the boundary zone.

Even the Earth's rotation may be affected

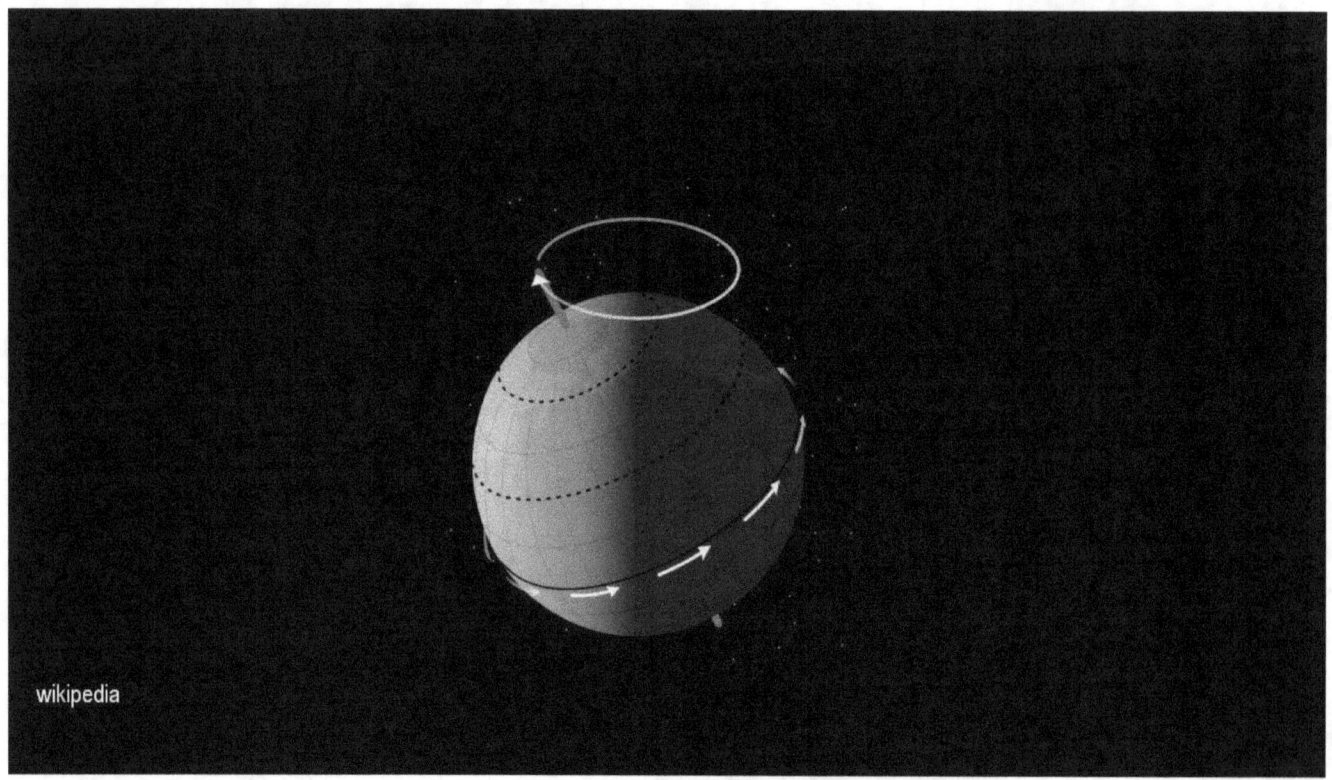

Even the Earth's rotation may be affected by the weakening of the solar system that began as far back as the interglacial optimum, 8,000 years ago. Analysis of historical astronomical records reveals that the Earth rotation has slowed 2.3 milliseconds per century during the last 2,800 years. This period covers a large portion of the asymmetric period in which the primer fields were diminishing more rapidly.

Gravitational drag by the moon on the Earth

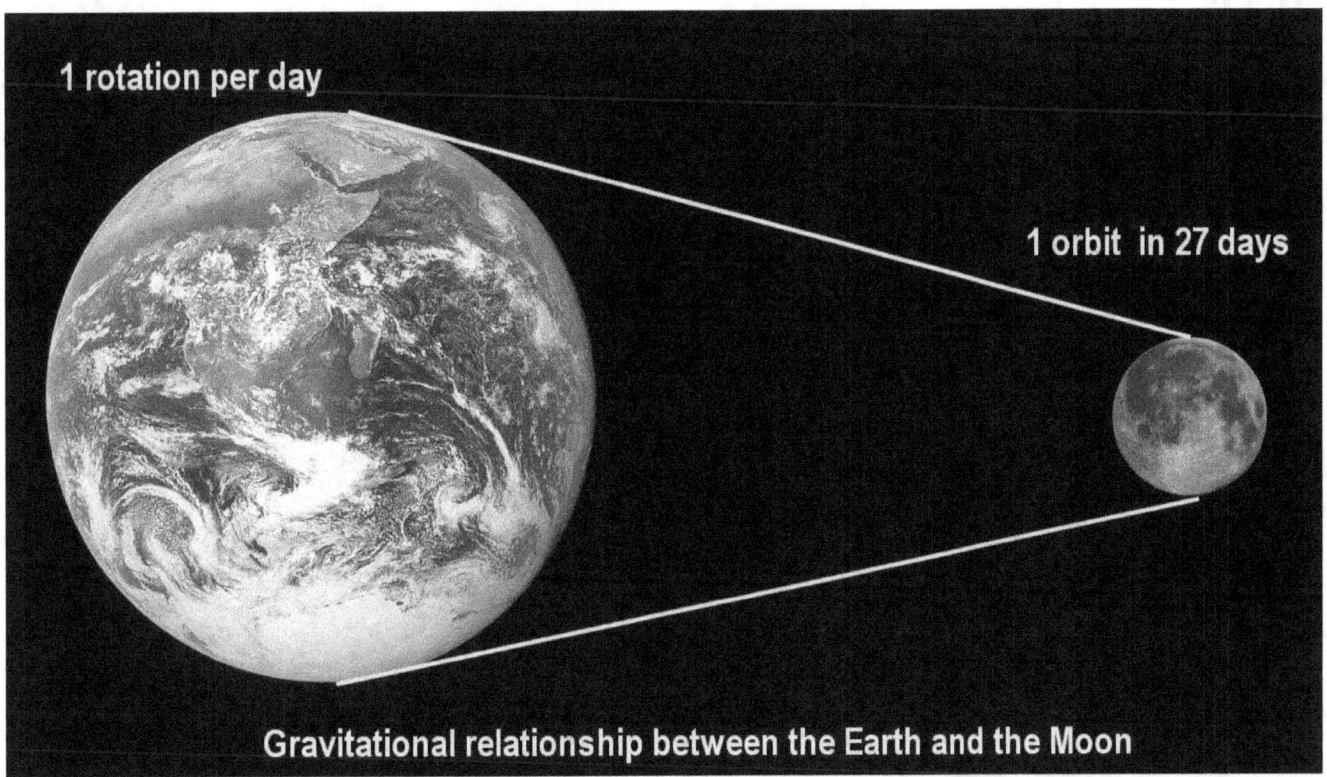

Some scientists say that the slowing is the result of gravitational drag by the moon on the Earth.

The weakening electrodynamics of the primer fields and their effects

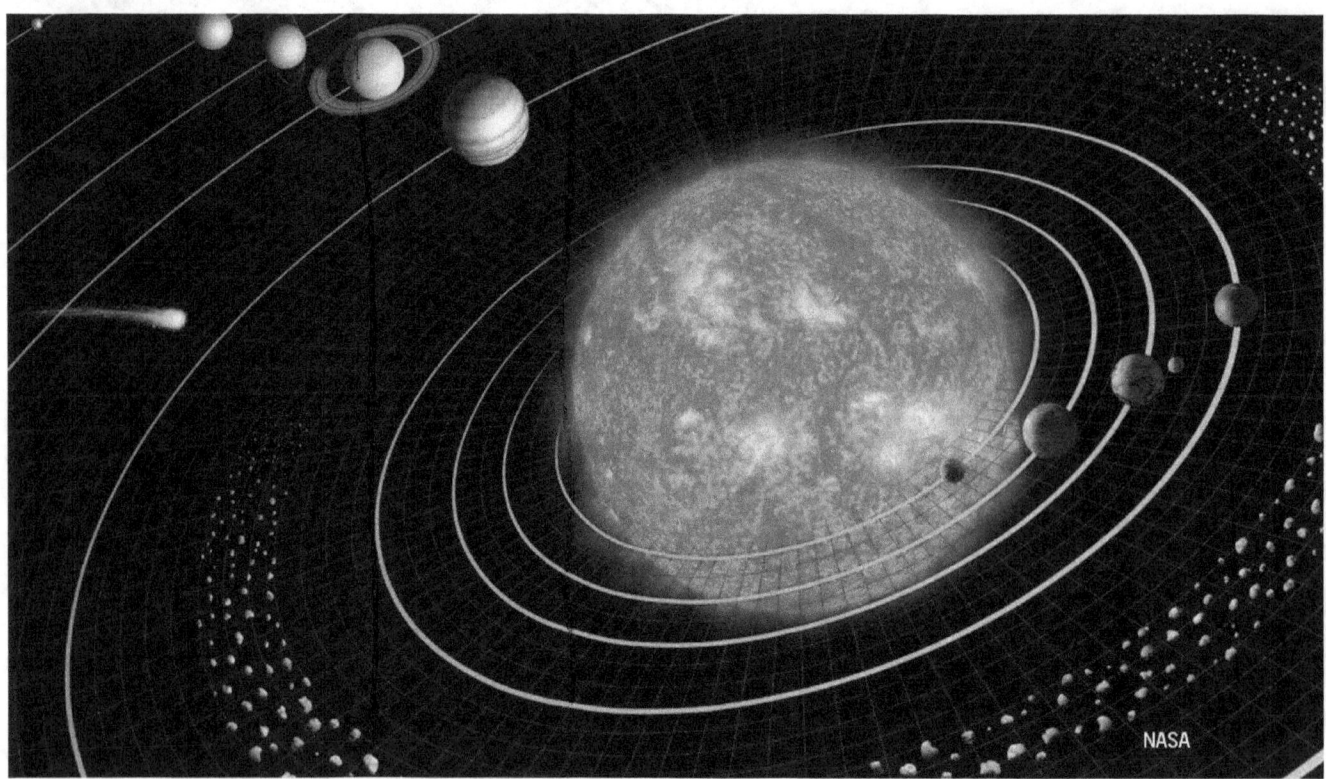

It is far more likely, however, that the slowing rotation reflects the weakening electrodynamics of the primer fields and their effects, which actively maintain the spinning of the Earth and of the planets, except for the innermost planets, Mercury and Venus, that are dominated by the Sun rather than the primer fields. Naturally, the large mass of the Earth, would keep the rotational variation at a low ratio.

Dramatic magnetic pole drift has been measured since the 1800s

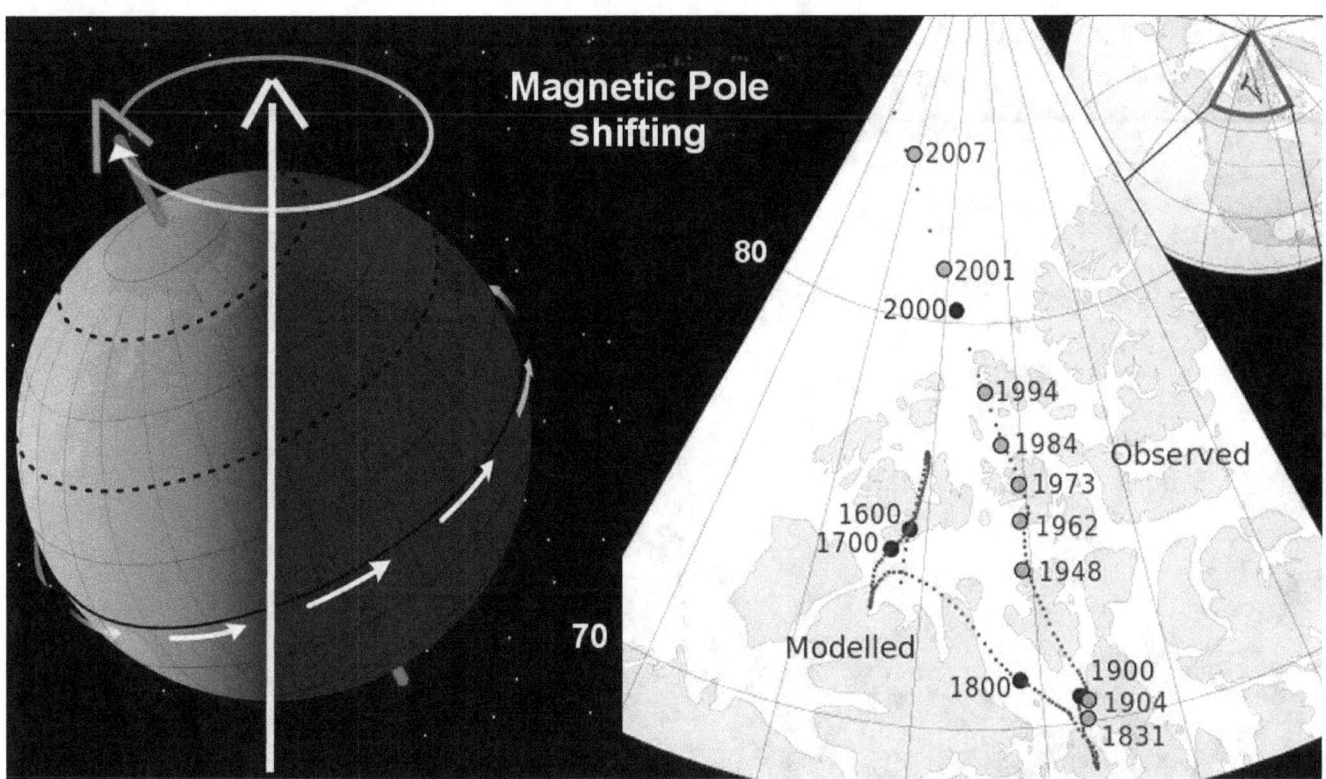

That the primer fields are getting weaker is also evident in other ways. It is most dramatically apparent in the weakening of the Earth's magnetic field, and in the dramatic magnetic pole drift that has been measured since the 1800s.

A part of the Earth's magnetic field is generated in line with the Earth spin axis, by an internal dynamo effect. This means that if it wasn't for the magnetic effects of the primer fields, the magnetic North Pole, and the geographic North Pole would be at the same location. But this is not the case. The magnetic orientation of the primer fields, which acts perpendicular to the planetary ecliptic, bends the Earth's magnetic field towards it, and away from its spin axis orientation, to a maximum of 23 degrees, according to the inclination of the spin-axis of the Earth.

With the effect of the primer fields now getting weaker, the deflection of the Earth's magnetic field away from its spin axis has been dramatically reduced, so that the Earth magnetic field, measured on the ground, began to approximate evermore the spin-axis orientation. The diminishing deflection away from it gives us a measurable indicator of the weakening of the solar primer fields. Of course at high altitudes, high above the Earth where the primer fields are the dominant force, does the deflection of the magnetic pole remain in the 20 degrees range.

Evident on the Sun itself

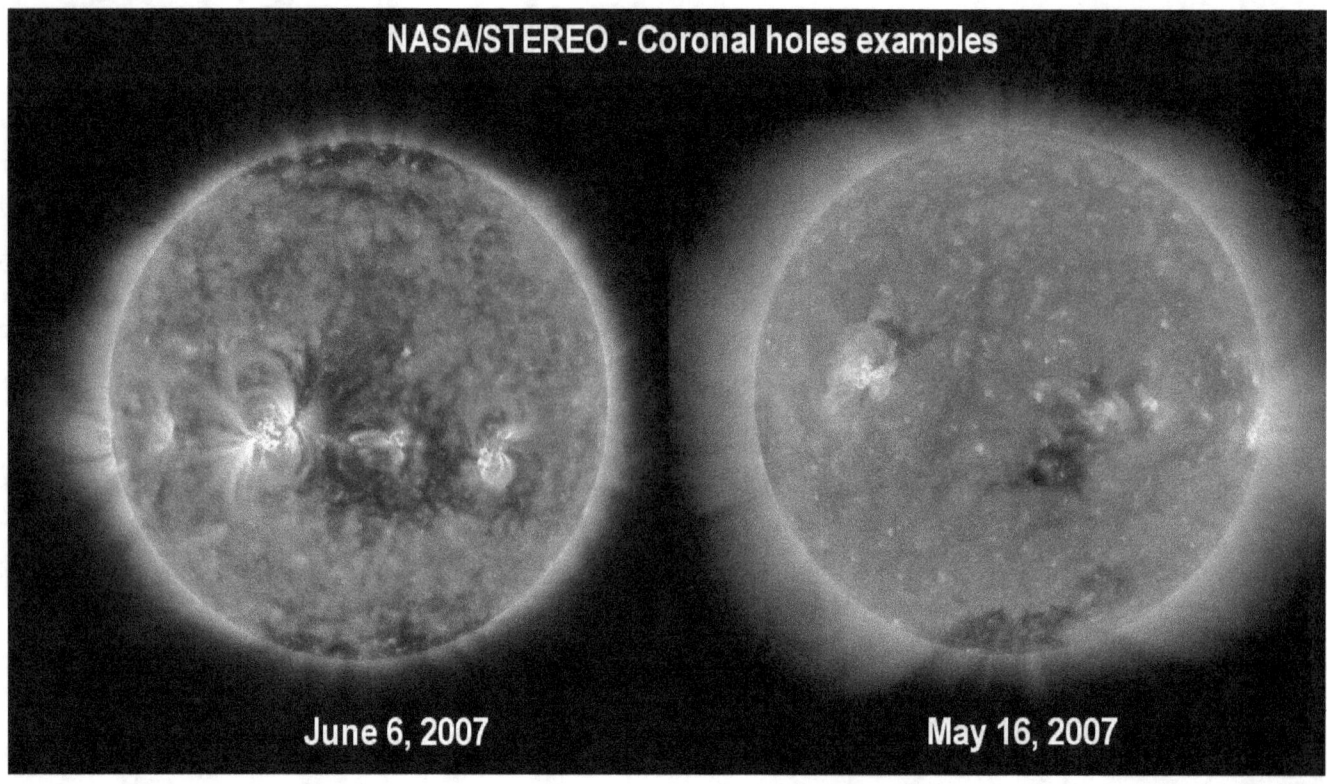

That the system that powers the Sun is getting weaker, is also evident on the Sun itself. A few years ago, around 2007, a feature on the Sun that had not been noticed before, suddenly became prominent. The forming of coronal holes began. When the holes were first noticed in 2007. the discovery made headlines.

Now, 11 years later, when the entire face of the Sun is peppered with holes

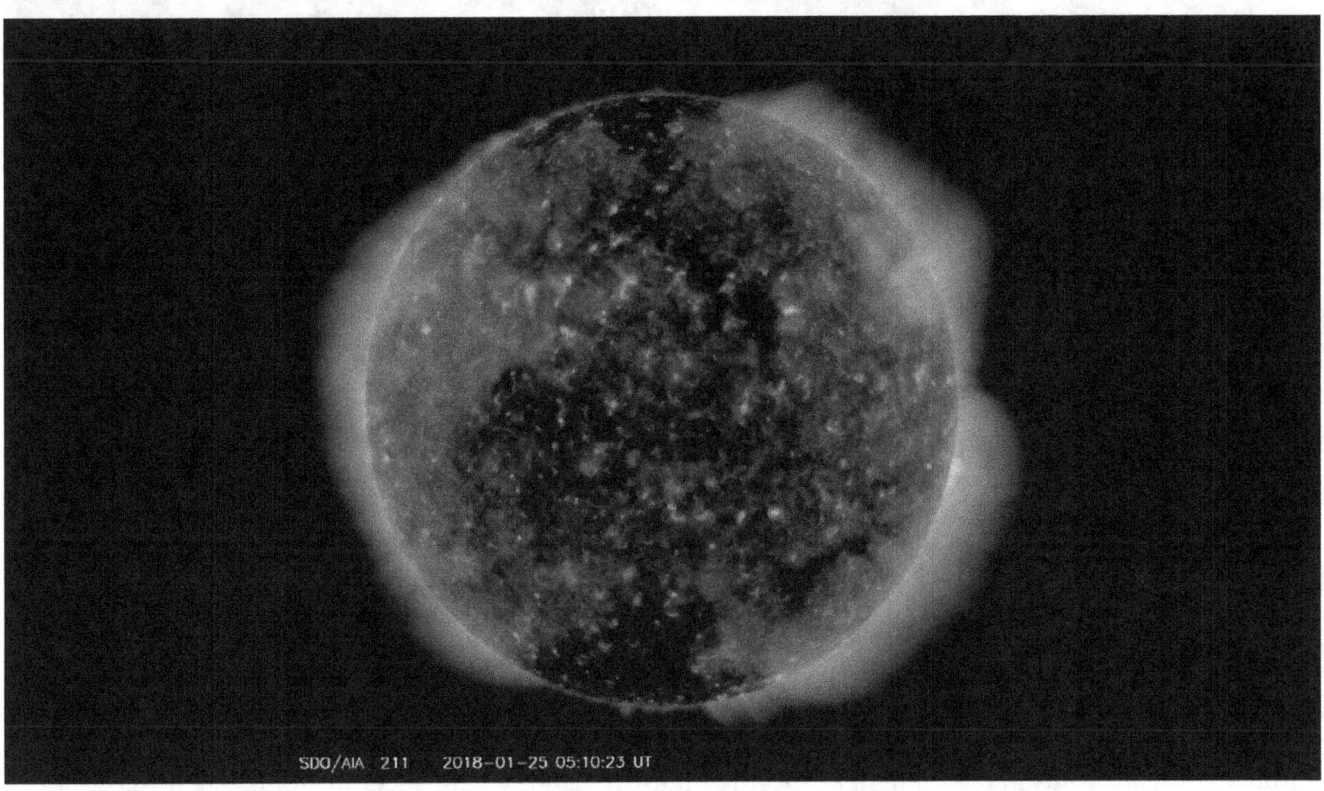

Now, 11 years later, when the entire face of the Sun is routinely peppered with holes, the phenomenon raises no eyebrows anymore. While the climate effects in terms of floods are often devastating, the link of them to the coronal holes are simply ignored.

Coronal holes are warning signs that we should heed

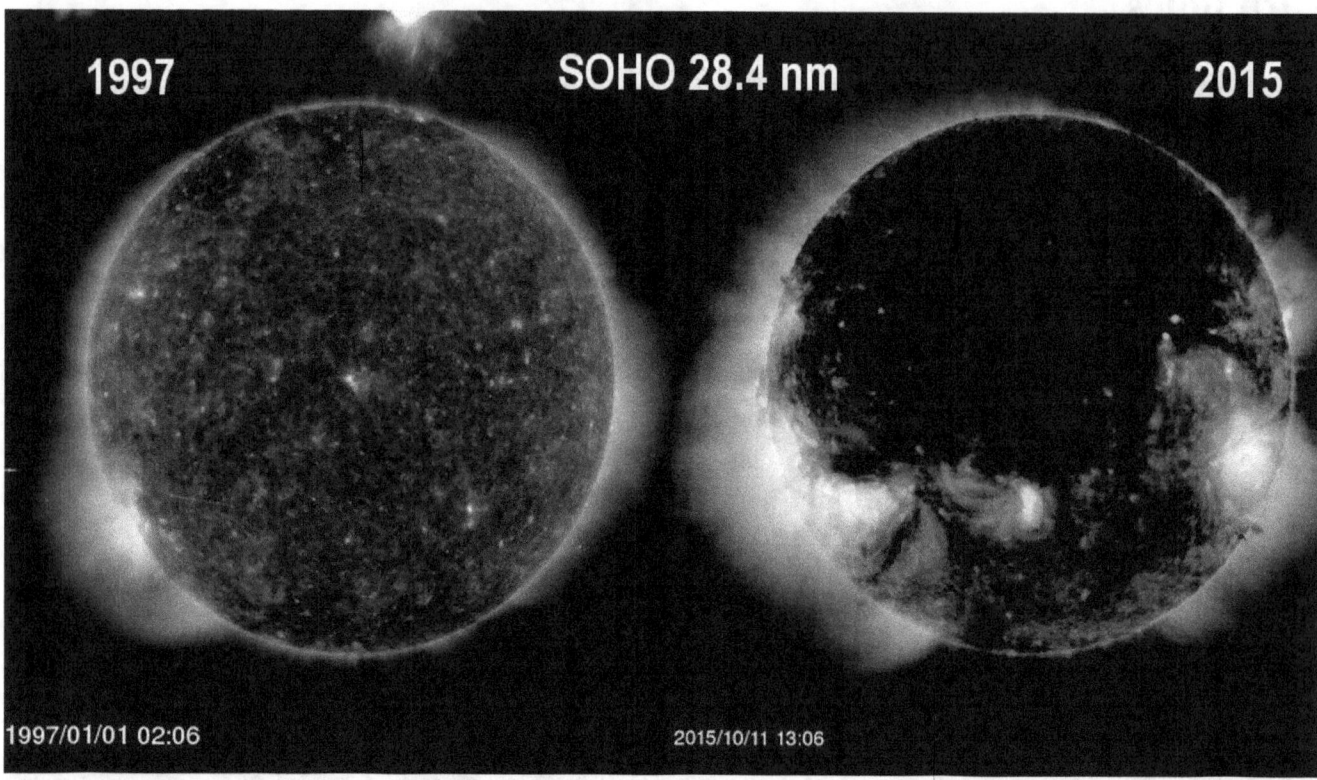

Coronal holes are voids in the sunlight at specific high-UV bands where specific atomic elements emit light. The voids appear to be physical voids, or density voids, in the plasma corona around the Sun, because these voids allow larger volumes of solar cosmic-ray flux to penetrate the corona and affect the climate on Earth with increased cloud nucleation. Are this effects all warning signs that we should heed that something big is in the making?

The giant flash flood event in 2015

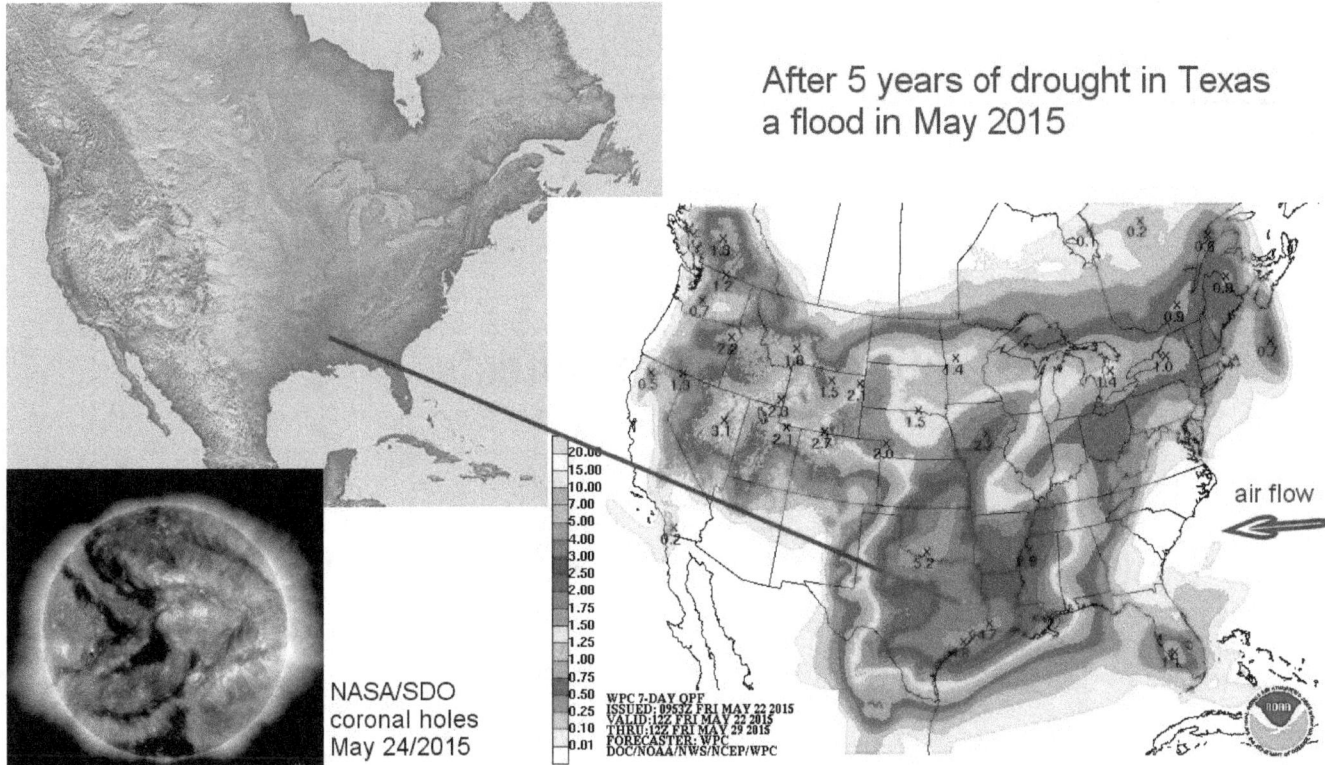

A case in point is the giant flash flood event in 2015. After years of drought, a flash flood event spread from Texas all the way to Canada. This happened in the shadow of a rather substantial coronal-hole event.

The weakening Sun has caused many such sudden climate effects. Some of the effects are short, but dramatic. Others are gradual and long in duration, and even more dramatic.

Distracting climate effects

Distracting climate effects

Distracting climate effects

Concerns are raised over the thinning of Arctic sea ice

Some of the big climate effects also have big secondary effects. One of these is Arctic warming. Concerns are raised over the thinning of Arctic sea ice, or over glaciers melting away on Elsmere island in the far North.

Arctic-warming the result of global cooling

While the Arctic ocean friezes over each winter, increasingly larger areas have been melting during the Arctic summers. The increased melting is cited as proof of manmade global warming, while the Arctic-warming phenomenon is in fact the result of global cooling.

Cold air is heavier than warm air

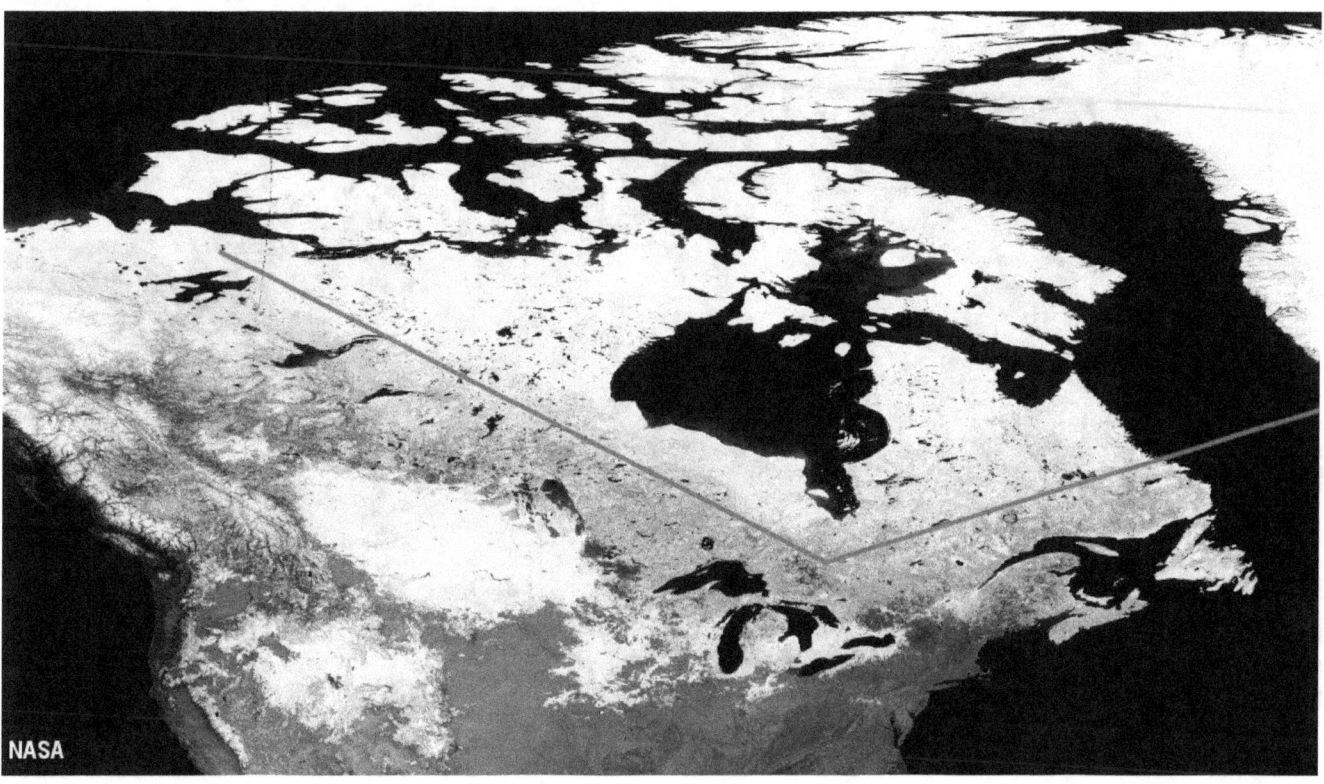

Cold air is heavier than warm air. This means that the cold air mass over the northern Canadian shield, is forced southward by the centrifugal force of the rotation of the Earth.

The colder the temperature is, the heavier is the air

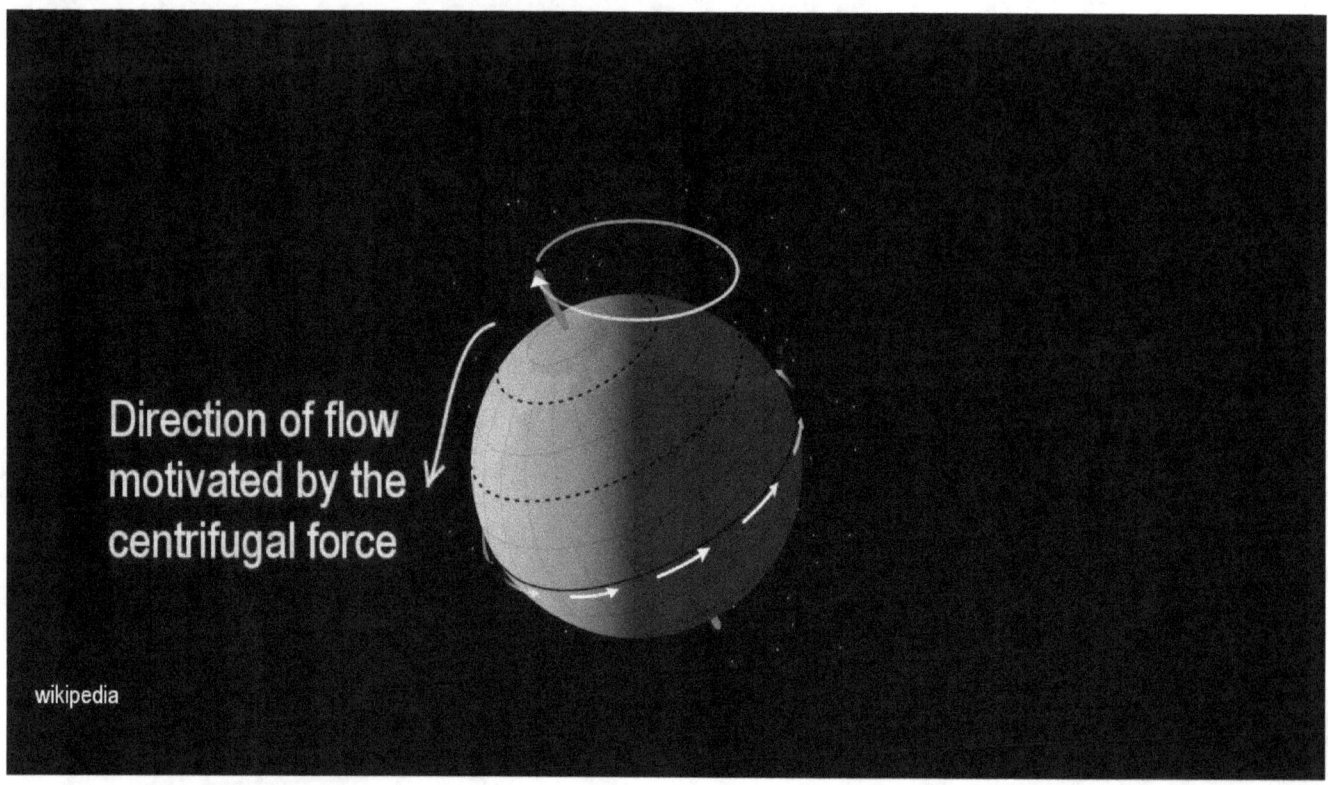

While the rotation of the Earth remains constant, the centrifugal air movement does vary. It varies with temperature. The colder the temperature is, the heavier is the air, and the stronger it is forced southward by the centrifugal force of the rotation of the Earth.

The circulation is termed 'The Polar Mobile Anti-Cyclone'

The cold air typically flows across the central U.S.A. between the Rocky Mountains in the West, and Appalachian Mountains in the East, into the Gulf of Mexico, where the moving air-mass warms up and begins to flow back into the Arctic. A giant air circulation pattern results by this process that carries warm air into the Arctic.

The circulation is termed 'The Polar Mobile Anti-Cyclone'. It is driven by cold air. This means that the colder the northern regions over Canada become, the stronger the circulation system operates, and the warmer the Arctic becomes. The return path sweeps across both the East and West coasts of Greenland, and further north, across Elsmere Island.

Thus, the resulting Arctic warming actually stands as measured evidence that the Earth is getting colder. Here is where the danger lies.

Part 4 - Our Near Future - The next 15 years, potentially

The stronger the Polar Mobile Anti-Cyclone operates

The stronger the Polar Mobile Anti-Cyclone operates, the greater is the danger that the cold-flow from the North endangers the Canadian and American agriculture, like this winter blizzard in April that spread snow and sub-zero temperature across the North American grain belt at a time when planting should be in progress.

While agriculture recovered from blizzard Xanto, it may not recover when such blizzards strike al late as May, June, or July. The current climate collapse is moving us in this direction. We will face these crisis increasingly over the next 15 years.

This systemic cold outflow that is now getting colder, year after year, will in due course overwhelm agriculture in the affected regions, and ultimately disable the affected nations. Without food production a nation ceases to exist.

In Russia and Europe the cold flow, named the 'Beast from the East,'

The ultimately overwhelming effect on agriculture, that the fast-cooling climate of the Earth invariably has, is already being felt worldwide with different effects in different regions, such as in Russia and Europe where the cold flow, named the 'Beast from the East,' is mainly westward oriented by the latency effect of the rotation of the Earth, flowing out of the cold northern regions in the East.

The southward component of the cold flow

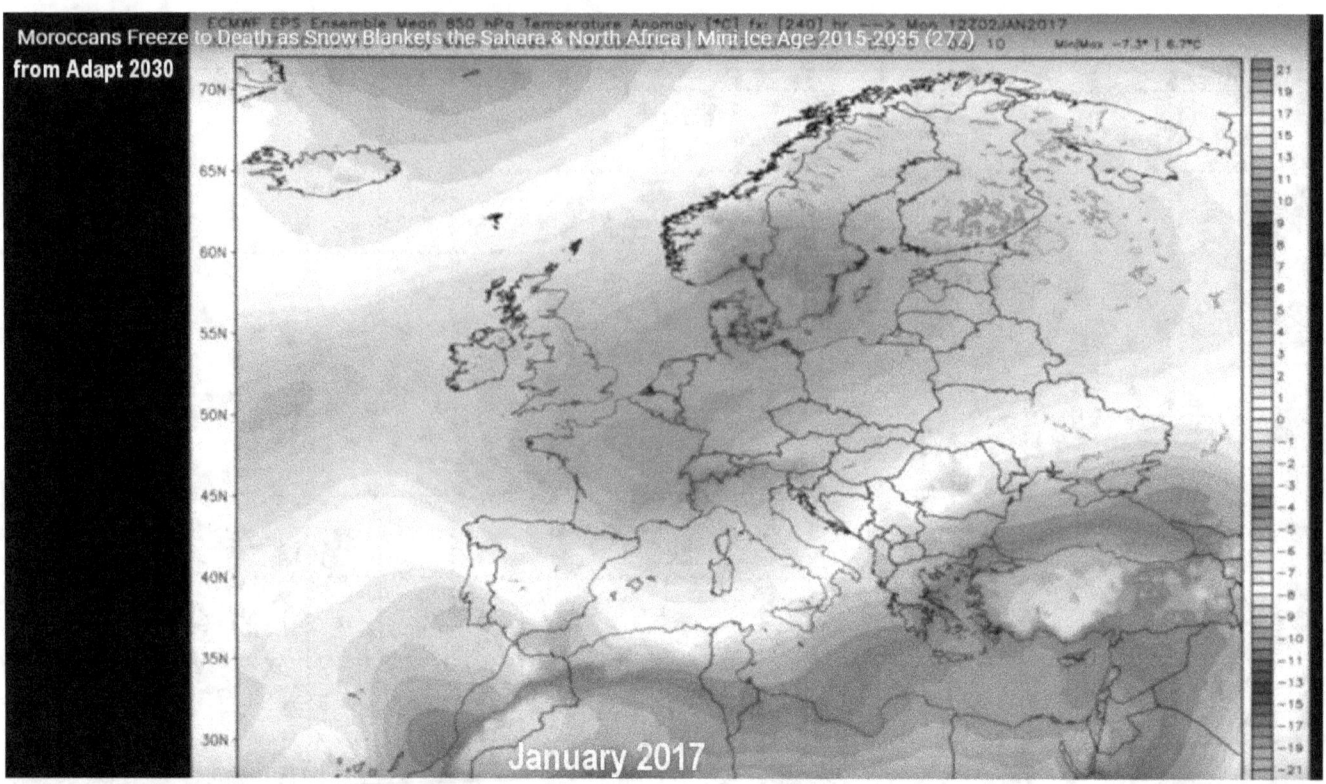

The southward component of the cold flow, when it flows stronger, also has the potential to sweep the cold air mass as far south as North Africa. This is already happening, and it is a part of the Polar Mobile Anti-Cyclone operating more strongly.

The anticyclone flow sweeps across Europe and predominantly Norway

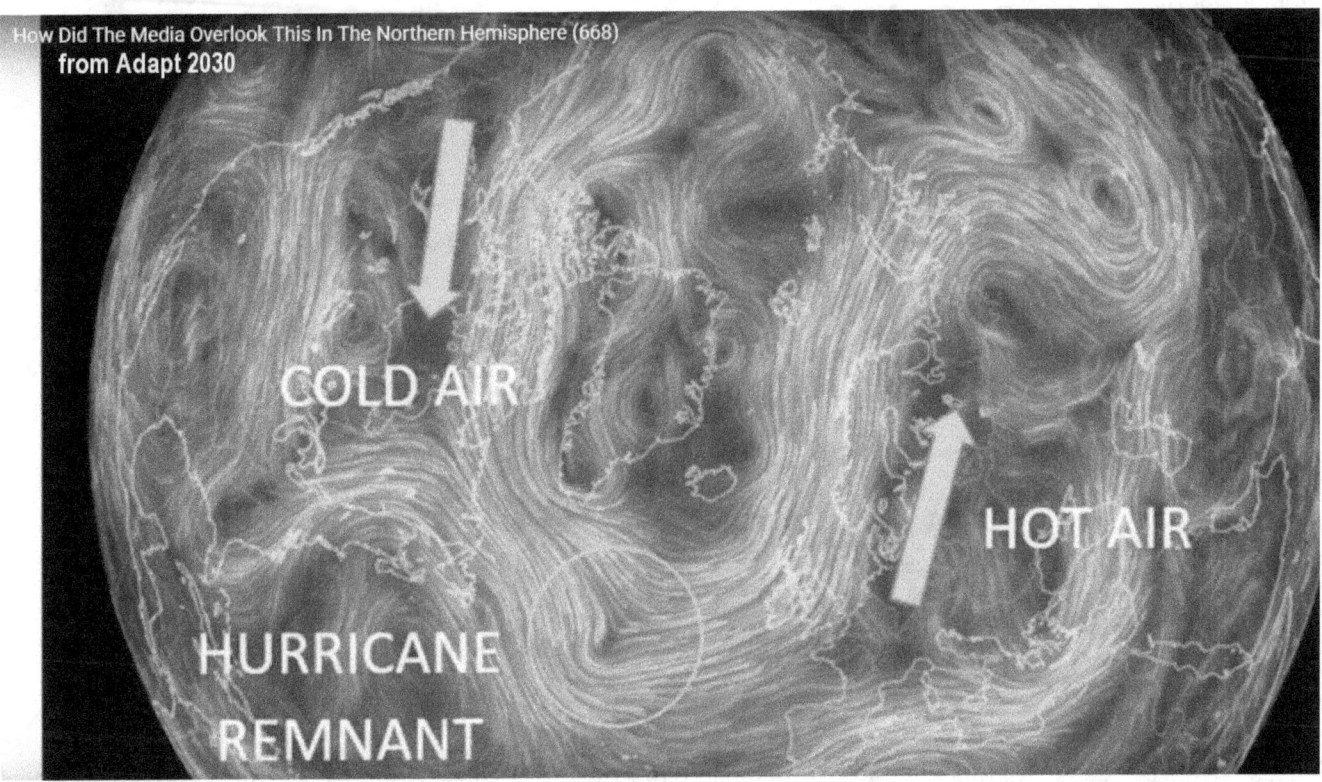

When a strong cold flow from the North gets warmed up over the subtropical Atlantic, the anticyclone flow sweeps across Europe and predominantly Norway. From there it flows in the Arctic region, where it cools and returns flowing south, then west, and sweeps across northern Africa.

Part 4 - Our Near Future - The next 15 years, potentially

Just the beginning, with the climate collapse now accelerating

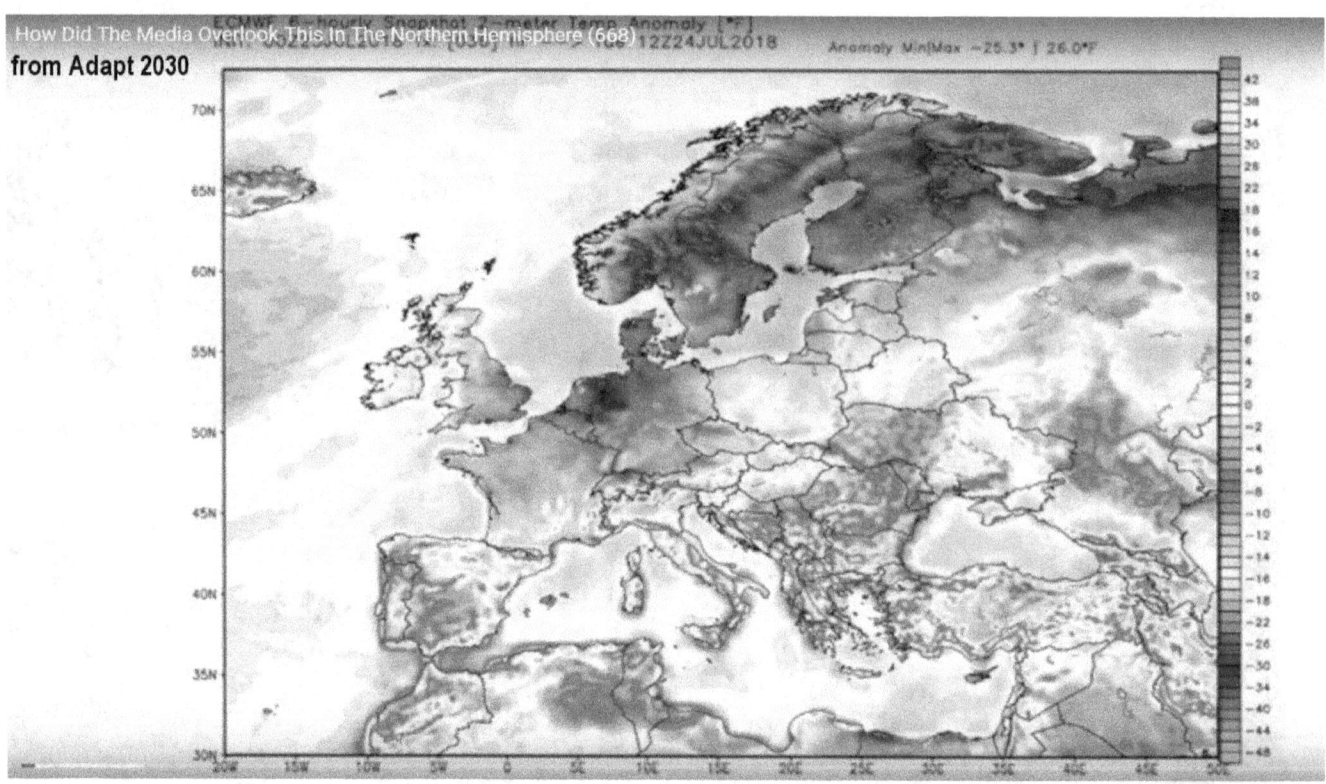

That's why we see such paradoxical phenomena as Norway getting unseasonably warmer, and the southern regions getting colder.

And all of this is apparently just the beginning, with the climate collapse now accelerating.

The Earth is getting colder year after year, until the Sun falls back into glacial hibernation

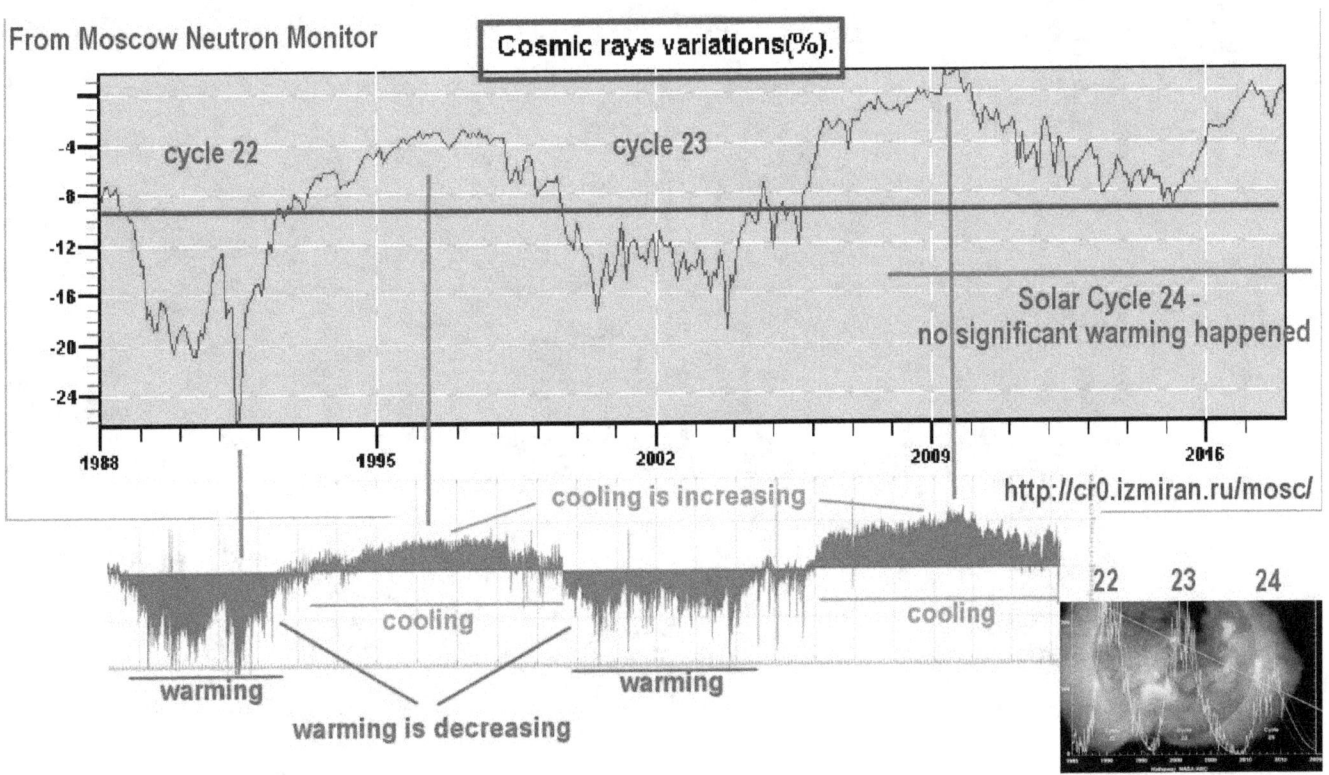

We are already past the point where the peak periods of solar cycles had a re-warming effect to offset the cooling caused by increased solar cosmic-ray flux. The last of the mid-cycle warming occurred during cycle 23, and it was already weak then. After that, no more mid-cycle warming happened, and won't be happening. Instead, the cosmic-ray cooling is increasing, with the result that the Earth is getting colder year after year, until the Sun drops its interglacial 'shoe' and falls back into glacial hibernation.

Part 4 - Our Near Future - The next 15 years, potentially

The world is already beginning to experience

The world is already beginning to experience some of the global cooling in many different ways.

Part 4 - Our Near Future - The next 15 years, potentially

To build a new world that the collapsing climate cannot touch

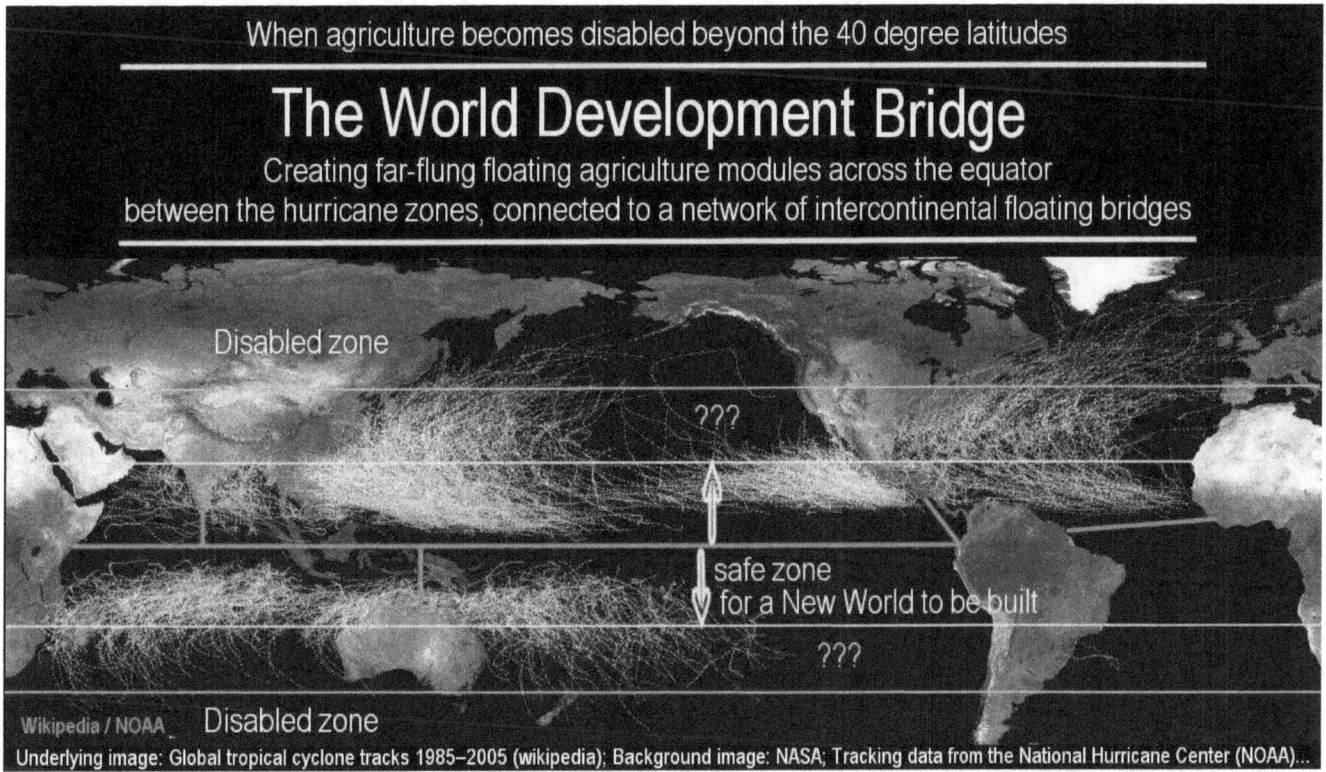

There is no escape possible from this trend, except to build a new world that the collapsing climate cannot touch.

This issue is so real and so big

This issue is so real and so big, that if it is honestly addressed, will sweep all the lesser issues off the table.

Bigger than all the silly games that politicians, bankers, and media play

The ongoing climate collapse is definitely bigger than all the silly games that the greedy politicians, bankers, and media of the world, play in their dancing at the edge of a precipice that they are not even aware of exists.

Only the effects can be prevented - not the cause

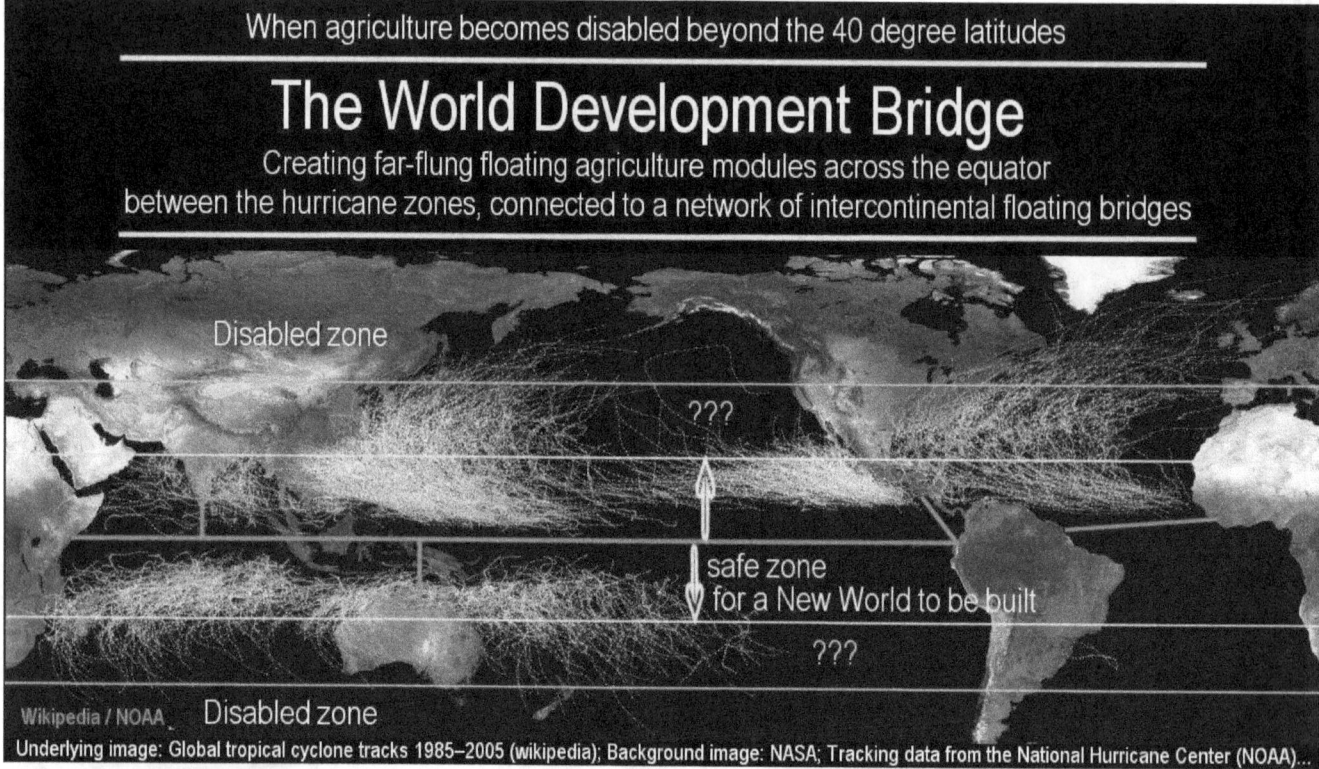

Only the effects can be prevented - not the cause - of the ongoing climate collapse that the global cooling has on our living on the Earth. The effects can be prevented by us building us a new world with technological infrastructures that the cooling of the Earth cannot affect.

Fear over Arctic warming should ring as warning bells to rouse society

Fear over Arctic warming, together with the hushed up crop losses in the path of the expanding cold, should ring as warning bells to rouse society out of its slumber, alerting it that the need for a new world has not yet been recognized, much less is it considered to be built.

Part 4 - Our Near Future - The next 15 years, potentially

Future winter blizzards in July or August, instead of just in April?

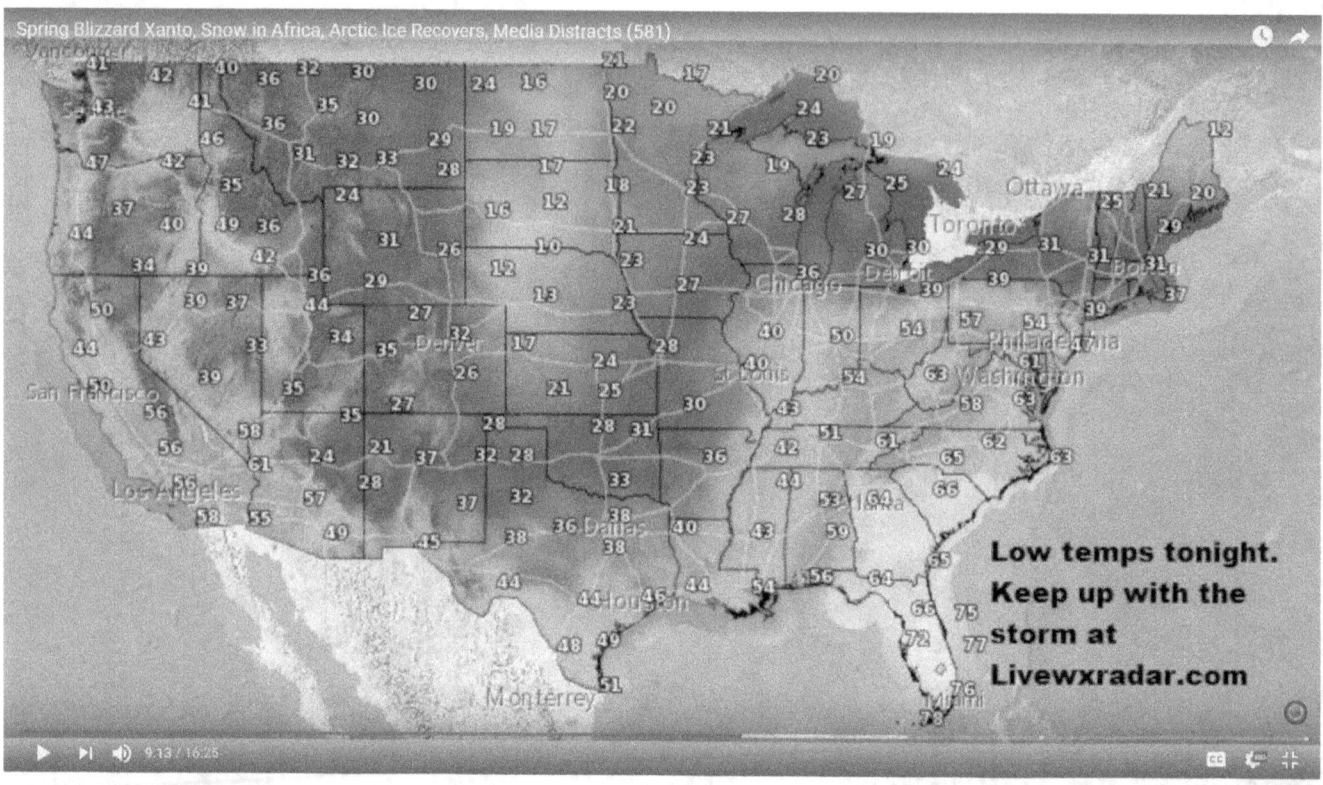

How soon will it be, when we experience future winter blizzards in July or August, instead of just in April?

Plan-B anyone?

Plan-B anyone?

Part 4 - Our Near Future - The next 15 years, potentially

Nobody in living experience has ever faced Ice Age conditions

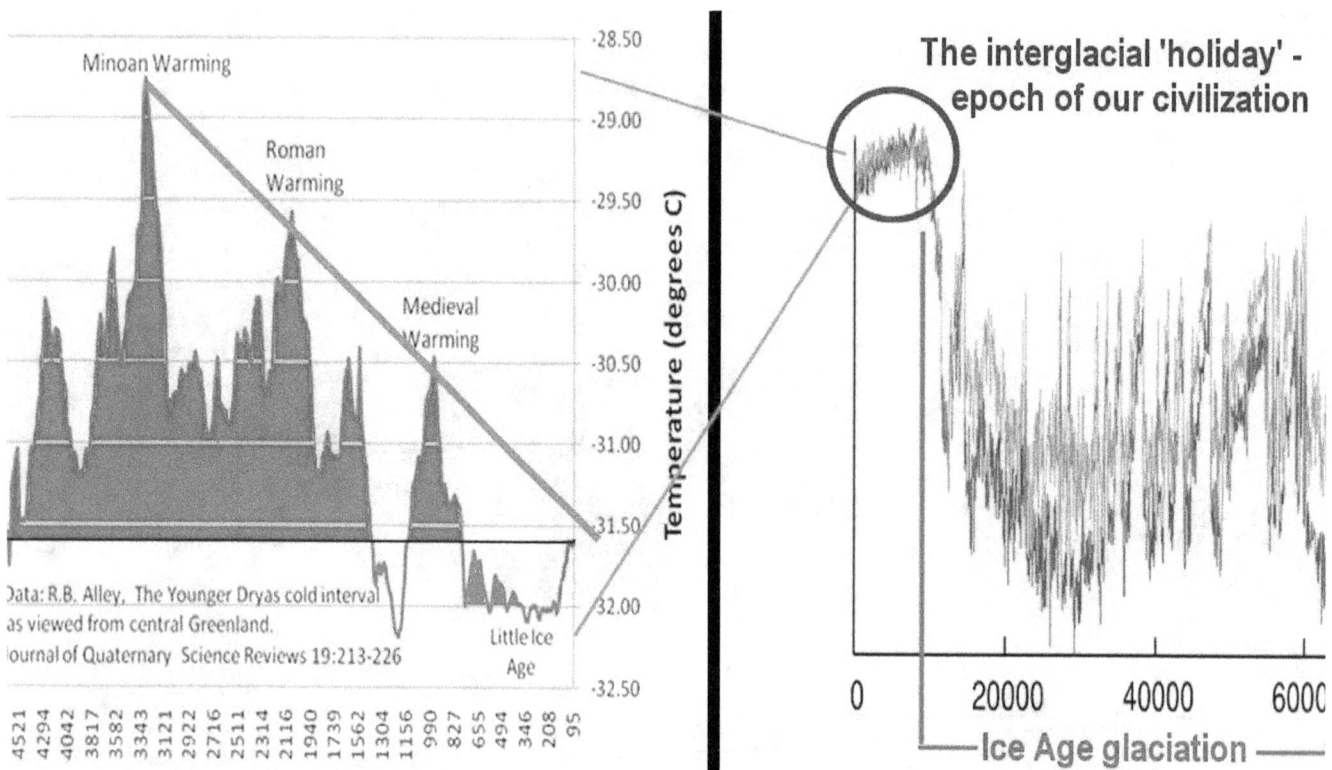

The reason that it is hard to get people interested in building the Plan-B infrastructures that enable the whole of humanity to live through the next Ice Age, is that nobody in living experience has ever faced Ice Age conditions. The entire development of civilization occurred during the climate anomaly that is our current interglacial period, and most of that development occurred during the last half of it.

When the phase shift happens that takes us back to normal

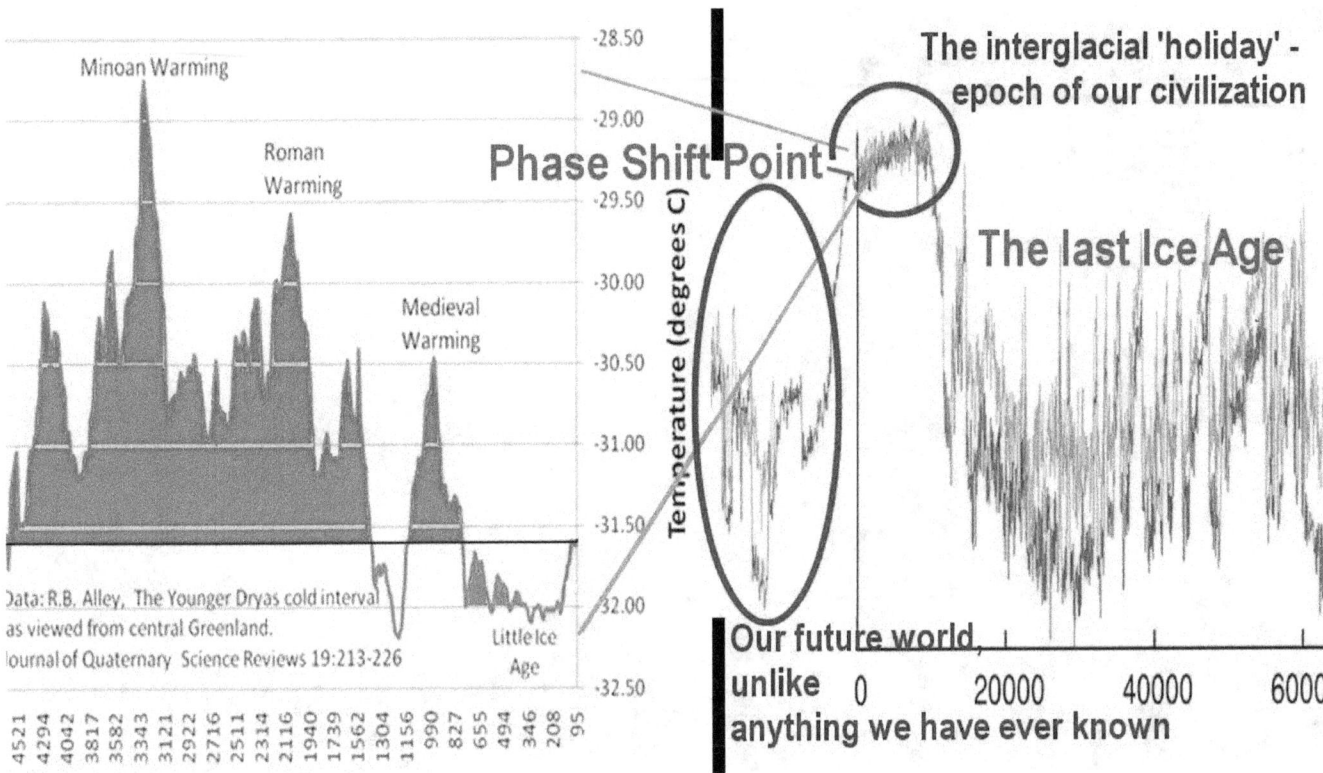

When the phase shift happens that takes us back to what has been the normal climate for the last half-million years or longer, which is the glacial climate that we have discovered in ice core samples, we enter a totally different world that is dramatically unlike anything we have ever experienced. Science has enabled us to to look into the future and explore that 'normal' world, and to create the infrastructures for us to live in that future 'normal' world that no one in our time has experienced before.

Ice core records speak to us of a 70% less-radiant Sun

Ice core records speak to us of a 70% less-radiant Sun. That's what we have measured in ice. That's what we will face from the 2050s on, after a rapid transition back to 'normal,' to glaciation. But who heeds the voice of science? The song is, "let's keep on dreaming that this won't happen"

Ice core records also speak of world with 80% less rain

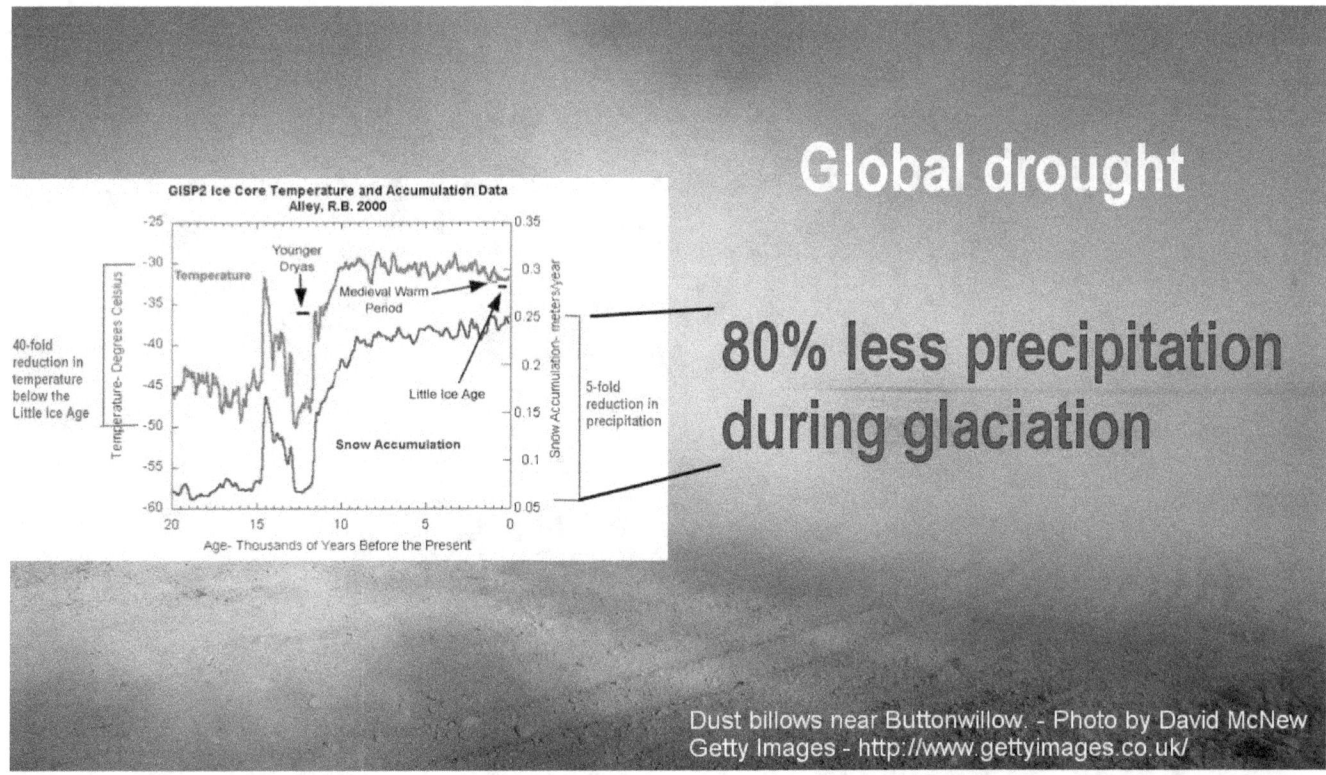

Ice core records also speak to us of a dryer world with 80% less rain. Has anybody any idea how this affects agriculture, and how it affects us if we don't prepare us for it with compensating technologies, which are presently not even considered? We should be singing the songs of science. But how does one inspire such songs?

Part 5 To rouse the living, wake the dead

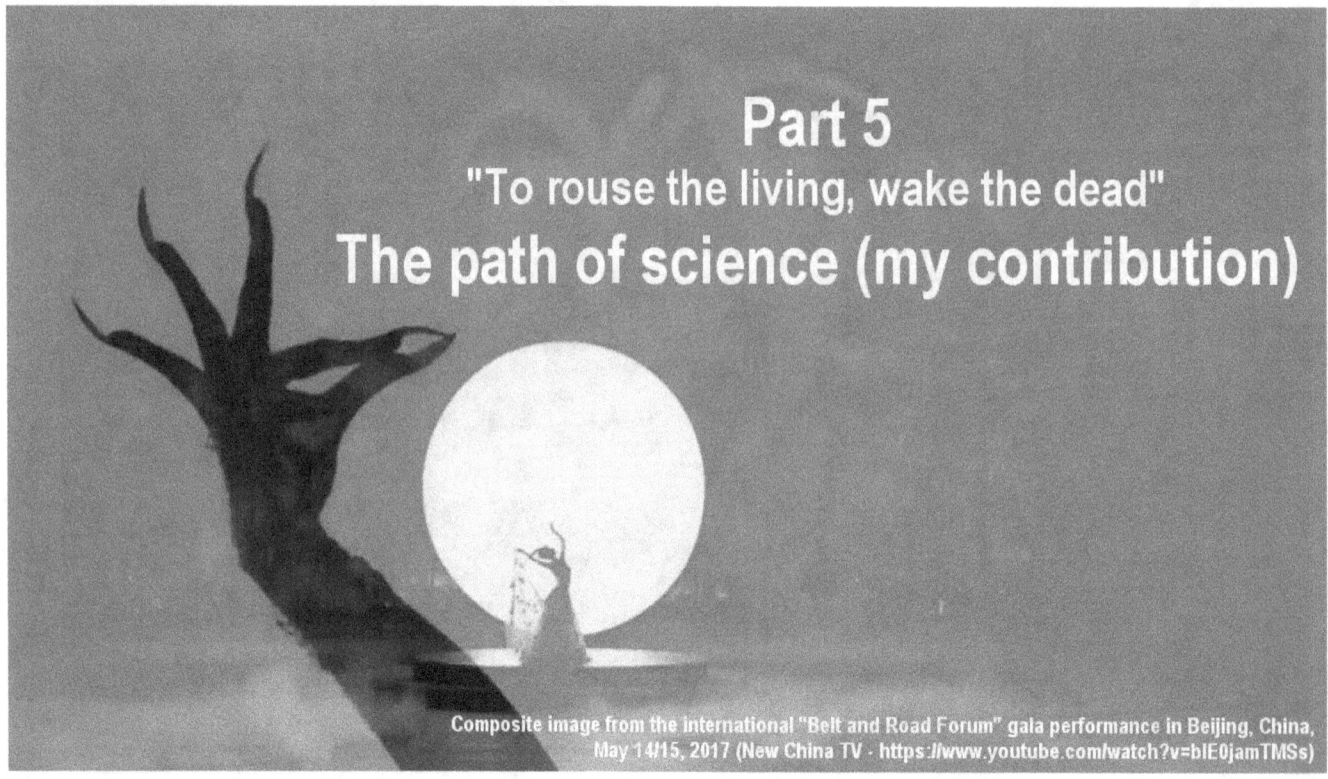

Part 5 - "To rouse the living, wake the dead" - The path of science (my contribution)

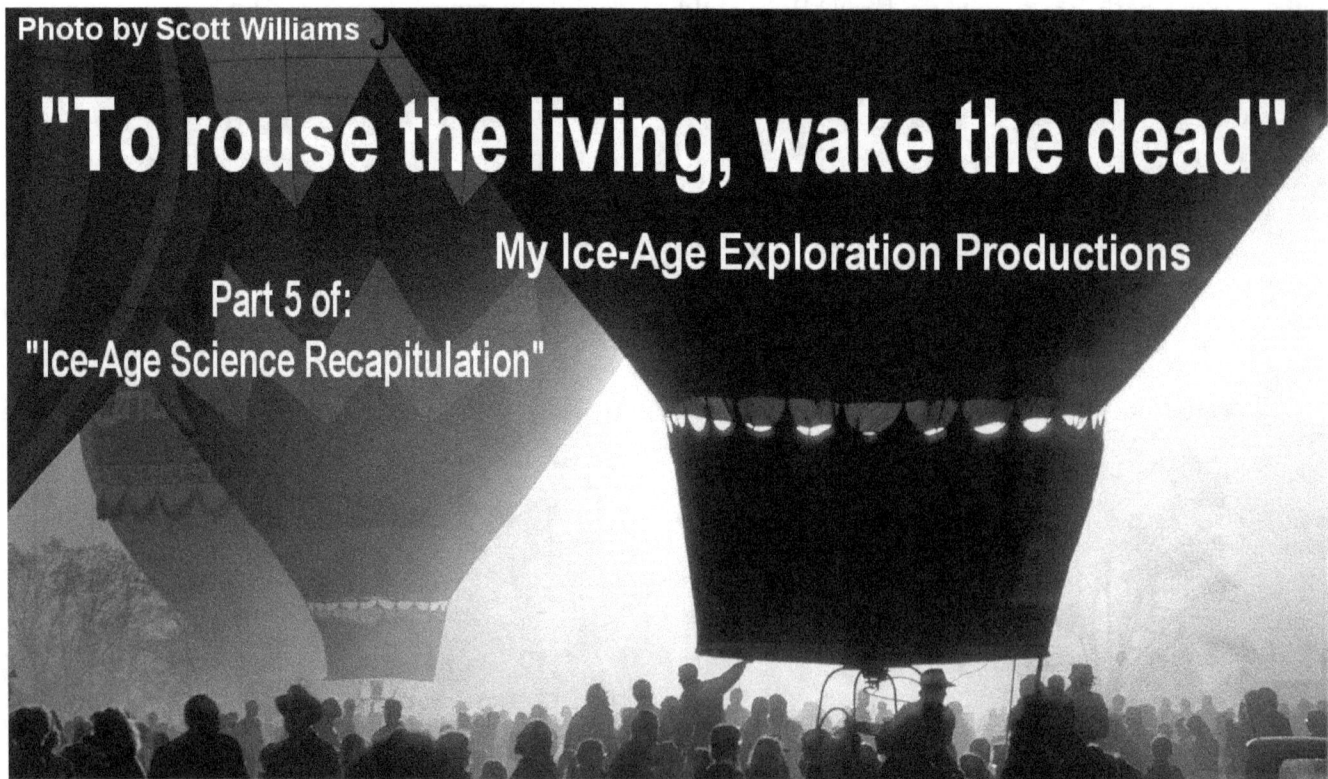

How does one rouse society to the truth ?

How does one rouse society to the truth, when the rulers cry, "there is no truth - all is opinion"? The measured evidence should take us beyond opinions. We are in the boundary zone to the near Ice Age, of which science presents to us measured data of what to expect.

Great science projects have measured the Sun and its effects

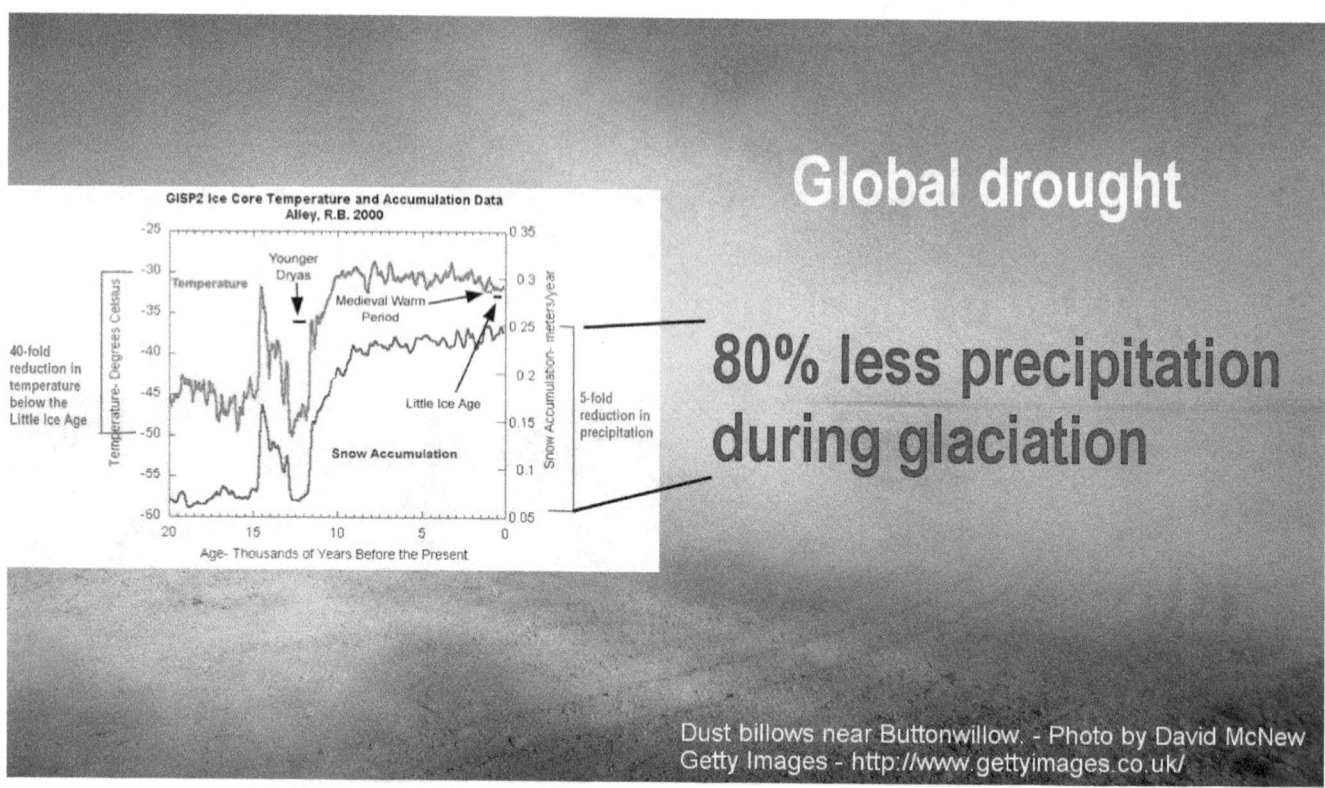

As I said before, great science projects have measured the Sun and its effects on the Earth, intensively, over many years, with amazing results. The results all tell us the same story, that the Sun is getting progressively weaker and the climate on Earth is getting colder and dryer.

The measurements also tell us that the final phase shift is near

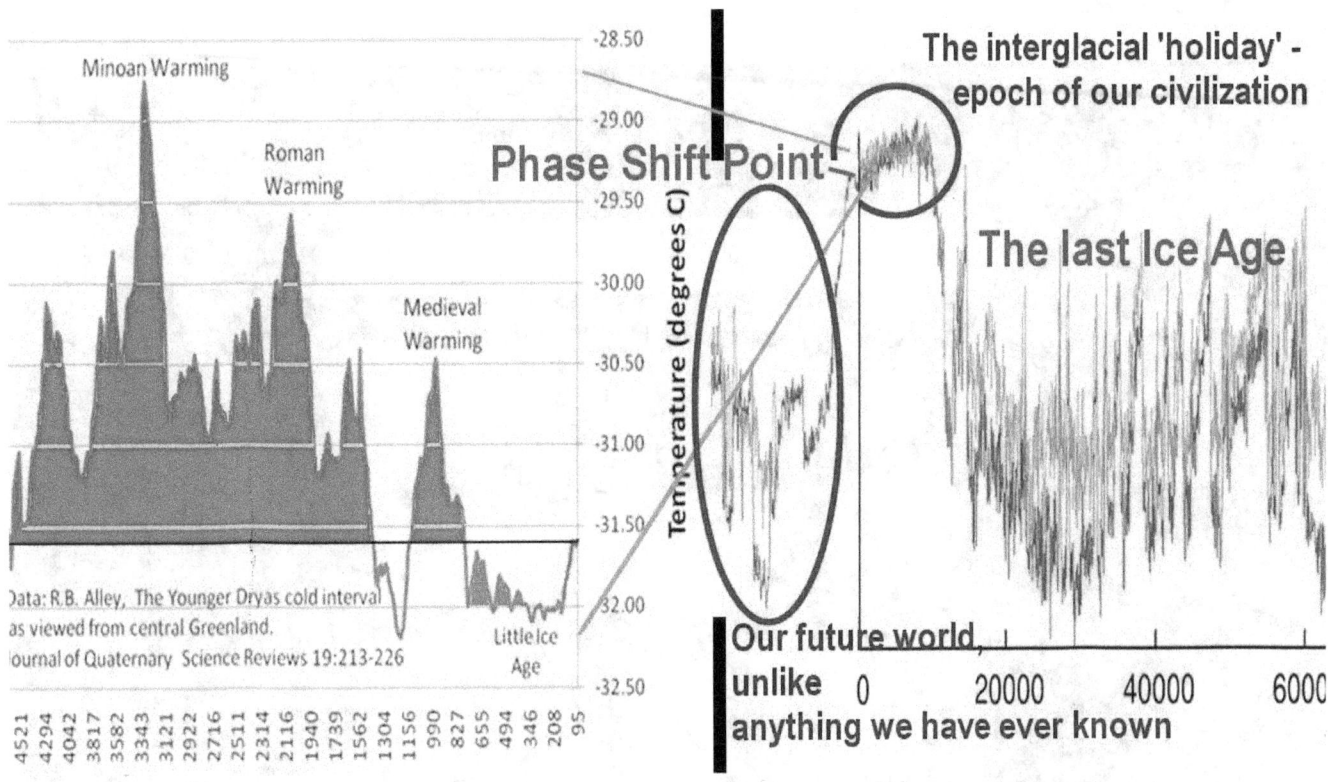

The measurements also tell us that the final phase shift to the full Ice Age is near, and that the full Ice Age pales anything that we experience so far.

The looming Ice Age certainly should not be ignored, as is presently the case, even if we don't fully understand yet the absolute of it all.

I have presented the scientific discoveries that have been made over the years

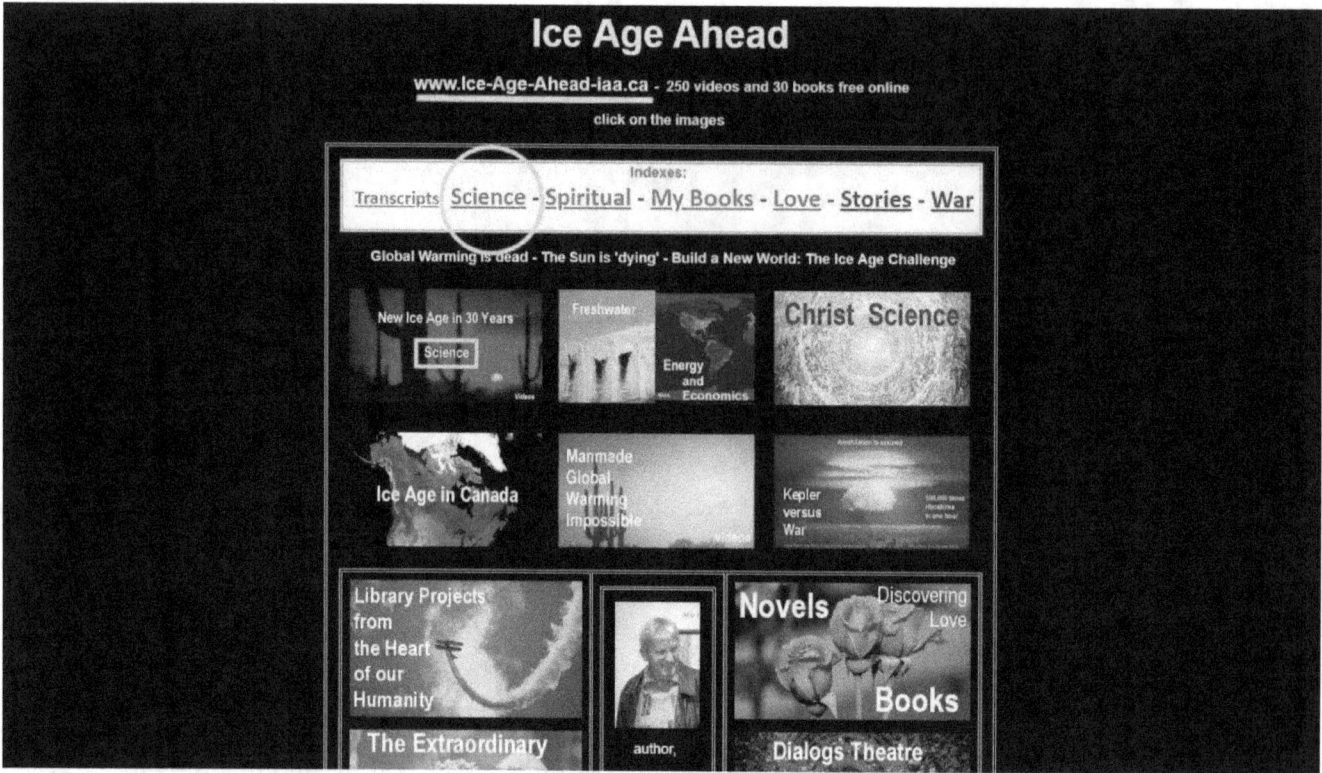

Towards this end I have presented the scientific discoveries that have been made over the years. I have presented these discoveries in my numerous exploration video productions that I have hosted on my website, Ice Age Ahead -(http://www.ice-age-ahead-iaa.ca/,) Click on the Science tab.

To alert society of the great danger that Canada, Europe, Russia, and the USA are facing

The videos have been produced to alert society of the great danger that Canada, Europe, Russia, and the USA are facing to their existence, by the incredible speed of the on-going collapse of the solar system, and with it the collapse of the climate on Earth.

The videos hosted on my website are categorized

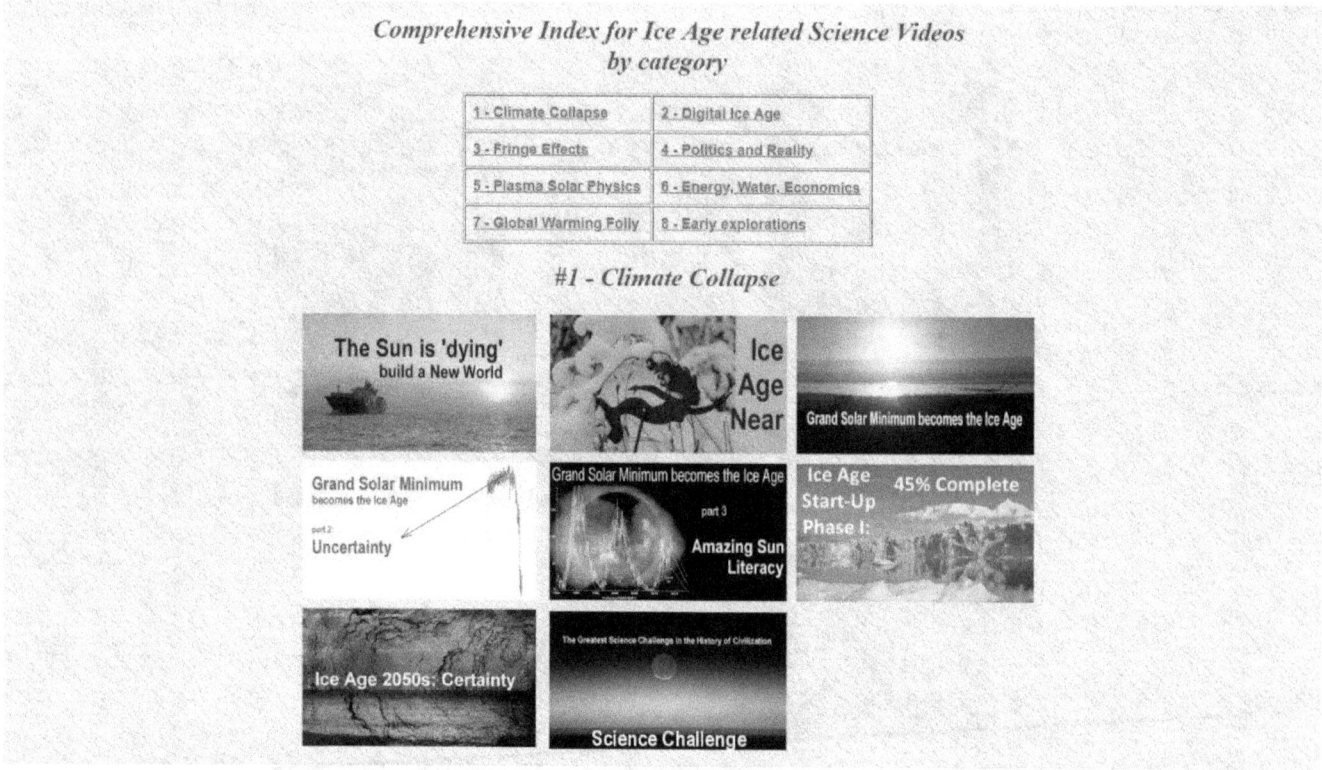

The videos hosted on my website are categorized into several types.

Just click on an image within the categories. It will link you to its start page.

The videos are also hosted on YouTube

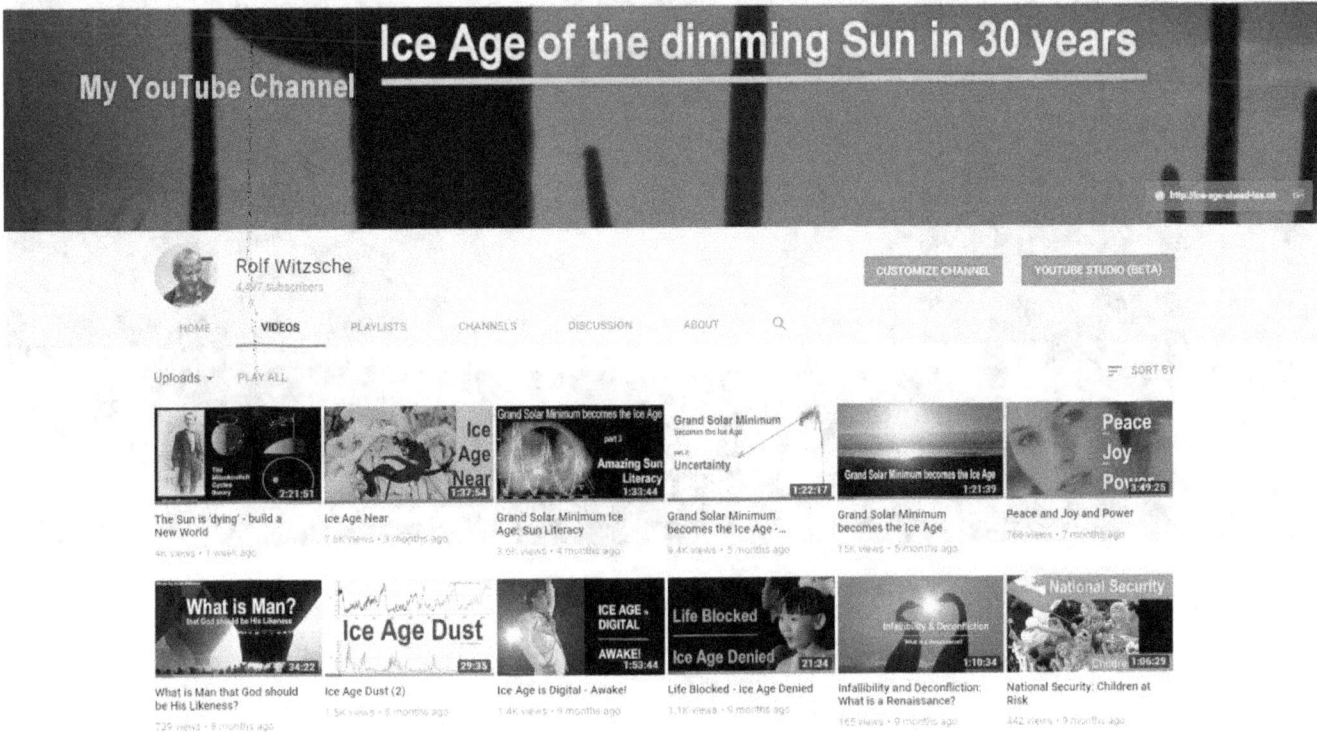

The videos are also hosted on YouTube on my channel: 'Ice Age of the dimming Sun in 30 years.'

Transcripts of the videos, complete with images, are also available

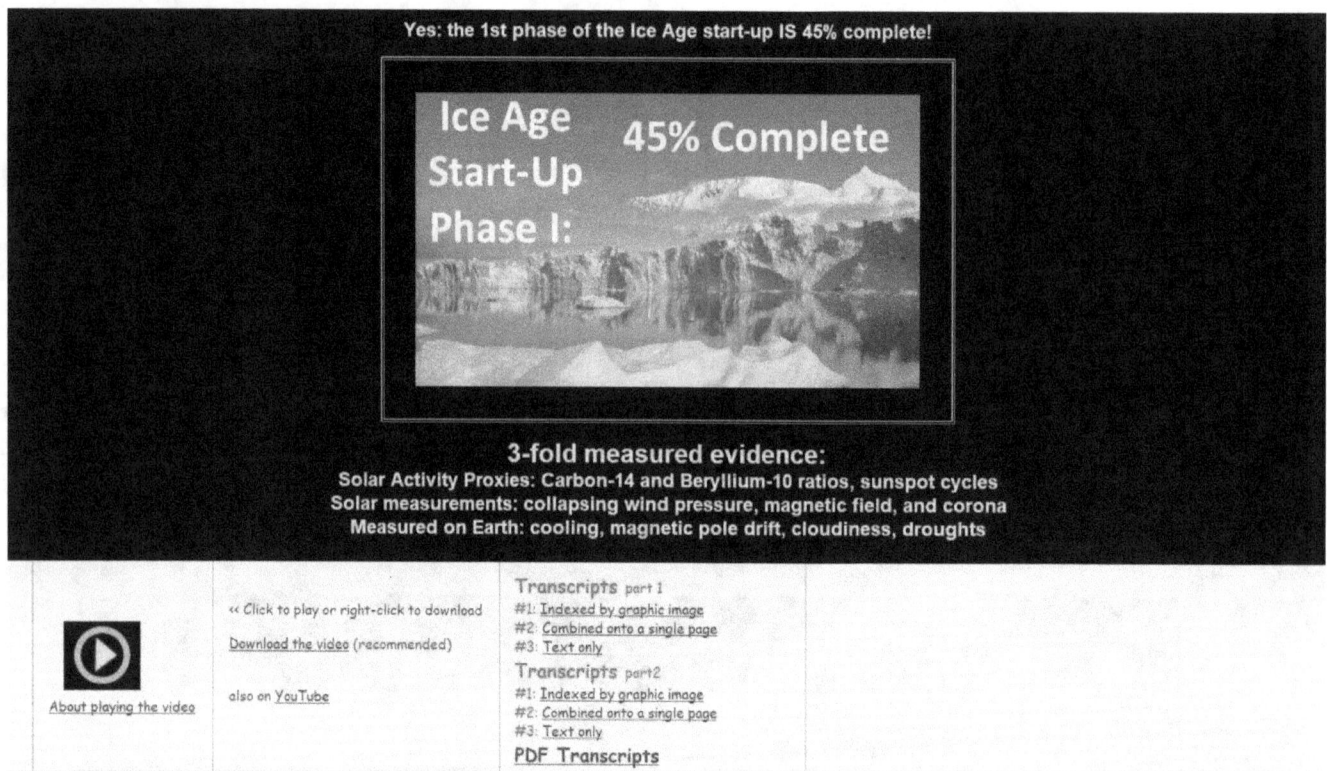

Links on the start pages on my website, enable the videos to be played online, and facilitate free downloading.

Transcripts of the videos, complete with images, are also available from the start pages in several different versions.

One form of the transcripts is interactive

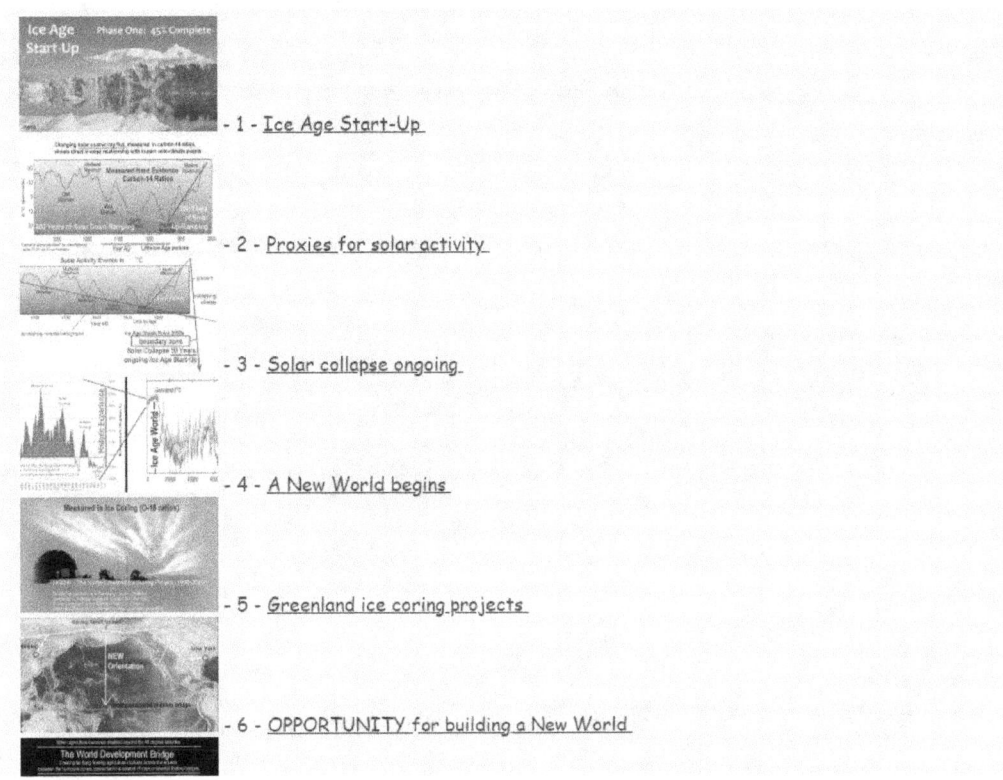

One form of the transcripts is interactive. Individual parts of the video transcript can be selected vial thumbnail images and descriptive title. - The interactive process selects specific pages, for the quick access of a topic within the transcripts of the videos.

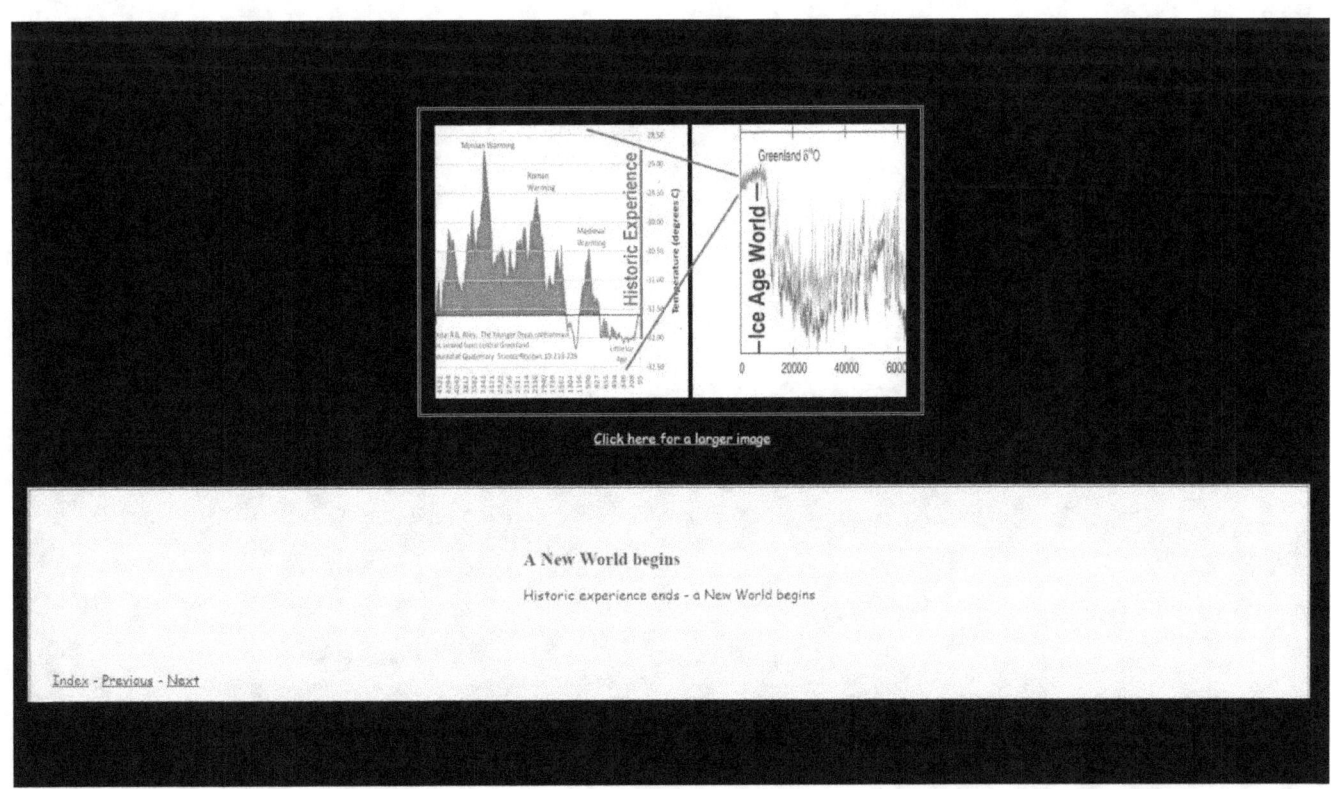

The selected page displays a larger image and the text that goes with the image. A link on this page also enables the full size of the image to be displayed and be downloaded.

All text and the images on a single page also as a PDF file

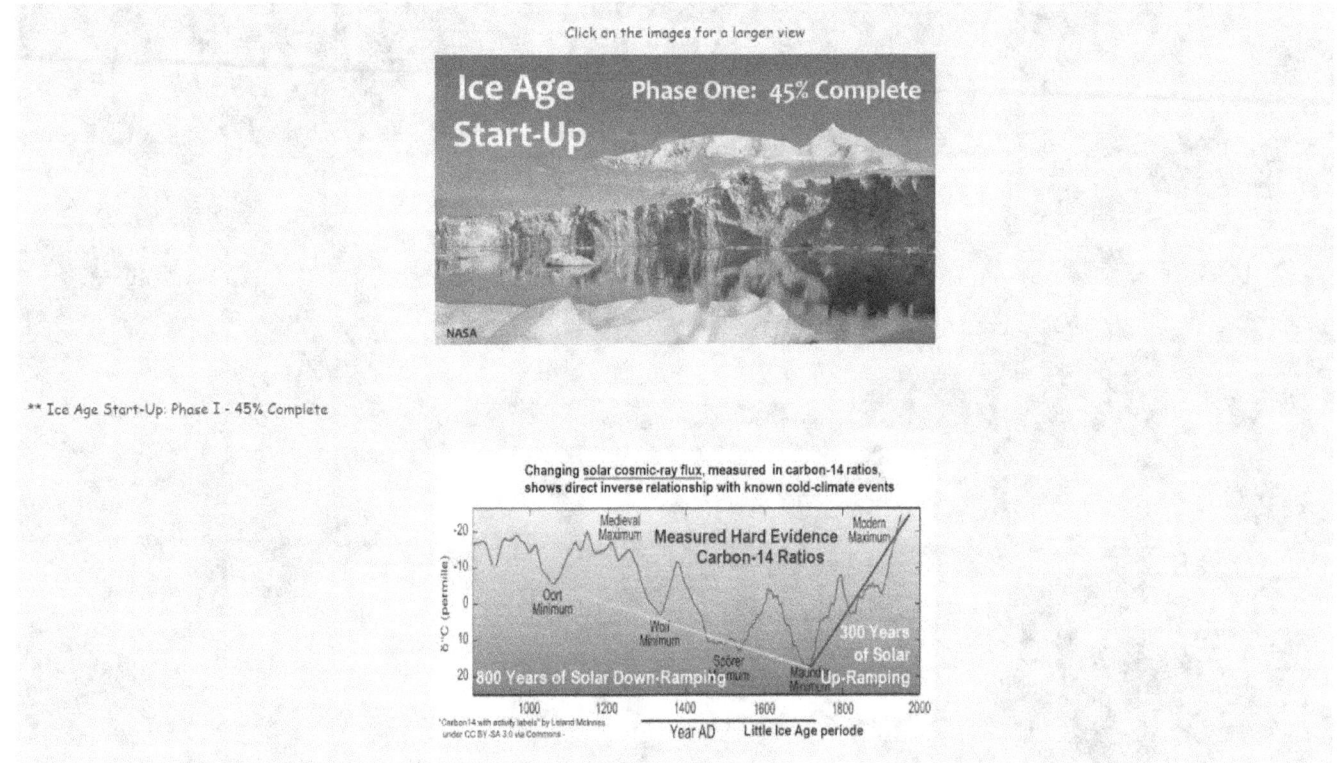

Another form of the transcript displays all text and the images on a single page. This combined version is also available as a PDF file.

The PDF transcripts contain medium-size images with the text interspersed. Clicking on an image causes the full size of the image to be displayed, dynamically loaded from the internet server.

Transcripts also available in printed form

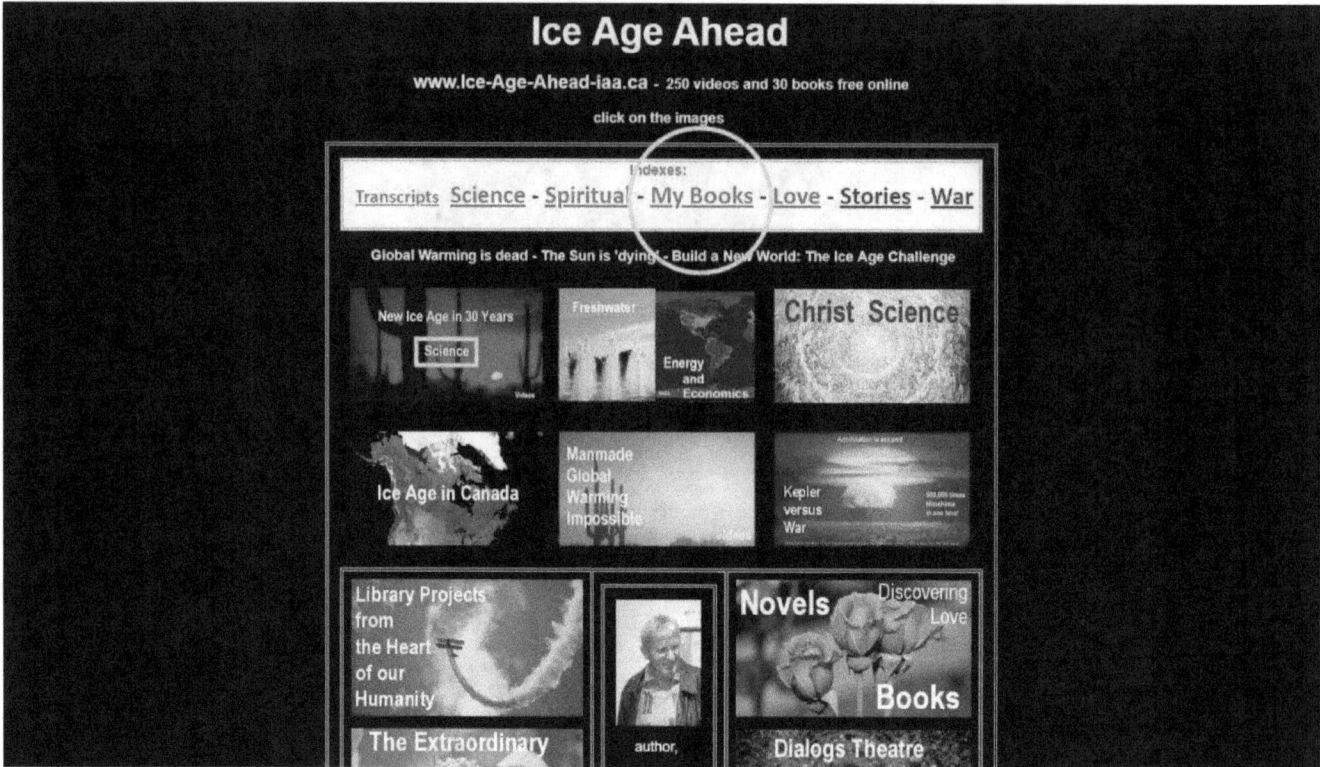

In some cases I have made the transcripts also available in printed form, as 6x9 paperbacks for the earlier versions, and in 8.5x11 format for the later ones. Click on the "My Books" tab.

The transcript books are indexed by their cover image

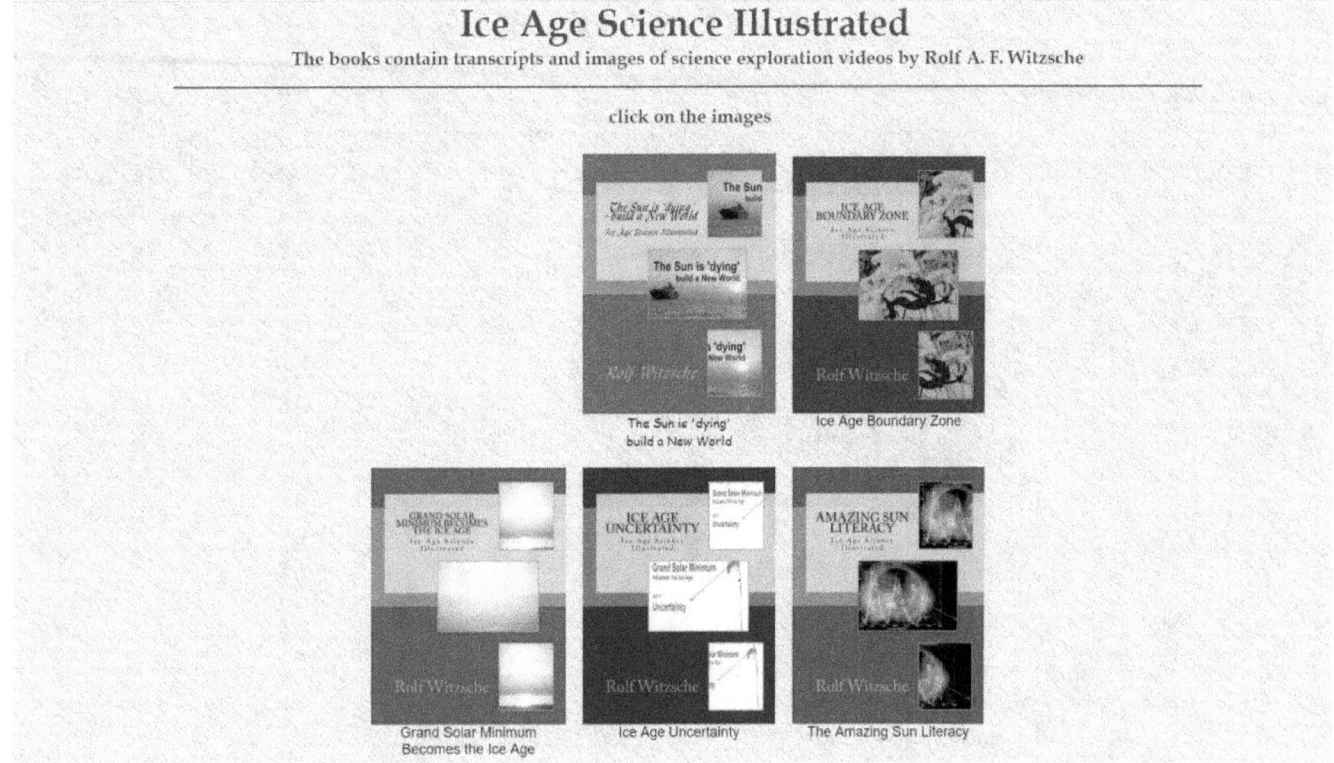

The transcript books are indexed by their cover image. Click on the image to link to the detail page.

The book detail pages include descriptions

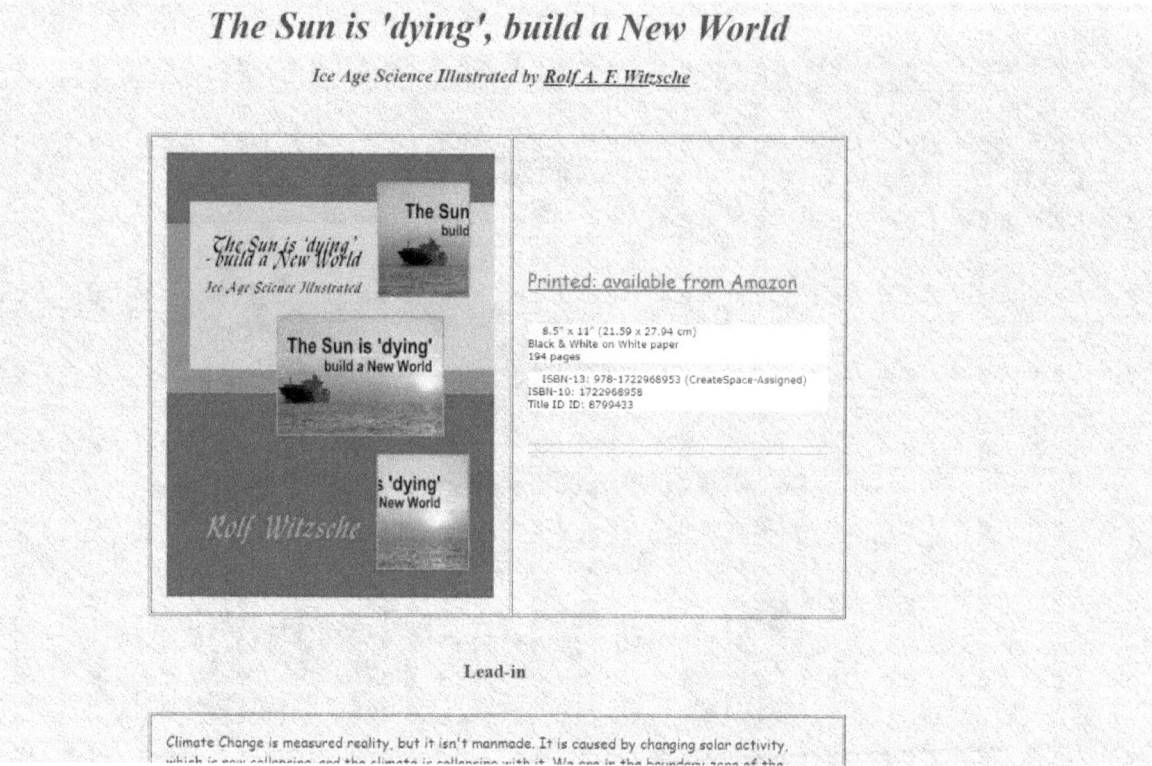

The book detail pages include descriptions, ordering information, and in some cases PDF versions, and front and back cover images.

Also available on the transcripts index page

Links to the transcripts, in all forms, are also available on the transcripts index page.

The index page lists all my videos and available transcripts

Transcripts for the videos and books on this stite

File Format	IDX= indexed (by images) - SPG or PG = singe-page version - PDF = PDF version (downloadable) Book = (available as a printed book)
IDX - SPG - PDF IDX - SPG IDX - SPG - Book	#1 - Sun is 'dying', build a New World #2 #3
IDX - SPG - PDF IDX - SPG - Book	#1 - Ice Age Near (the Ice Age Boundary Zone) #2
IDX - SPG - PDF IDX - SPG - Book	#1 - Grand Solar Minimum becomes the Ice Age #2
IDX - SPG - PDF IDX - SPG - Book	#1 - Grand Solar Minimum becomes the Ice Age -Part2- Uncertainty #2
IDX - SPG - PDF IDX - SPG - Book	#1 - Grand Solar Minimum becomes the Ice Age - part 3 - Amazing Sun Literacy #2
IDX - SPG - PDF IDX - SPG - PDF IDX - SPG - PDF IDX - SPG - PDF IDX - SPG - PDF	To Be or Not to Be? (in an Ice Age World) Peace and Joy and Power Part 1 - Part 2 - Part 3 - Part 4 - Part 5a

The index page lists all my videos and available transcripts for them. The videos cover a wide range of subjects. Most of the exploration videos contain discoveries that were new to me at the time. With a few rare exceptions, they are discoveries presented to me, or publically presented by leading institutions, individuals, and researchers within these institutions. Some of the material presented includes data published by international science research projects, some spanning many years, such as the ESO and NASA Ulysses project. that took 36-years to complete, or the multi-year ice coring projects in Greenland and Antarctica.

A major focus on physical science is evidently needed

A major focus on physical science is evidently needed, because society has lost its connection with what is affecting it. My science videos are presented to bridge this gap that has developed over time.

Society has become lost in a dream where nothing is real

Society has become lost in a dream where nothing is real. It still speaks the words of science and promotes science, but has blinded itself to the momentous discoveries that science has already made. The dreaming has tragic consequences for its future.

Consider the Bering Strait tunnel project

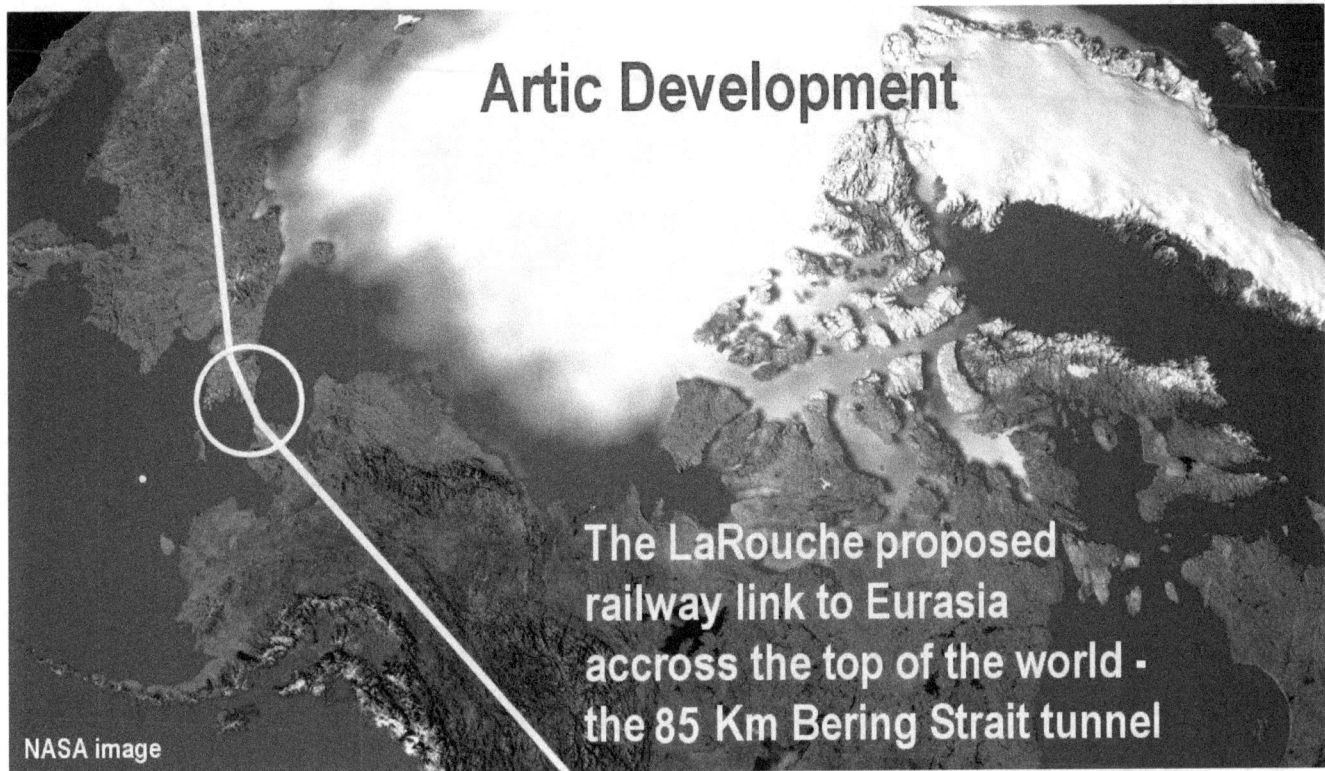

Consider the Bering Strait tunnel project as an example. The project has been long proposed and is still being promoted. It would connect the American continent and the Eurasian continent across the top of the world via 50 mile long tunnel, crossing the Bering Strait below the sea.

However, the project has two fundamental flaws. One flaw is that the top-of-the-world path would become disabled by the ongoing climate collapse before the project would be completed.

The other flaw is that the tunnel isn't designed to connect with anything

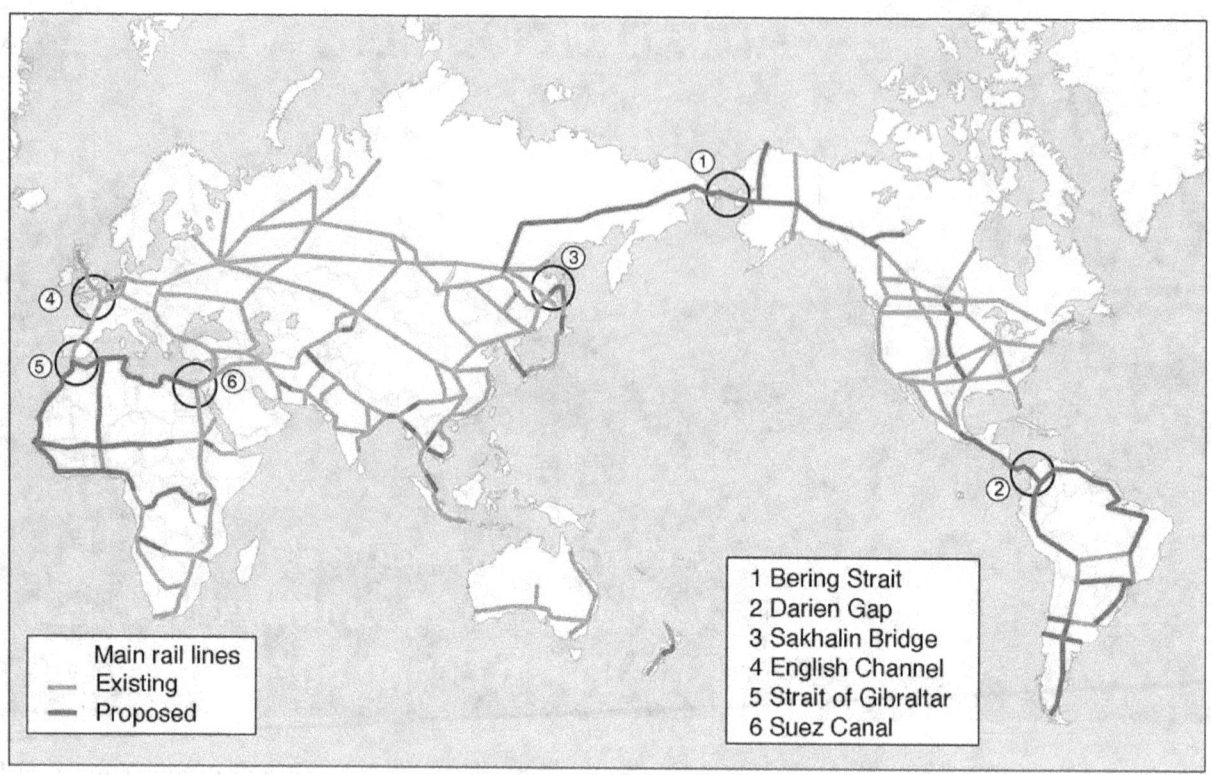

The other flaw is that the tunnel isn't designed to connect with anything. Thousands of kilometers of railways would have to be constructed, across difficult terrain, to link the tunnel project up with major population centers.

Now compare the tunnel link with my World Bridge infrastructure

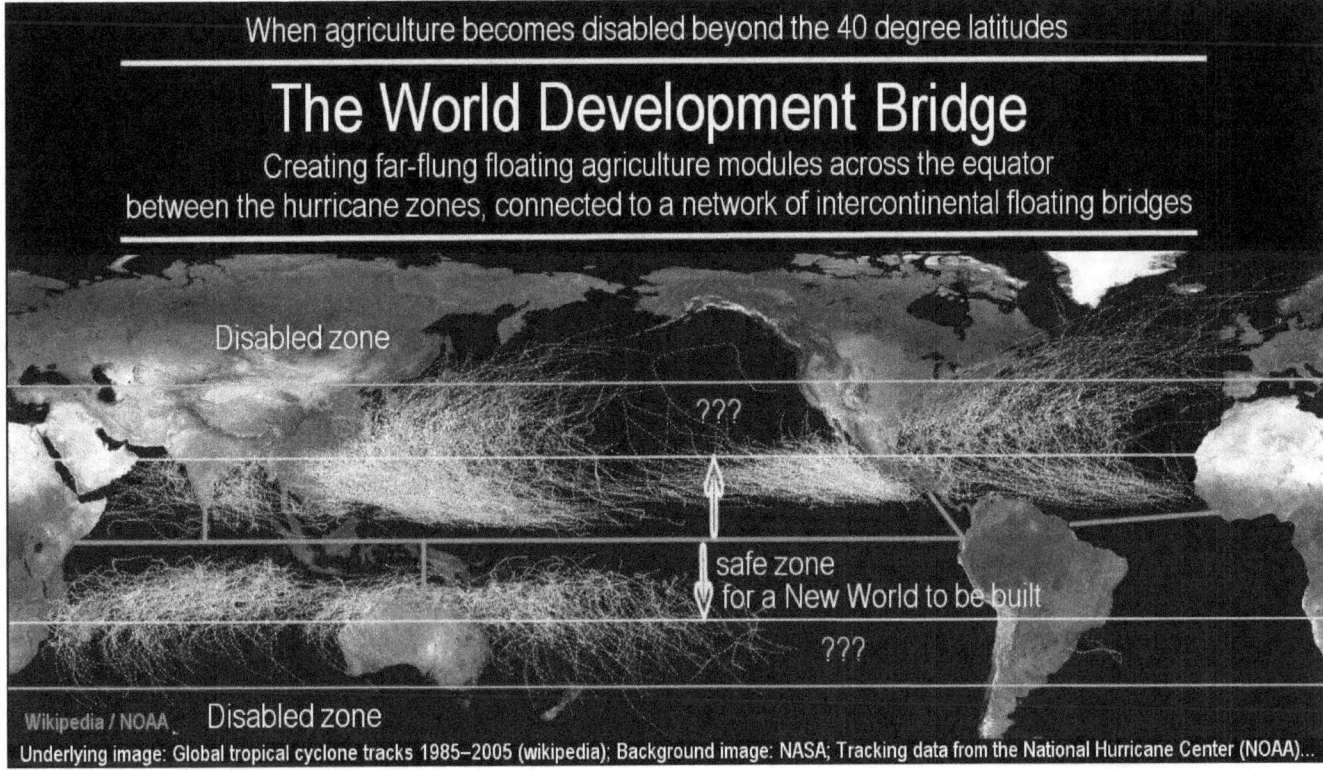

Now compare the tunnel link with my proposed floating World Bridge infrastructure that is designed to connect all the continents along the Equator. The project has none of the basic flaws that the tunnel project has. It wouldn't become obsolete by the ongoing climate collapse, even by the Ice Age when it begins. In fact it would be essential for it.

In addition, the World Bridge would link all the major population centers of the world in the most efficient manner possible, with links extending to Africa, India, Indonesia, and links to Burma, China, Japan, Korea, the Philippines, Australia, and with further links to North and South America. The World Bridge would link up with Europe through Africa, and Russia through India.. The far-flung floating Word Bridge would thereby become a World Community Bridge, with all nations participating and benefitting.

Also the World Bridge project would become immediately beneficial while the construction was in progress. It would become the backbone for vast new forms of floating agriculture that would be attached to the floating World Bridge. The new agriculture would become operational in step with the advancing construction. It would start producing food almost from the start, while the traditional agriculture begins to diminish. The world Bridge would also have thousands of new cities attached, rent free, servicing the agriculture, complete with new industries and cultural centers.

This giant project will likely be more easily built with automated industrial processes, than the tunnel project would. The material and energy resources for the project exist in near infinite quantity. The material sits process-ready on the ground, in the form of basalt that melts at 1,400 degrees, and is lighter and stronger than steel and is none-corrosive. And the energy for the processing is available from high-temperature liquid-fluoride nuclear reactors that have been sitting on the shelf for 50 years. We lack nothing to build our New World, but the courage to do it, and of course the science awareness in society that inspires that courage. This is what my Ice-Age video series is aiming for.

My part in the video productions is focused on drawing the field together

My part in the video productions is focused on drawing the wide field of related science research and physical measurements together into a comprehensive whole, in order that the measured evidence may speak for itself in its numerous forms and contexts, and in their relationship to our living on this planet that is our common home. However, the exploration is still incomplete at this point.

Part 6 - Freshwater and Energy

Part 6 - Freshwater and Energy

Both, freshwater and energy are needed in abundance

Both, freshwater and energy are needed in abundance to maintain a civilization, and both are available in abundance.

Freshwater can be drawn from the outflow of rivers

Freshwater can be drawn from the outflow of the world's great rivers, especially the tropical rivers.

Distributed worldwide in large-volume arteries

The Atlantic part:

The Amazon (219,000 cm/s)
The Congo River (40,000 cm/s)
The Orinoco River - Venezuela (33,000 cm/s)
The Parana River - Brazil (25,000 cm/s)
The Mississippi - USA (12,700 cm/s)
The St. Lawrence River - Canada (9,800 cm/s)
The Fraser River - Canada (3,400 cm/s)
Total: 343,000 cm/s (cubic meters per second)

The Pacific part:

The Yangtze River (31,000 cm/s)
The Brahmaputra River - India (19,200 cm/s)
The Pearl River - China (13,600 cm/s)
The Ayeyarwady River - Myanmar (13,000 cm/s)
The Ganges River in India (12,300 cm/s)
The Xi River in China (7,400 cm/s)
The Indus River - Pakistan (7,100 cm/s)
The Yellow River - China (2,500 cm/s)
Total: 106,000 cm/s (cubic meters per second)

Combined total: 450,000 cm/s
Compare, a large nuclear desalination plant: 2 cm/s

The outflow of the rivers can be distributed worldwide in large-volume arteries placed in the oceans, connected to submerged reservoirs.

Freshwater can also be drawn directly from the oceans

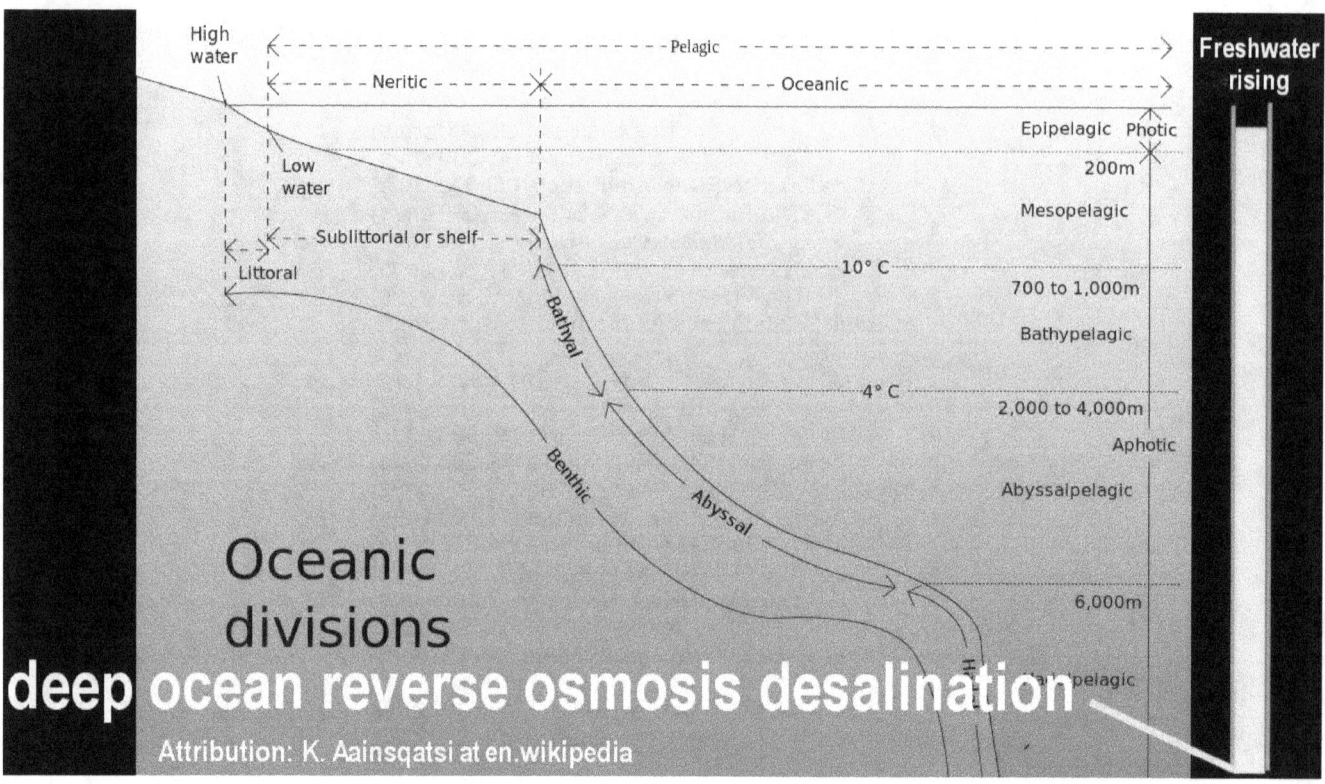

Freshwater can also be drawn directly from the oceans by deep-ocean reverse osmosis desalination.

For a boundless energy resource we need to look no further than the ionosphere

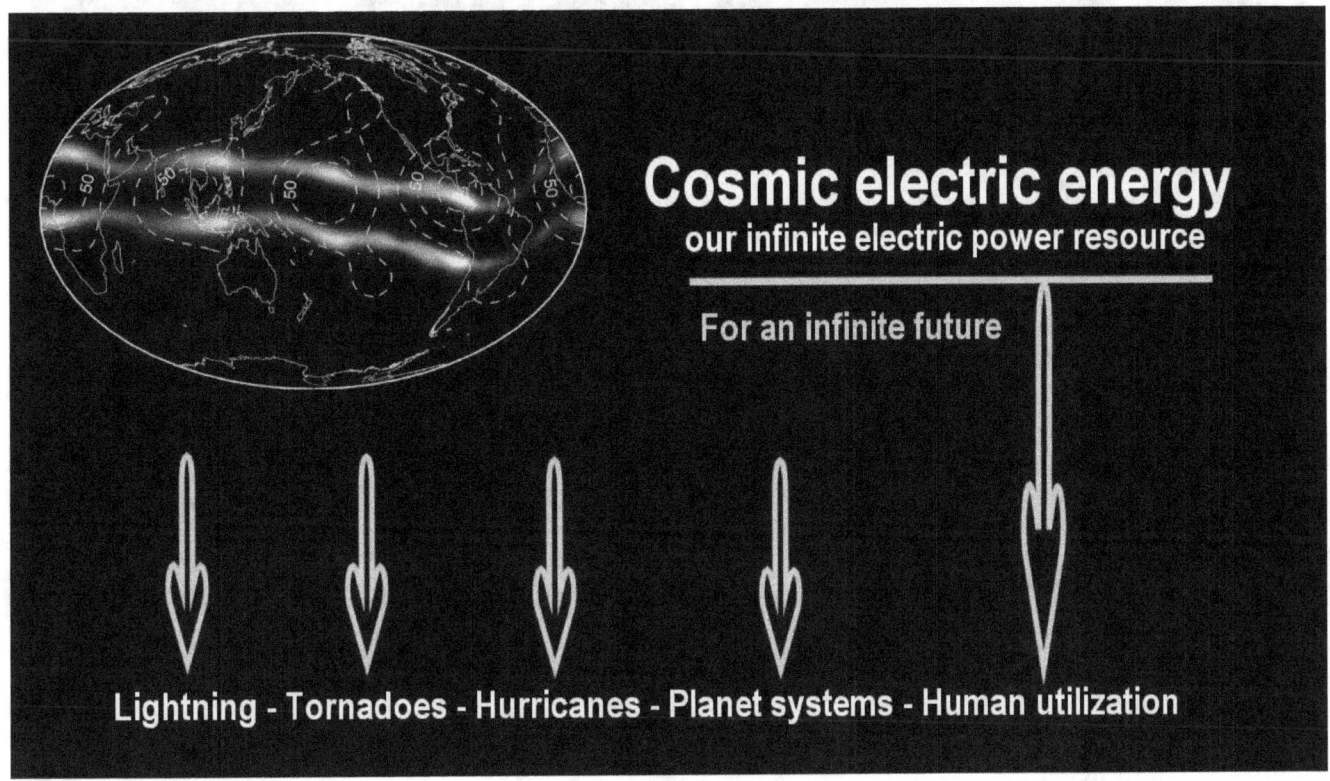

And for a boundless energy resource we need to look no further than the ionosphere that surrounds the Earth. It already serves as our interface to the electric plasma in space that powers the Sun, and also powers some of the large natural systems.

Freshwater Unlimited

Freshwater Unlimited

Freshwater from the Amazon River, the Orinoco River, the Parana River, the Congo River

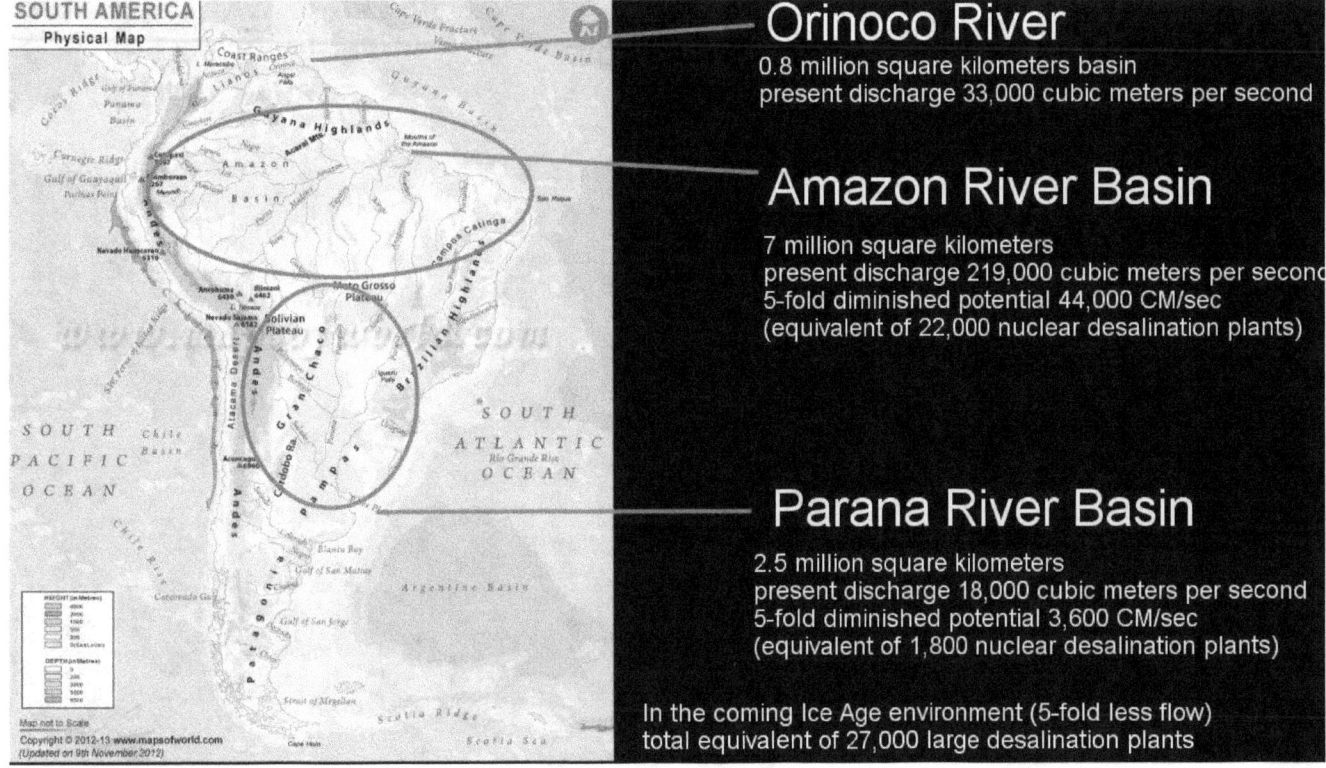

As for freshwater, it can be drawn directly from the outflow of the Amazon River, the Orinoco River, the Parana River, and also the Congo River, the four largest tropical rivers in flow volume. Together, they discharge roughly 300,000 cubic meters of freshwater per second, unused into the Pacific Ocean.

Worldwide water distribution critically important during Ice Age conditions

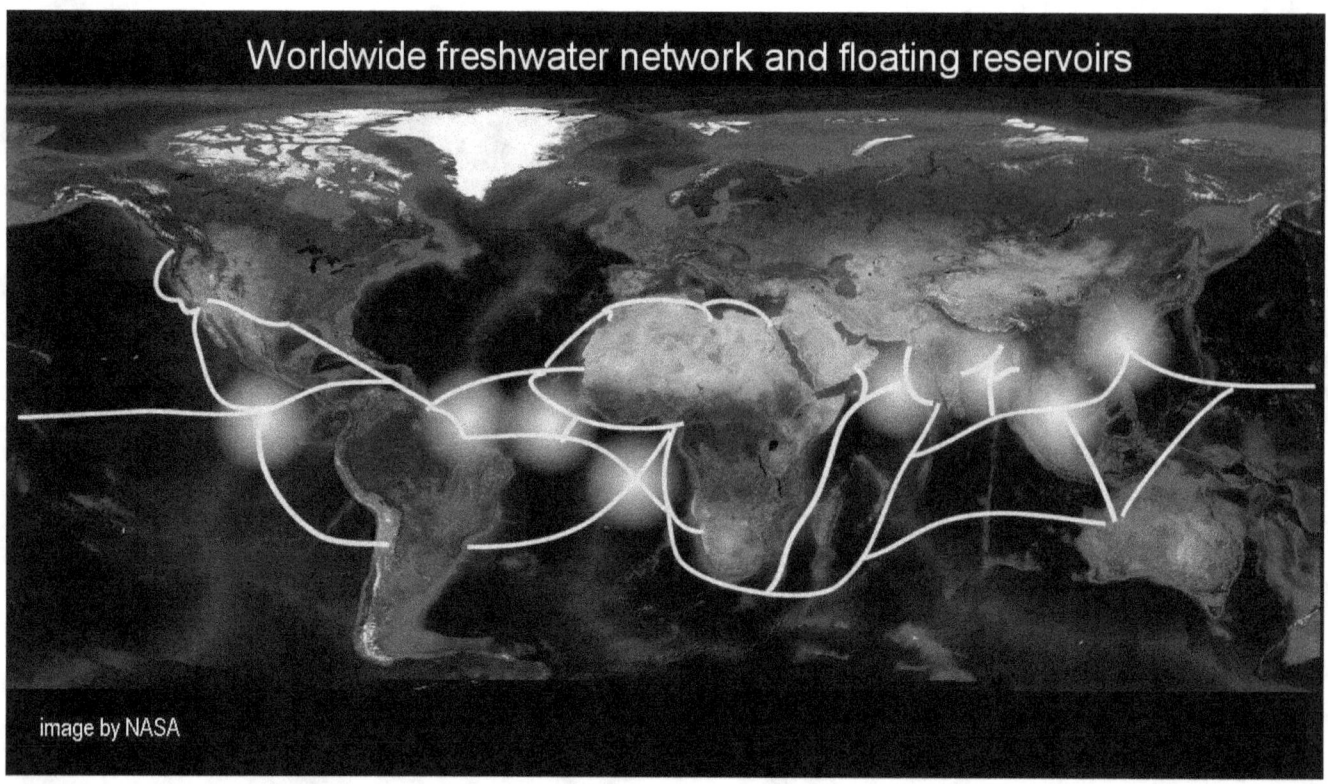

There is enough water flowing from these four rivers by themselves to feed a worldwide water distribution system, including the World Bridge. The movement of water through arteries floating in water, requires little effort as no elevation difference need to be overcome. This type of worldwide water management and distribution systems become critically important during Ice Age conditions that will be our future climate from the 2050s onward..

The climate will then be 40 times colder

Ice core measurements tell us that the climate will then be 40 times colder in relative terms that the Little Ice Age had been.

The colder Sun will produce 80% less precipitation

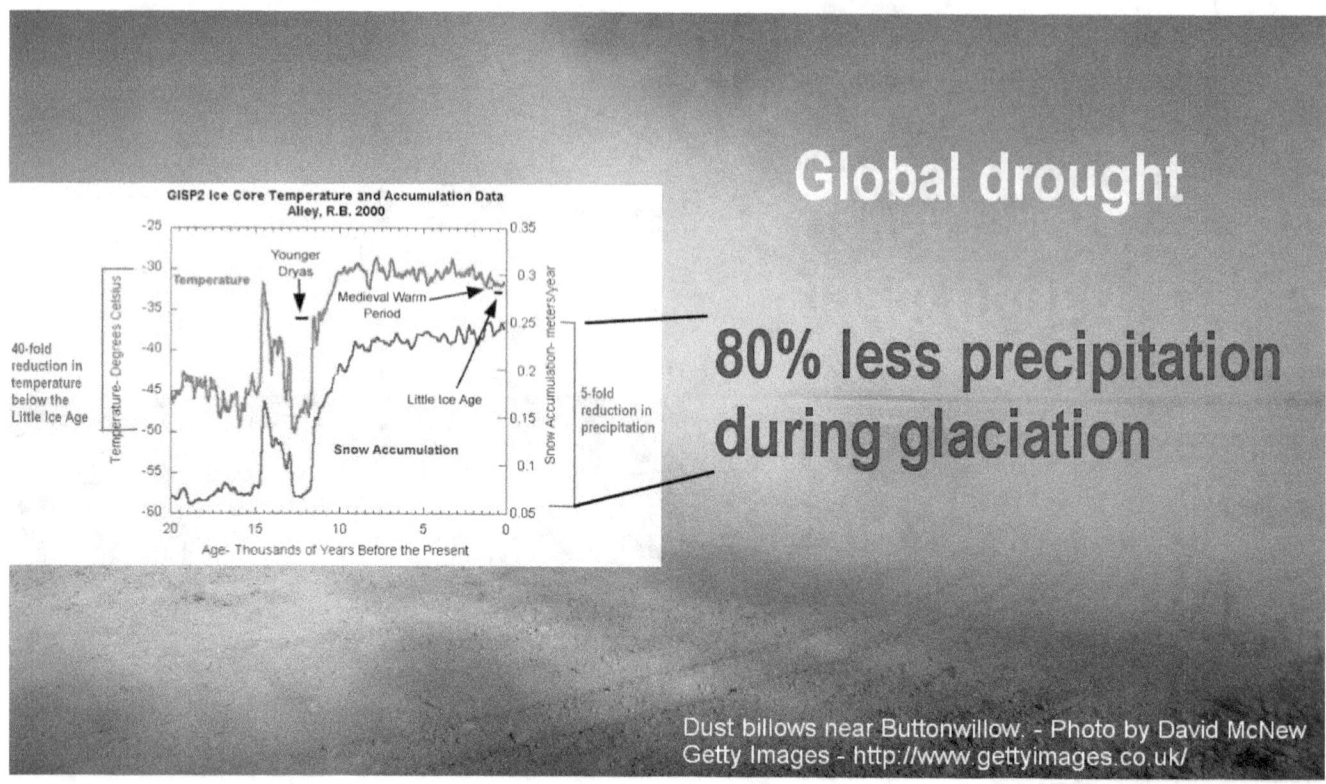

The same measurements also tell us that the colder Sun will produce 80% less precipitation, by which many regions become deserts, unless water can be brought to them, primarily from the tropics.

The tropics receive the strongest rainfall

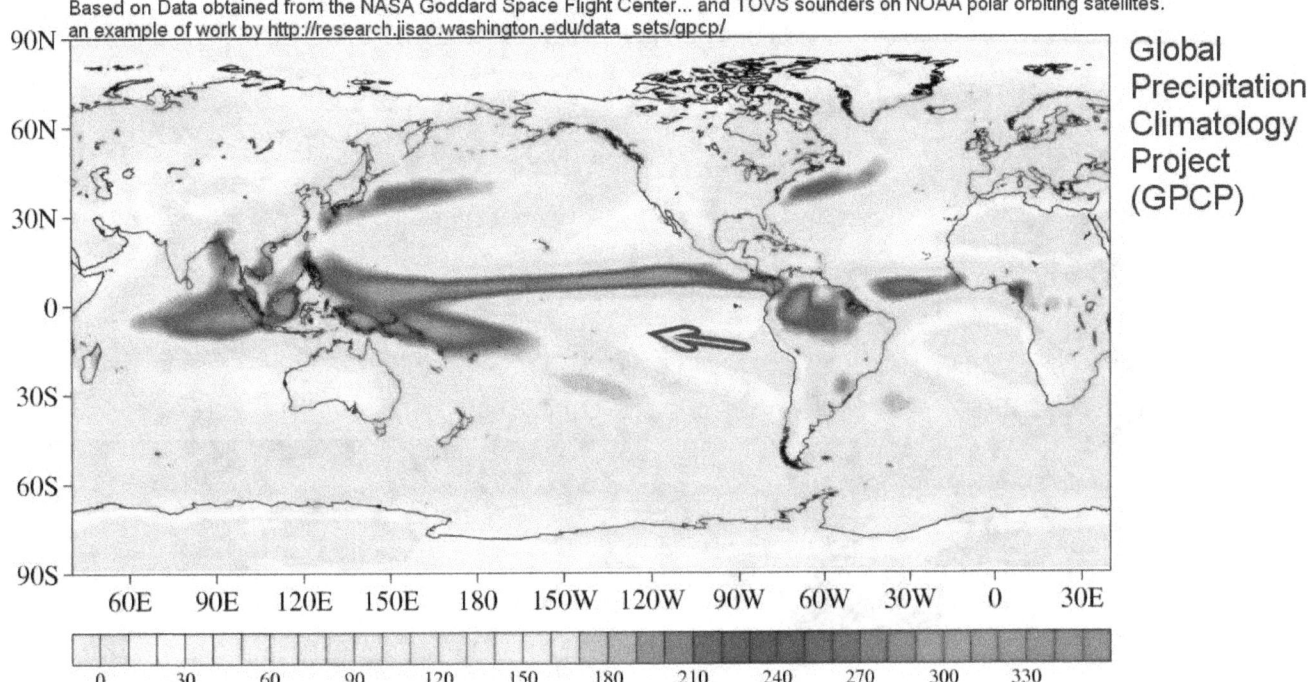

The tropics receive the strongest rainfall in the world. It is unknown, however, how much precipitation the tropics will loose during Ice Age conditions.

The loss may be far less in the tropics than the 80% reduction

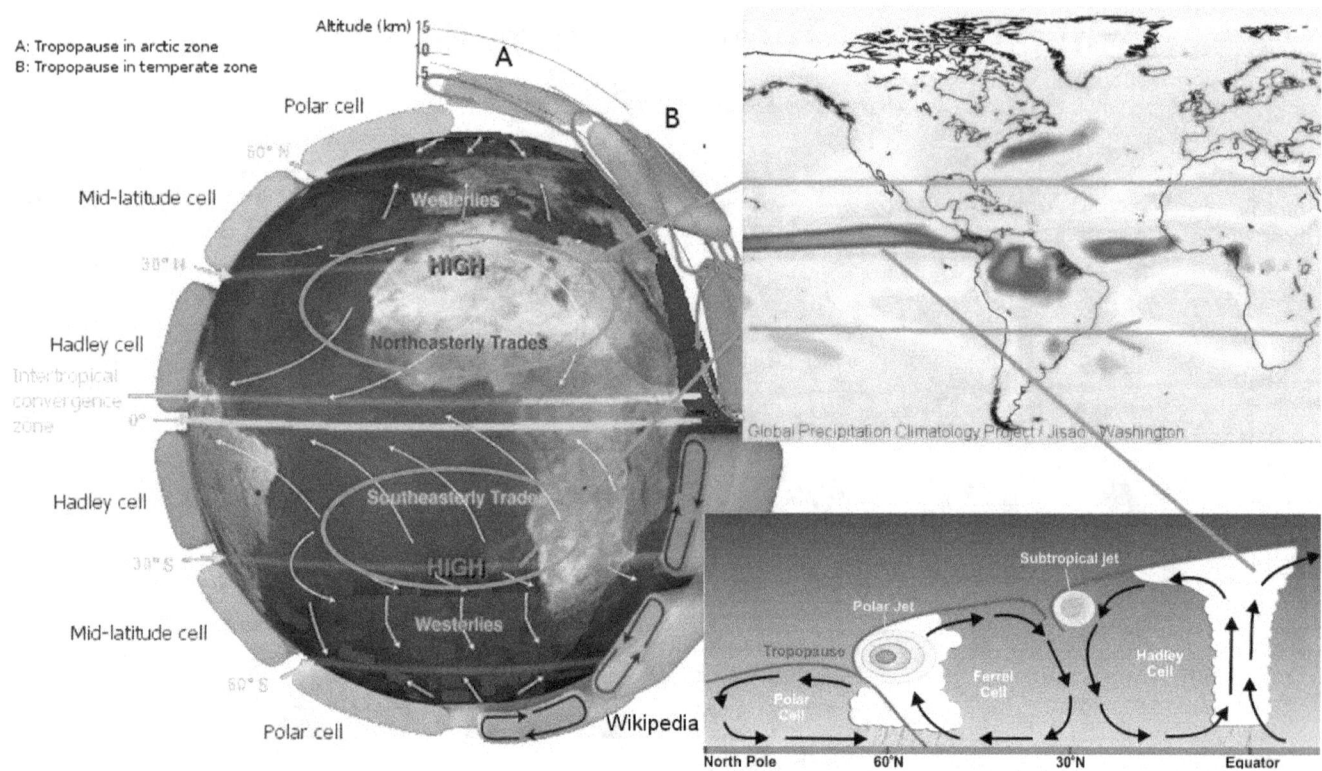

The loss of precipitation may be far less in the tropics than the 80% reduction measured in ice core samples from Greenland. Under present conditions, the worldwide air circulation system converges in the tropics, near the equator.

Here the northern and southern circulation systems flow into each other and are lifted up to higher altitudes whereby they loose their moisture as rain. It is highly unlikely that this principle will change during Ice Age conditions.

Tropical rivers will become the world's main source for freshwater

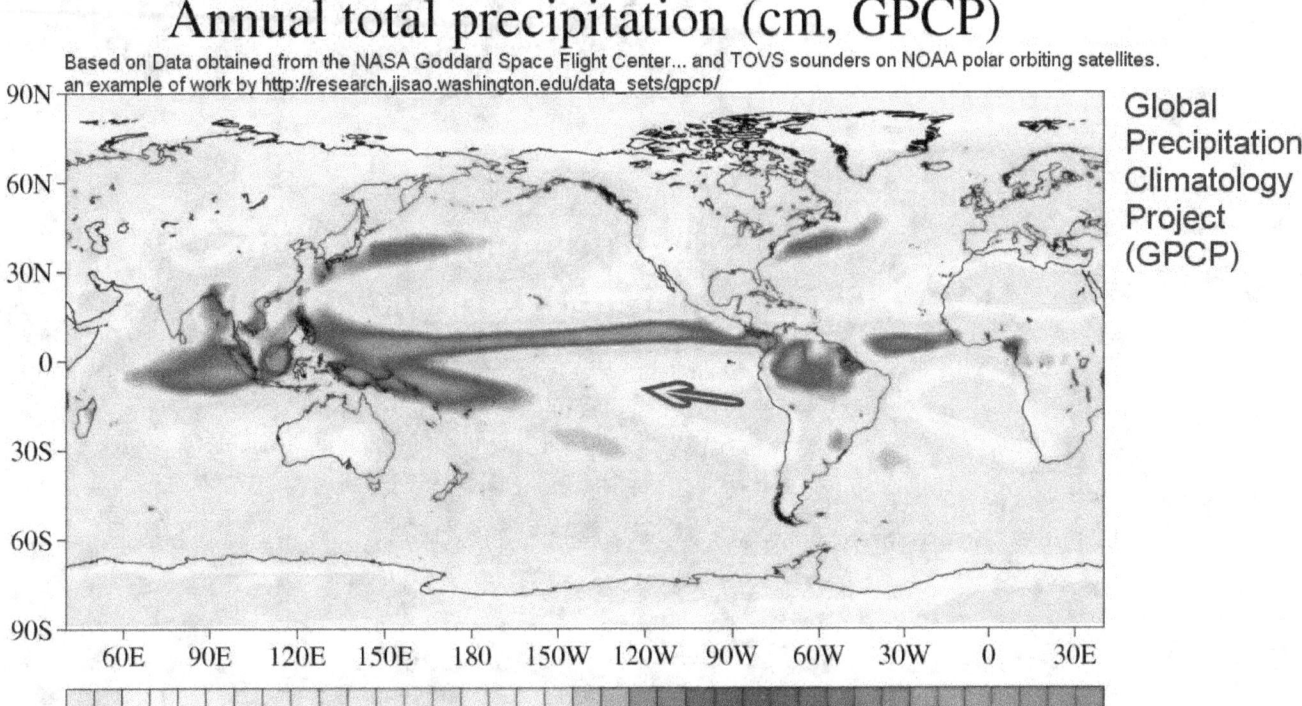

The tropical rivers will thereby become the world's main source for freshwater in the harsh times to come, even during times of reduced precipitation.

Large-scale ocean-water desalination offers a supplemental solution

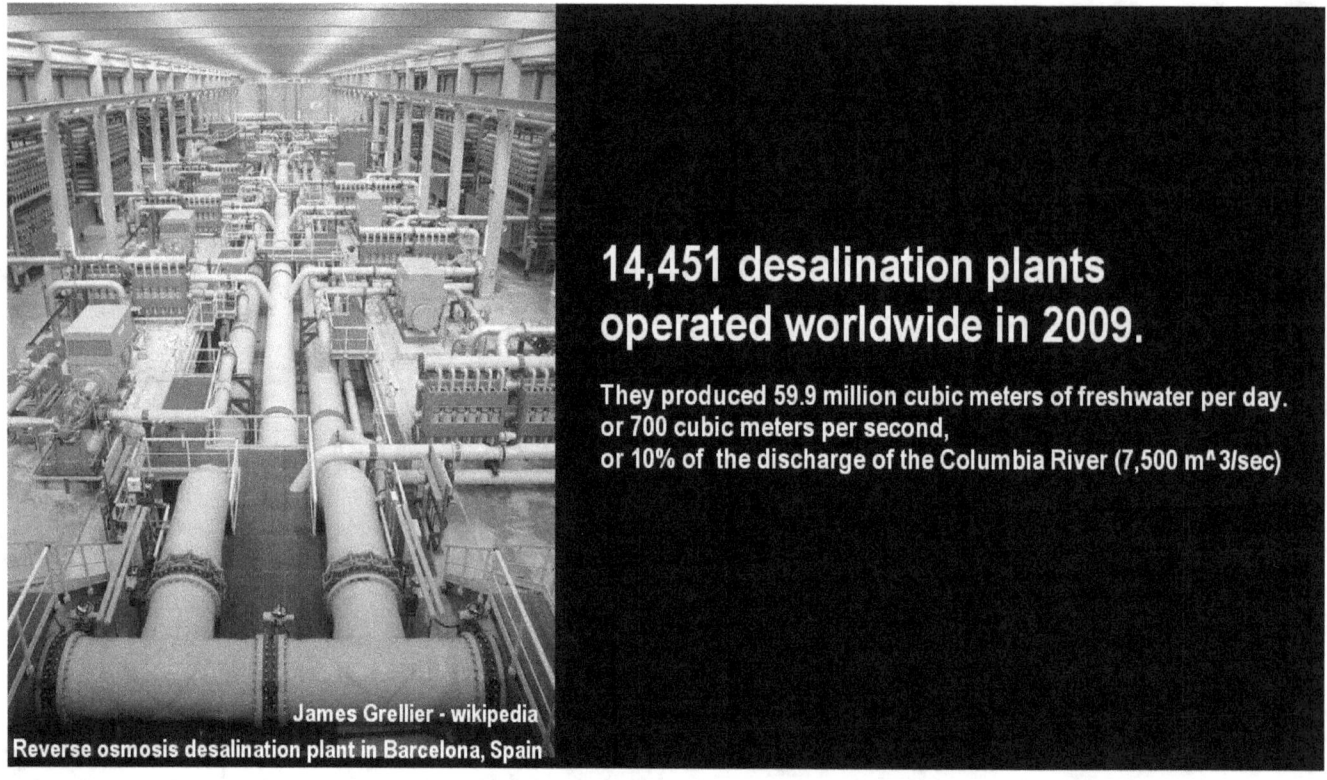

Reverse osmosis desalination plant in Barcelona, Spain. James Grellier - wikipedia

14,451 desalination plants operated worldwide in 2009.

They produced 59.9 million cubic meters of freshwater per day. or 700 cubic meters per second, or 10% of the discharge of the Columbia River (7,500 m^3/sec)

Should the loss of precipitation affect the tropics more extensively than expected, so that additional freshwater becomes needed, large-scale ocean-water desalination offers a supplemental solution. But for this, improvements are needed.

The modern desalination process is inefficient

China's largest seawater desalination project in Tianjin at the end of 2009

The modern desalination process is inefficient. It is highly energy intensive. The reason is that extremely high water pressure is required against the filter membranes, in order to overcome the seawater's inherent osmosis pressure. And still more pressure is needed to achieve a high rate of water throughput. The combination of the two makes the process energy costly and low in output volume.

The leading edge designs are expected to produce in the near future upwards to 15 cubic meters per second at over 600 psi pressure.

The produced volume is minuscule in comparison with major rivers

The produced volume is minuscule in comparison with the current flow-volume of major rivers, but may be big when the rivers diminish or dry up up, for which desalination would become essential.

When the desalination filters are located in deep oceans

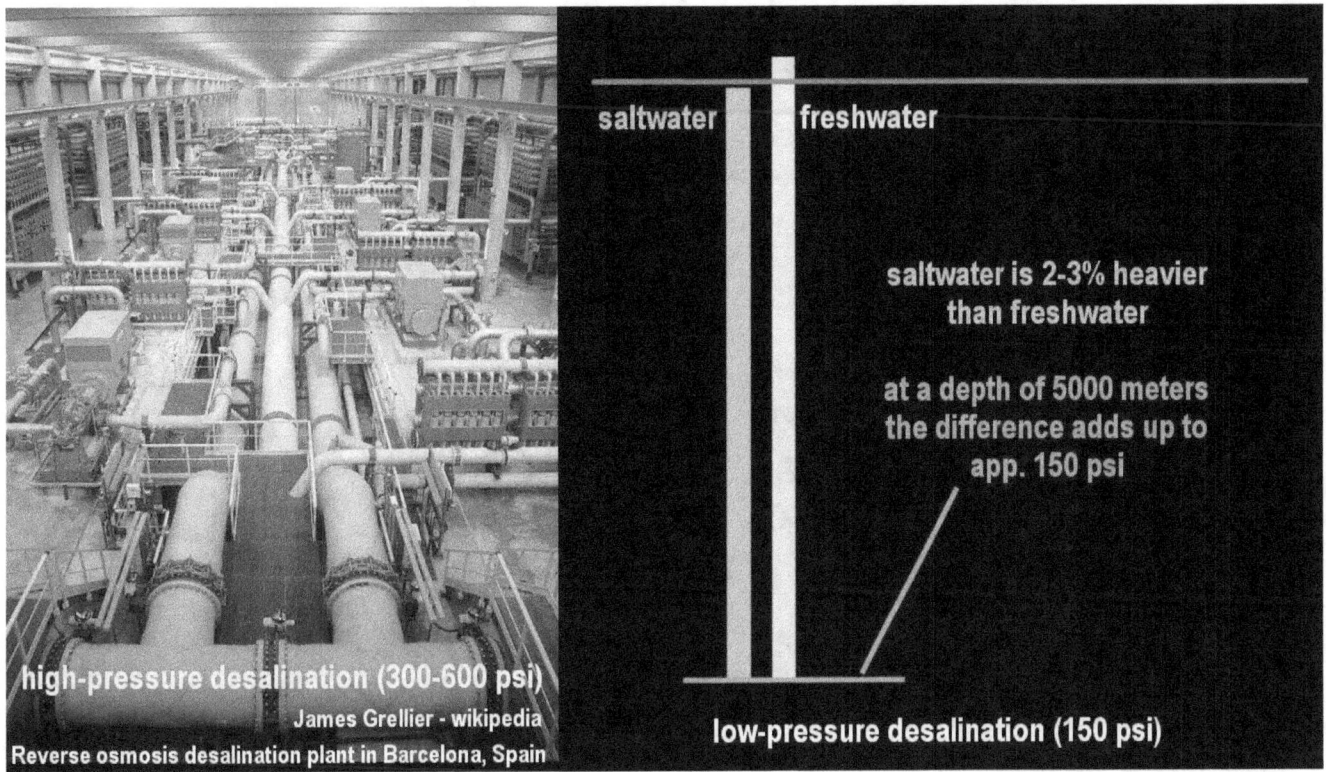

The energy cost for desalination is currently being reduced somewhat by energy-recovery technologies. It could be further reduced when the desalination filters are located in deep oceans, such as in the 5,000 meter range, where the 3% weight differential between saltwater and freshwater adds up to a pressure differential that helps the reverse osmosis significantly.

The first 270 feet would be required to overcome osmosis pressure

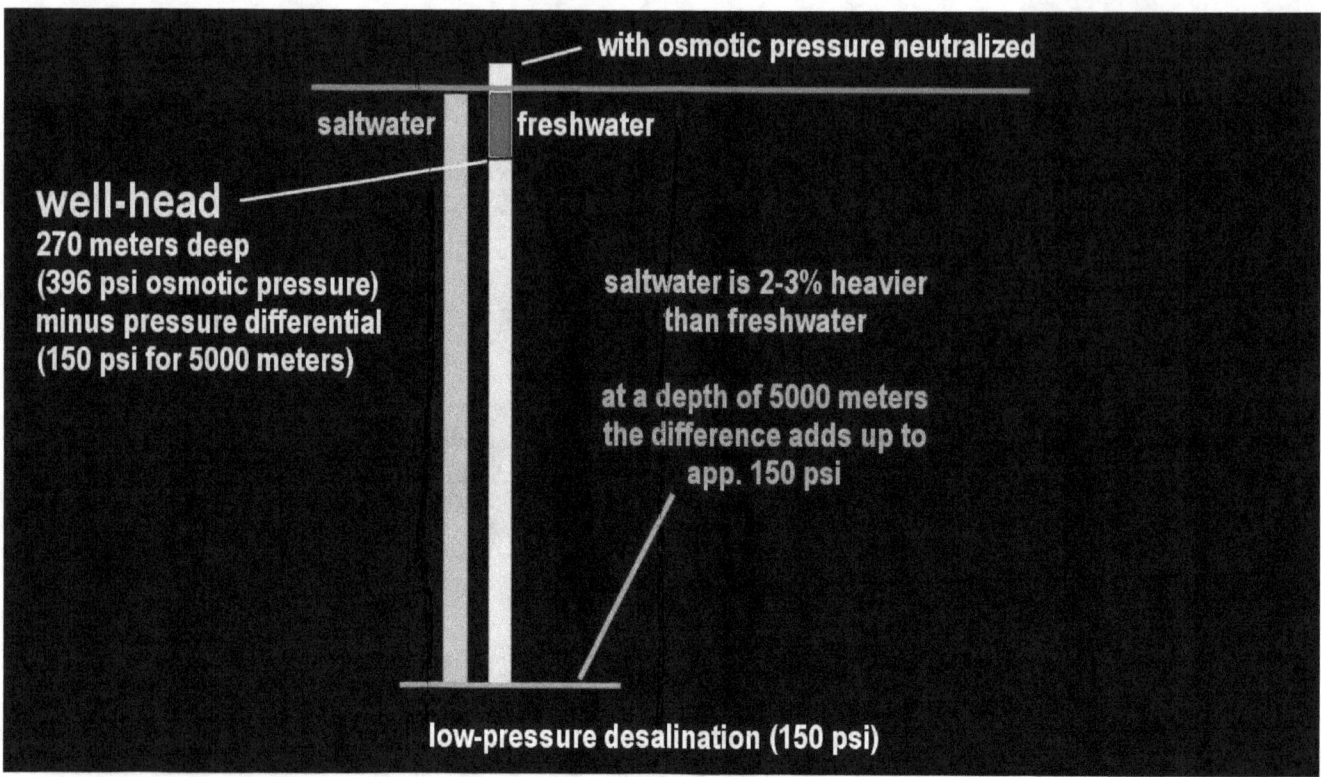

well-head

270 meters deep

(396 psi osmotic pressure)

minus pressure differential

(150 psi for 5000 meters)

The first 270 feet of the submerged filter's depth would be required to overcome the natural osmosis pressure of the seawater. In this case, a resulting freshwater well would be 270 meters deep, and the flow rate would be extremely low. However, if the filter membrane was placed 5,000 meters deep, so that the weight differential plays a role, the well depth would be raised 100 meters, to about 170 meters. Thus, the only energy cost that would remain, would be to draw the water from the effective 170 meter deep well, with high rate of flow.

If the filters were placed 9,000 meters deep

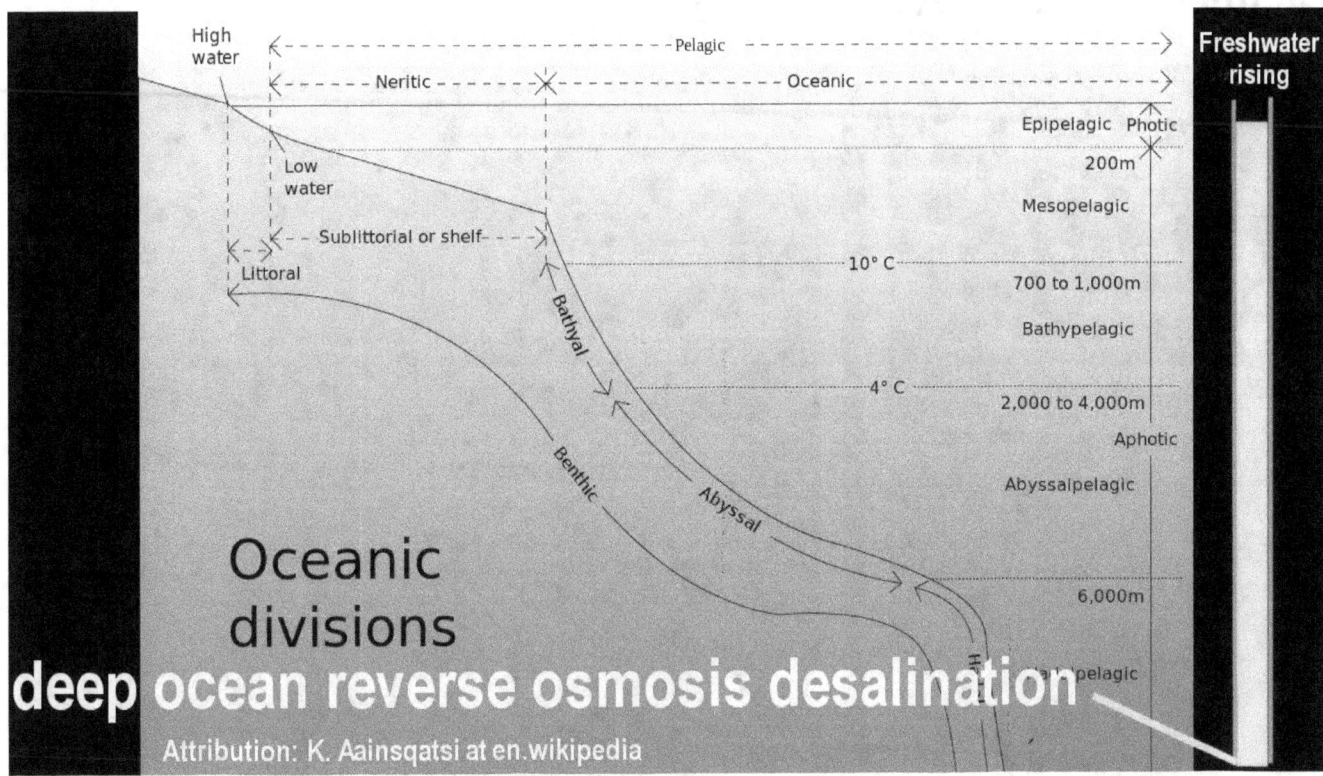

But if the filters were placed 9,000 meters deep, the resulting freshwater well would be raised by the increased weight differential, to only 80 meters below the ocean surface. It would be comparable to groundwater wells. The pump-lift would then require a mere 120 PSI pressure.

The potential that the inhibiting osmosis pressure can be completely neutralized, electrically

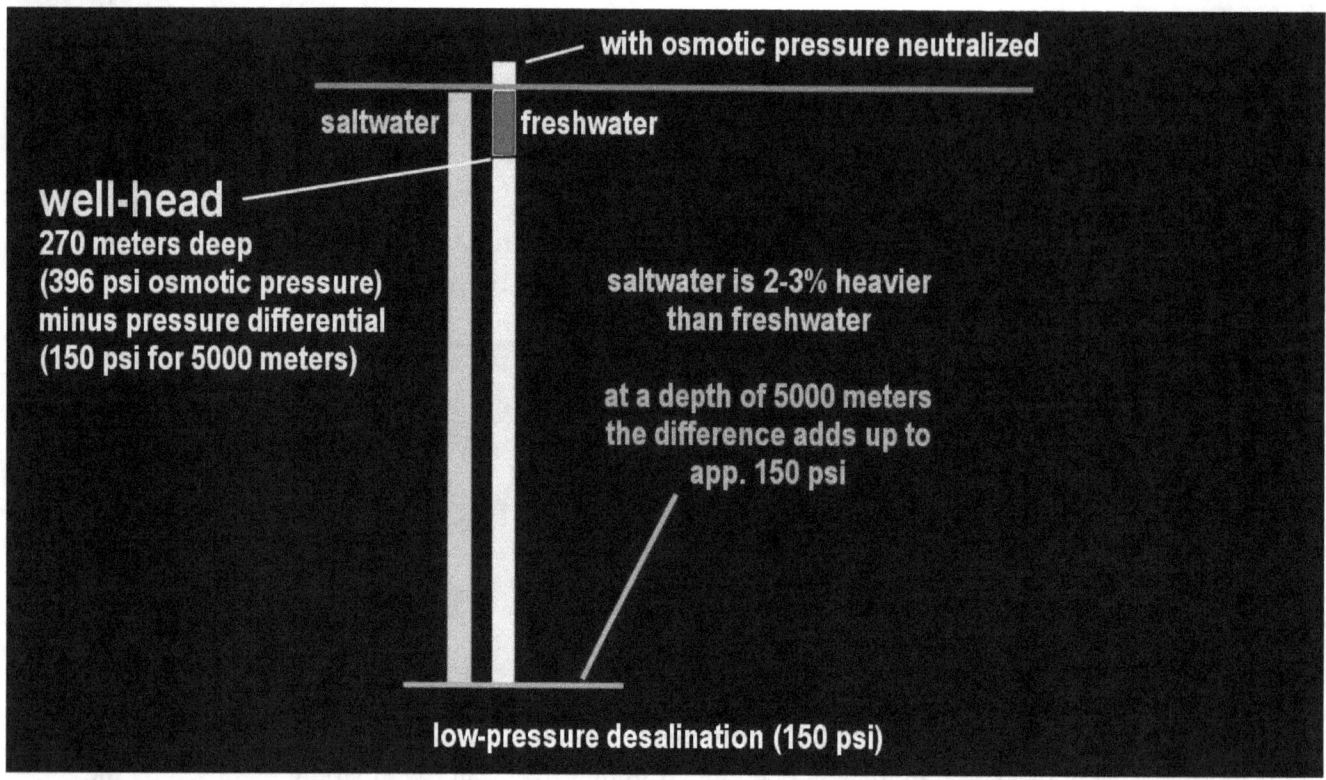

In addition, the potential does theoretically exist, that the inhibiting osmosis pressure in seawater can be completely neutralized, electrically, because the osmosis pressure is the result of an electric effect.

Water has a slight dipole electric property

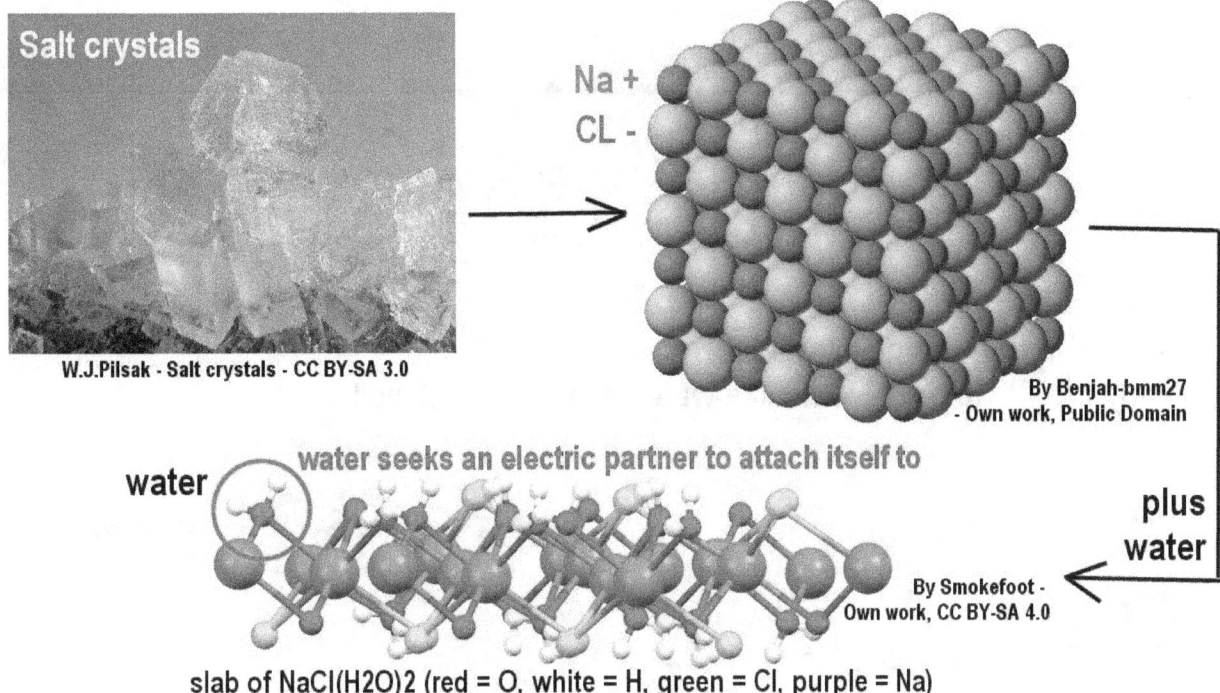

Water has a slight dipole electric property. By this electric property it becomes electrically attached to the natrium atoms of the salt dissolved in water.

By the electric effect, water is drawn into a salt solution, towards the salt

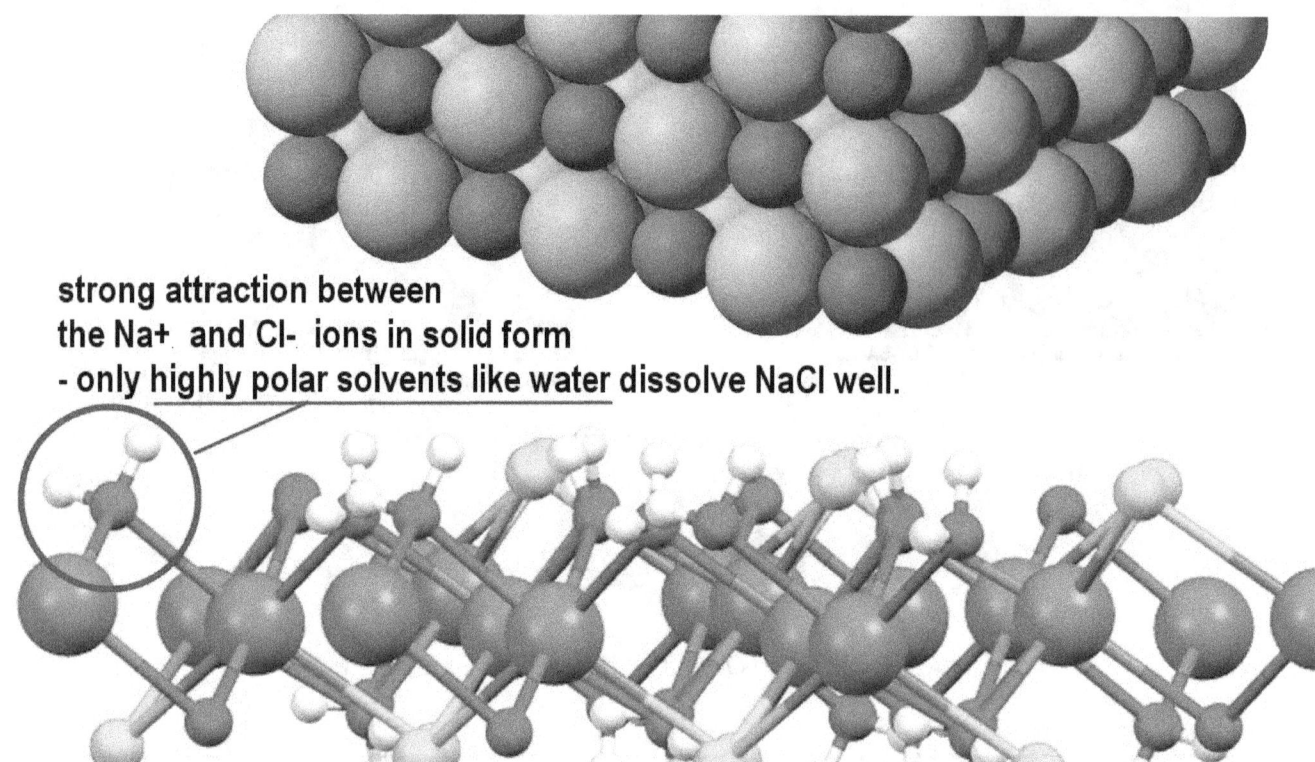

strong attraction between the Na+ and Cl- ions in solid form
- only highly polar solvents like water dissolve NaCl well.

By the electric effect, water is drawn into a salt solution, towards the salt. When dissolved in water, the sodium chloride framework disintegrates. The positive natrium or sodium ions, and the negative chlorine ions become surrounded in solution by the electrically polar water molecules, which, with their electric polarity compete with the ionic bonds of the sodium chloride. The resulting electric interaction breaks down the tight framework of the sodium chloride crystals and dissolves them.

The electric attraction lifts the salt column until equalization is achieved

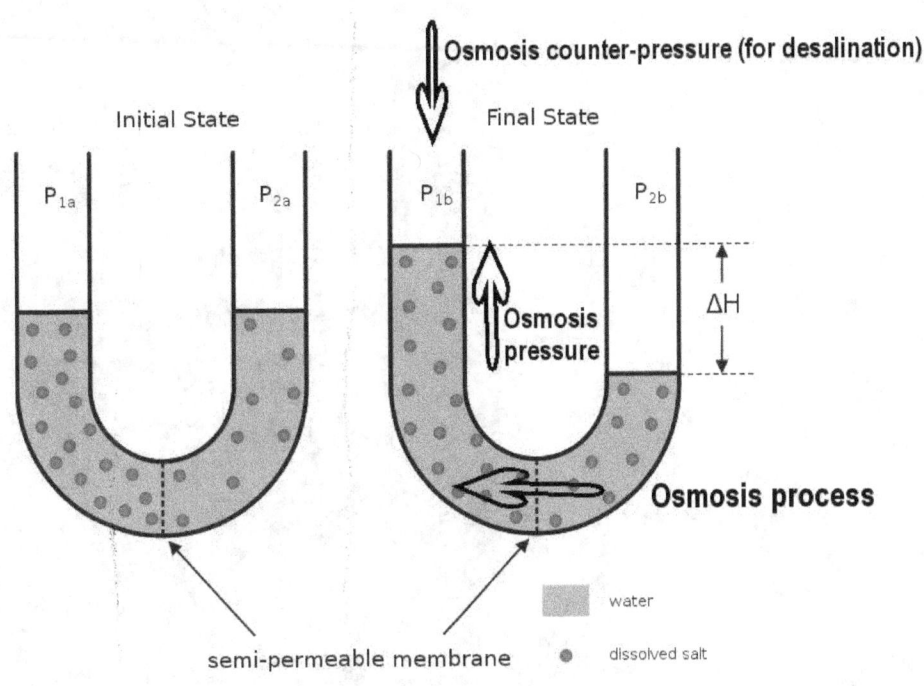

based on
by Hans Hillewaert / CC BY-SA 3.0

The electric attraction of the water (which is the solvent) towards the soluble (the salt), creates a tendency for attraction, and if unequal, for the attraction to equalize. If a barrier stands in the way that blocks the salt, but does not block the water, the attracted water is drawn through the filter towards the salt. It lifts the salt column until equalization is achieved.

In order for us to get the water to flow backwards, to flow out of the solution, a large amount of pressure needs to be applied for this reversal to happen. The needed reverse pressure for total separation, is termed the osmotic pressure. It forces the water to move contrary to its attraction. It forces it to break its ionic connection with the salt. Desalination is thereby, presently, a brute force process. And the force isn't trivial.

For seawater, the osmotic pressure is 396 PSI, or 27 atmospheres, the equivalent of a 270 meter tall column of water. This large osmotic pressure must be overcome before anything happens on the desalination front, unless the electric attraction can be neutralized at the point of the filter.

Since the electric neutralization has not yet been achieved

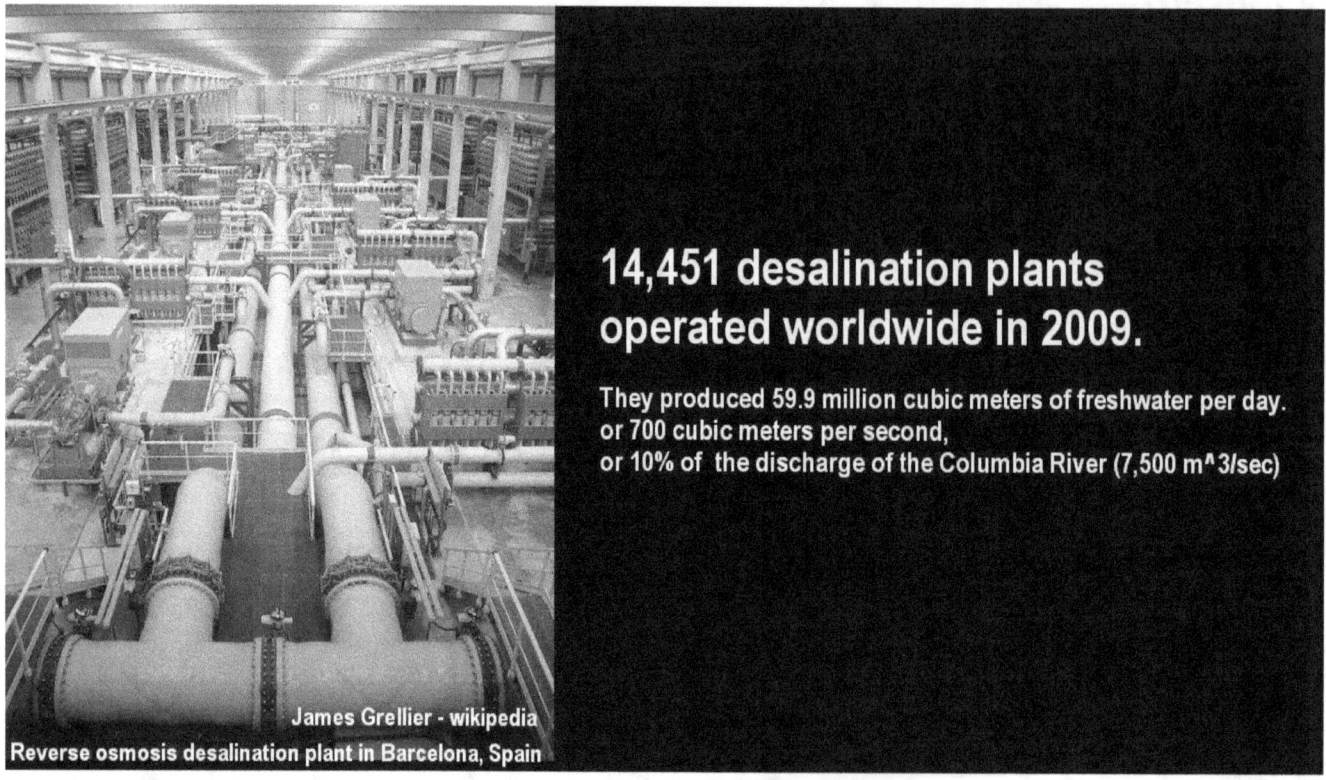

Since the electric neutralization has not yet been achieved, desalination plants typically operate with pressures from 600 PSI to 1,200 PSI to overcome the strong osmosis pressure and to achieve a large enough rate of flow.

The theoretically possible potential to break the osmotic connection is promising

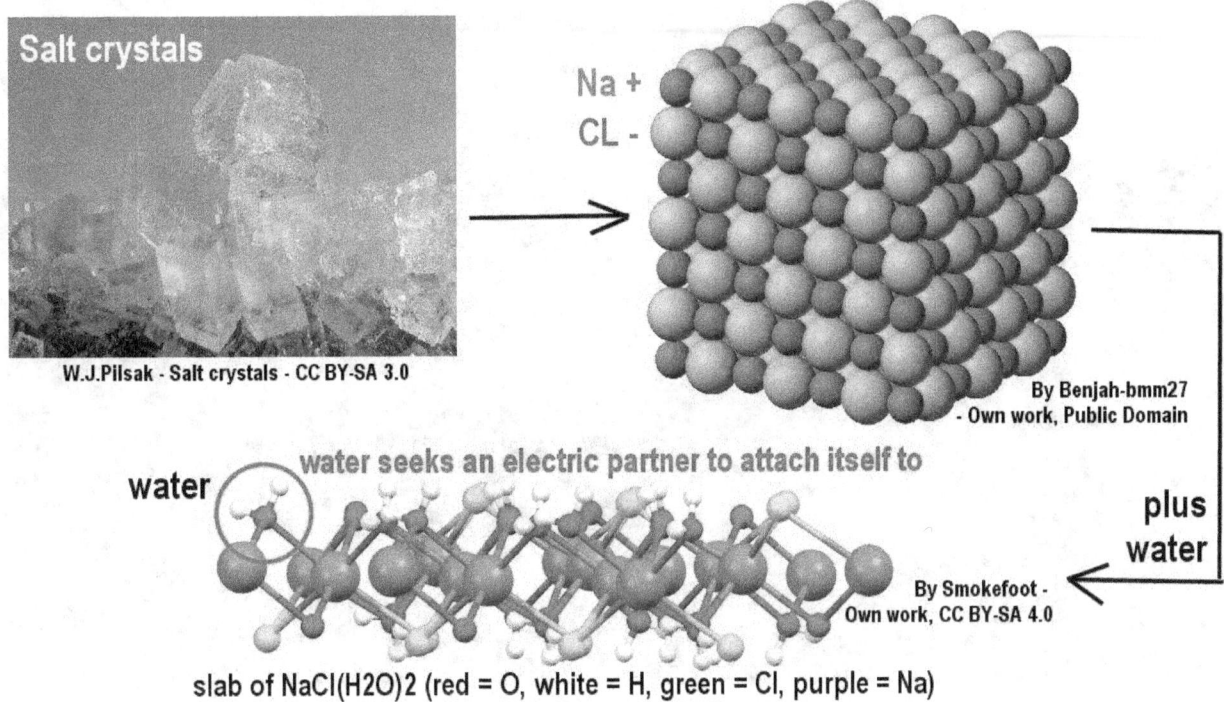

While the theoretically possible potential to break the osmotic connection, electrically, has not yet been achieved; research is ongoing and is promising.

The use of carbon nanotubes as filter elements is being explored

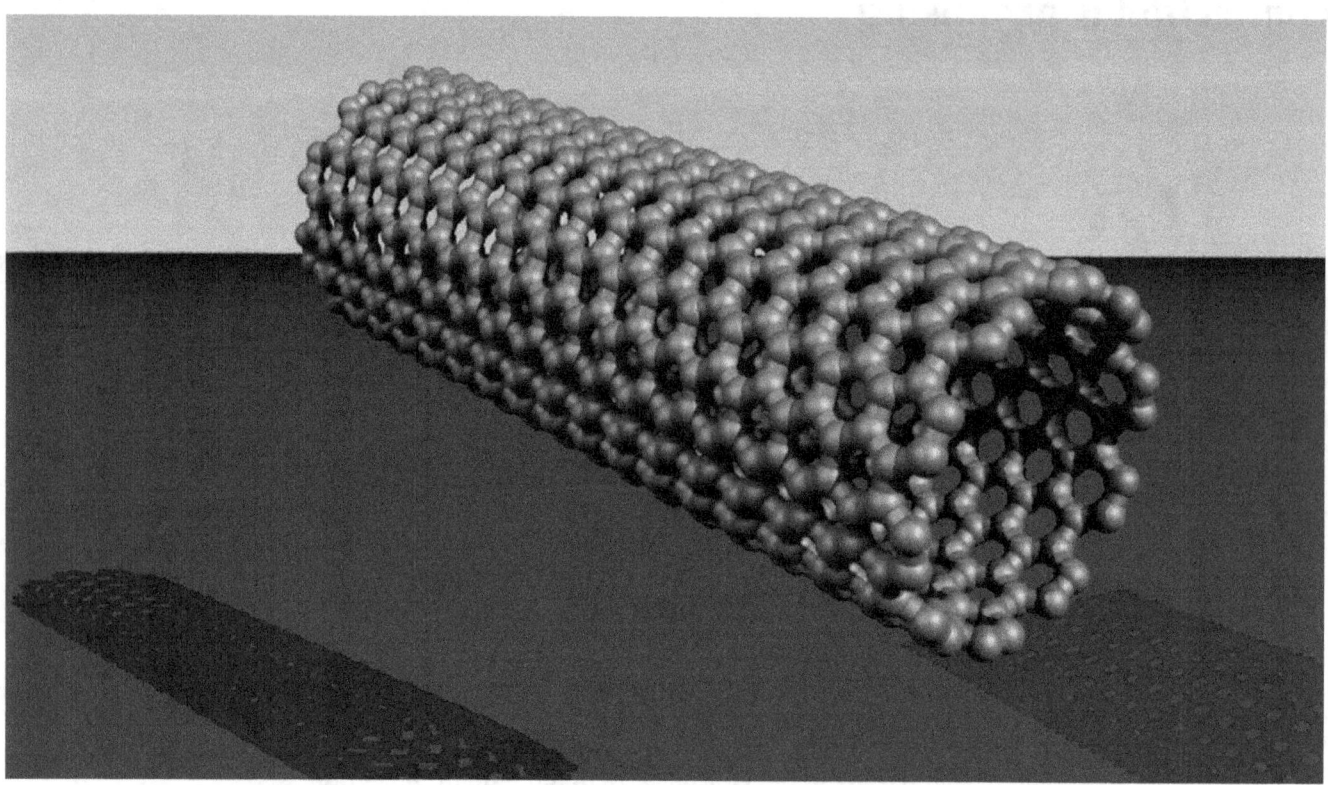

The use of carbon nanotubes as filter elements is being explored. The nanotubes have a unique electric quality.

Another promising new type of filter for desalination would utilize graphene sheets

Another promising new type of filter for desalination would utilize the recent advances in the manufacturing of graphene sheets. A graphene sheet is a sheet of graphite atoms linked together into a tight lateral lattice one atom thick. If the technology can be worked out that cuts the right size of holes through the sheet of graphene, a more than 100-fold increase in the desalination efficiency is deemed to be possible, according to research done at the Massachusetts Institute of Technology.

We could then have rivers of freshwater flowing out of the oceans

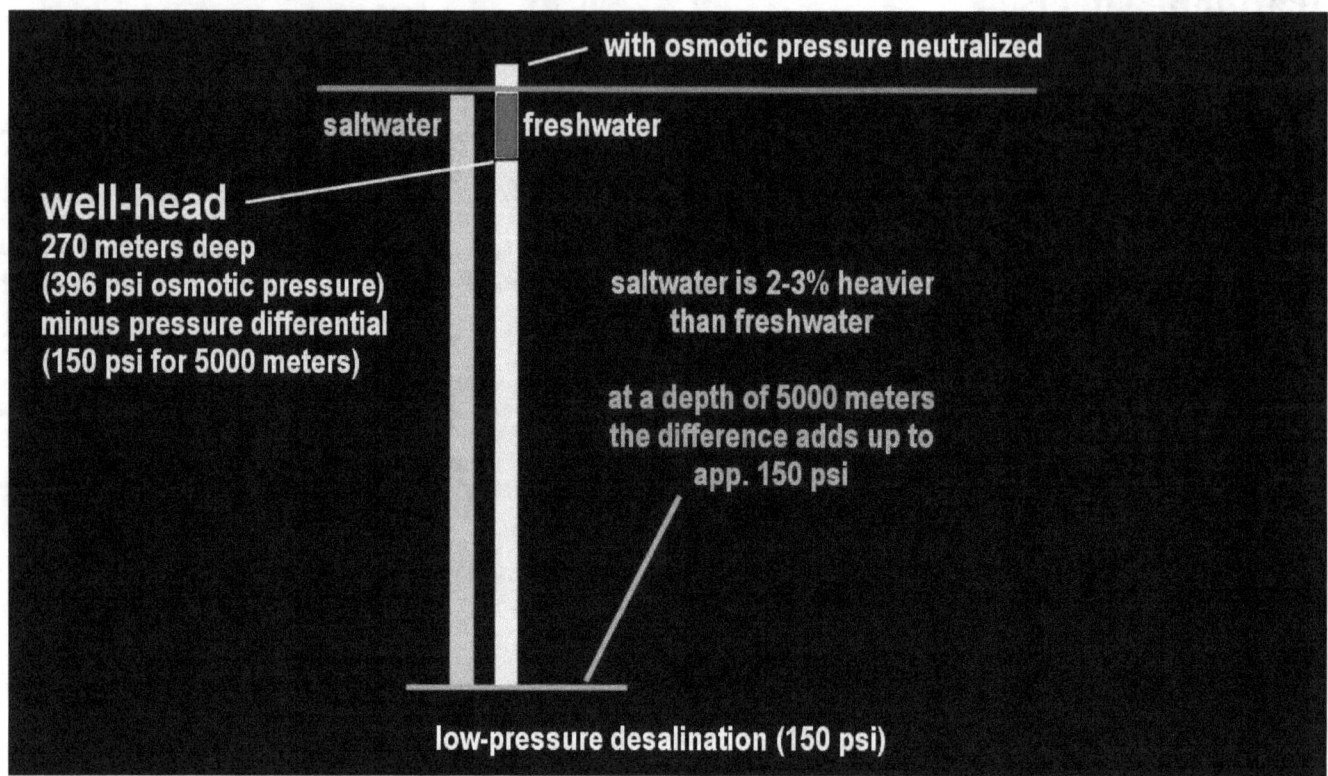

If the osmotic pressure could thereby be neutralized, the the deep-ocean-desalination well head would be 270 feet higher - high enough to reach the surface. We could then have rivers of freshwater flowing out of the oceans. The principle for this to be possible may exist.

Desalination shouldn't be needed at the present stage

China's largest seawater desalination project in Tianjin at the end of 2009

Ultimately, desalination shouldn't be needed at the present stage, and wouldn't have been pursued if the worldwide water distribution system had been built. Desalination is inefficient. A large modern desalination plant can produce a million cubic meters of freshwater a day. The Amazon River drains this volume into the Pacific Ocean very 5 seconds. Desalination would only become important when the rivers were to run dry during Ice Age conditions, or were reduced to low volumes of flow. For this potentially exceptional case, desalination technology needs to be developed to the utmost as a fall-back option, even if it won't ever be needed. We must make these types of extraordinary scientific, technological, and economic efforts without fail, because, if we don't make the efforts, and we find us later in need of their products in times of crisis, our failing to move today may cause the demise of civilization in future times.

The principles apparently do exist that make a secure future possible, and for those principles that are not yet known, we must surge ahead in scientific discovery and discover them, and not allow us to be deterred if we fail at times and chase after ghosts.

Energy Forever

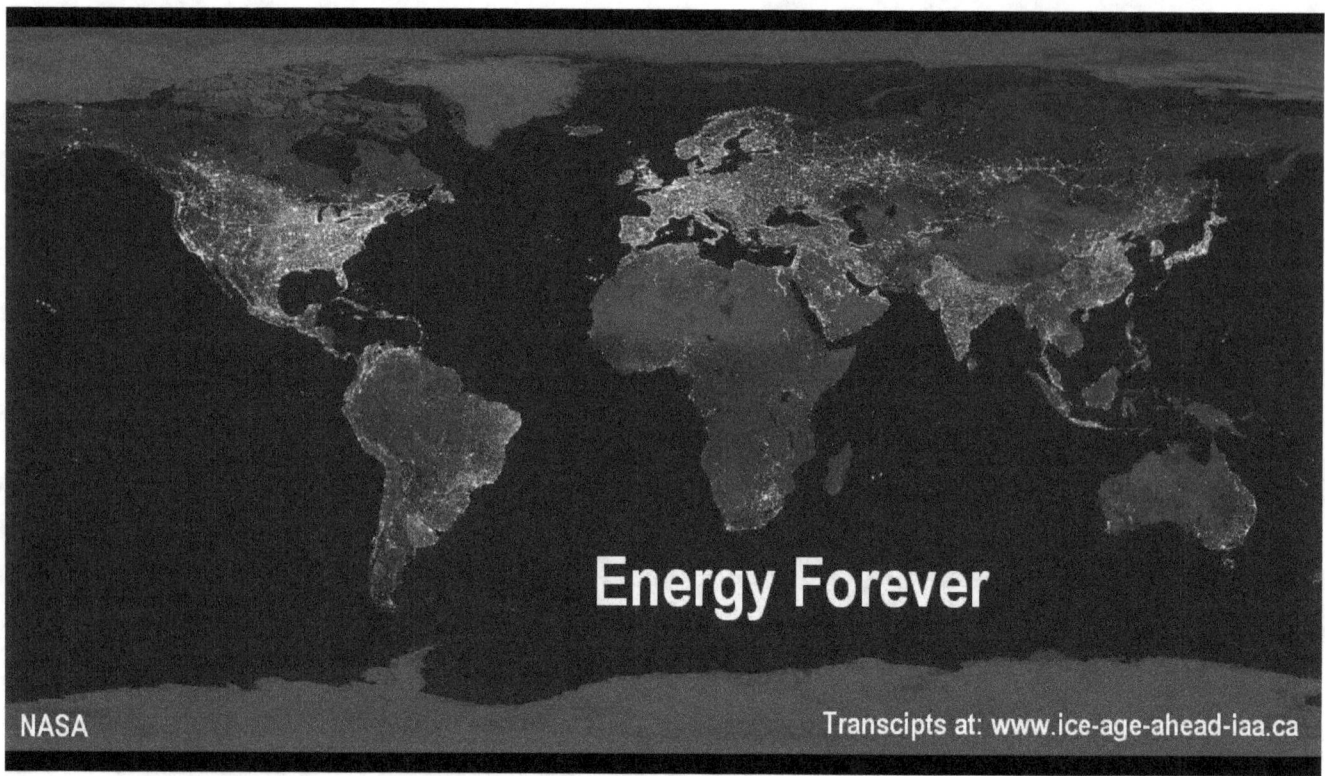

Energy Forever

One of the ghosts of pure illusion is nuclear-fusion power

One of the ghosts of pure illusion, that has been pursued as real, is nuclear-fusion power. The dream of harvesting nuclear fusion energy is a dream that will never come true. This is so, for the simple reason that no principle exists that would make this possible. It doesn't happen, not anywhere in the universe. It can't happen for the simple reason that nuclear fusion is an energy consuming process. The Joint European Tokomak experiment achieved fusion for one second with a 10-fold energy loss..

It took scientists decades of dreaming, and billions of dollars

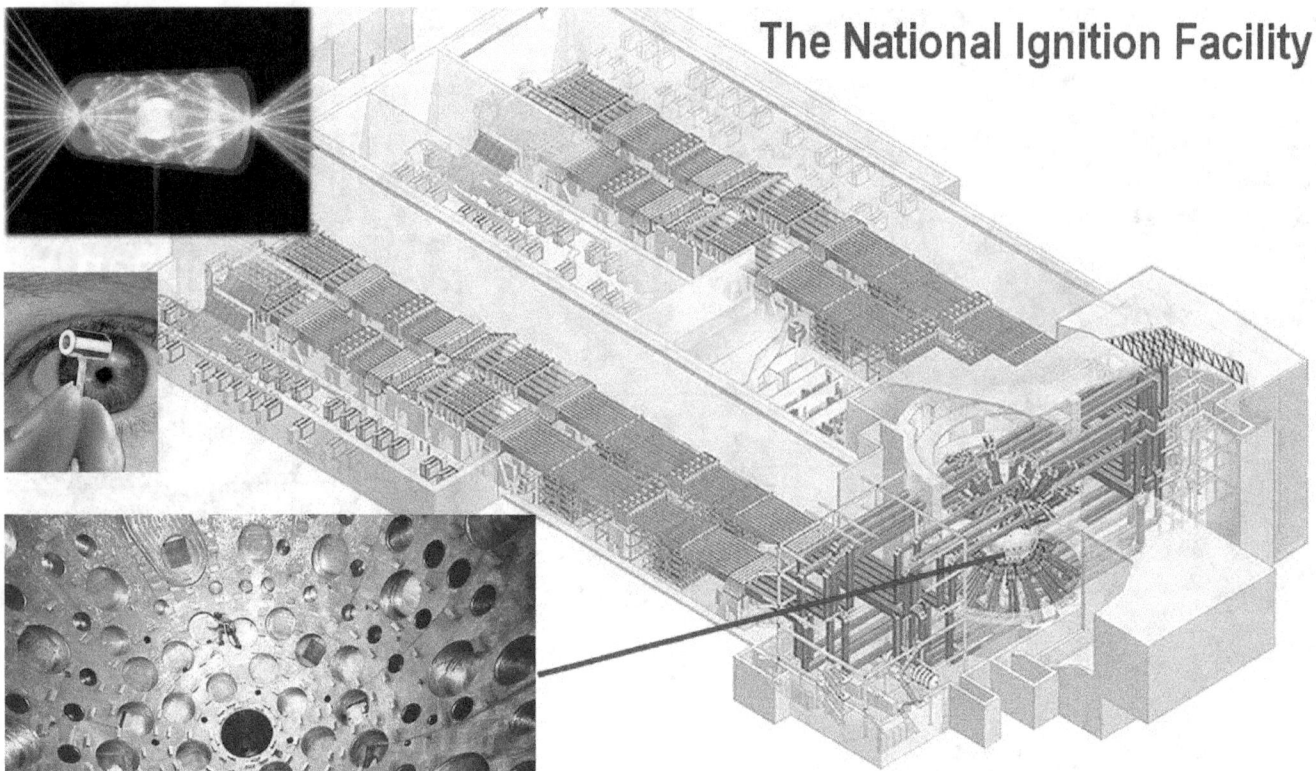

The National Ignition Facility

It took scientists decades of dreaming, and billions of dollars for building experiments, struggling to make the impossible happen, to recognize that the process isn't possible. Scientists had said to each other that if the Sun can do it, so can we. Except, they didn't realize that the Sun doesn't work the way they had imagined; that it isn't powered by nuclear fusion.

The giant National Ignition Facility, a facility the size of a stadium, has never, in all its years, achieved the once hoped-for fusion ignition. It now serves a different purpose. To judge the effort by its size, note the worker in the target chamber, in the lower-left image.

So, where do we go from here, after the dream has failed?

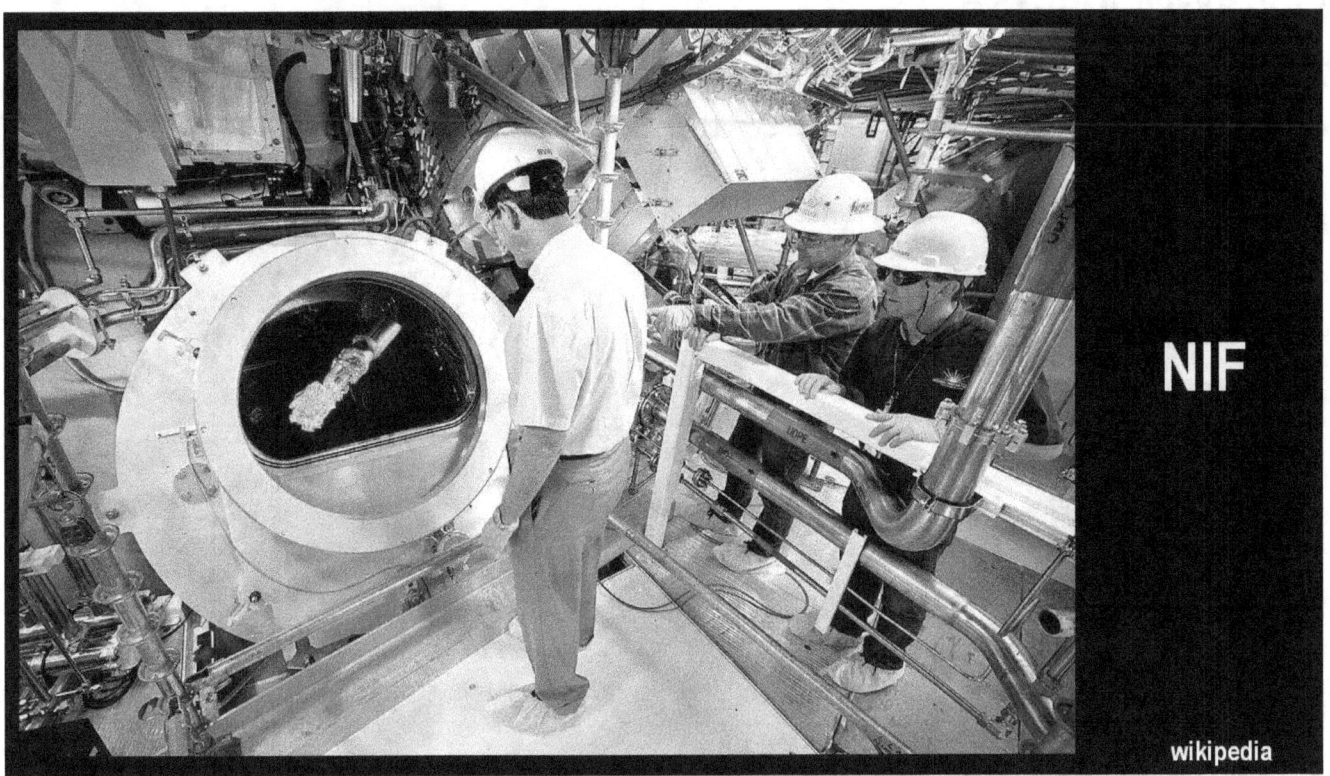

So, where do we go from here, after the dream has failed? Should we have never pursued advanced energy systems?

Scientists had an idea that this tiny pellet of fuel could unlock an energy-rich future

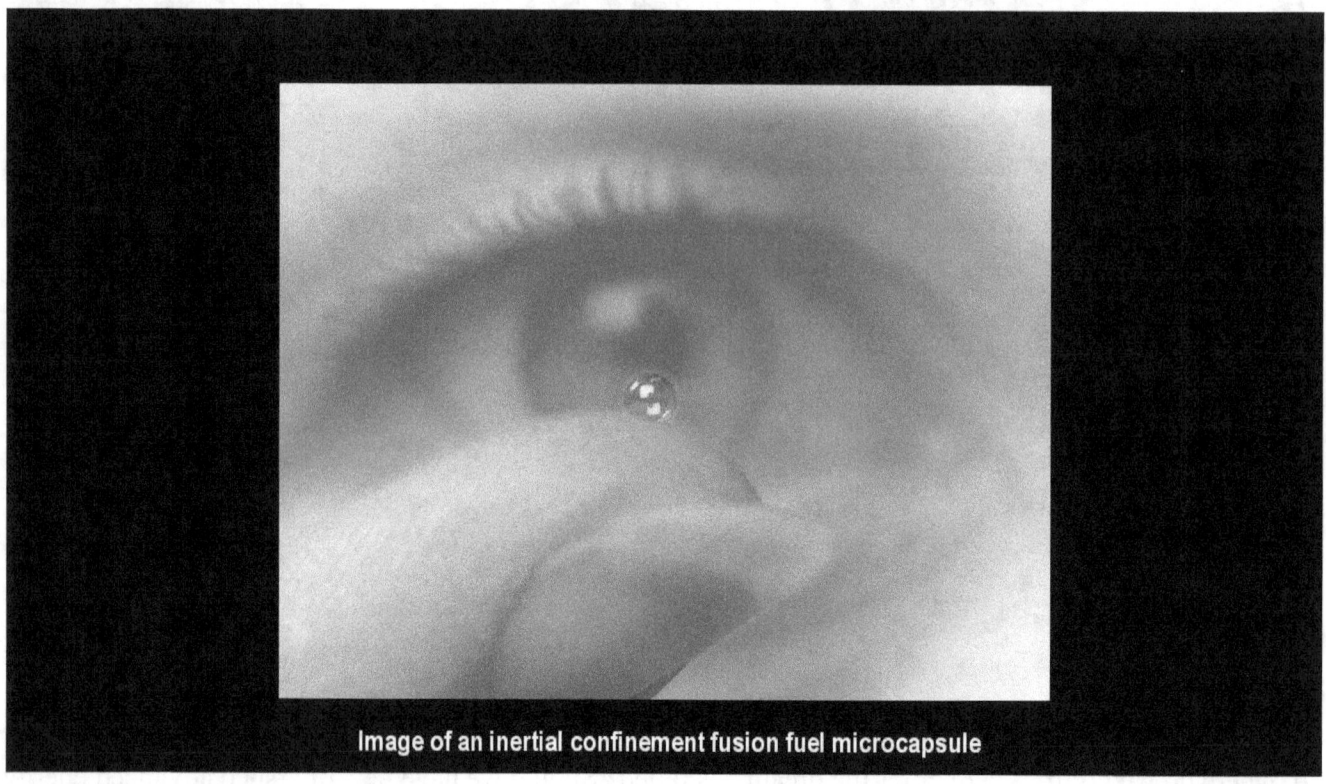

Image of an inertial confinement fusion fuel microcapsule

Scientists had an idea that this tiny pellet of fuel forced into fusion could unlock an energy-rich future for mankind. This energy-rich goal is still valid. Without an energy rich potential, we have no hope to meet the Ice Age Challenge.

We have depleted our energy resources so intensively that we have only 60 to 200 years left

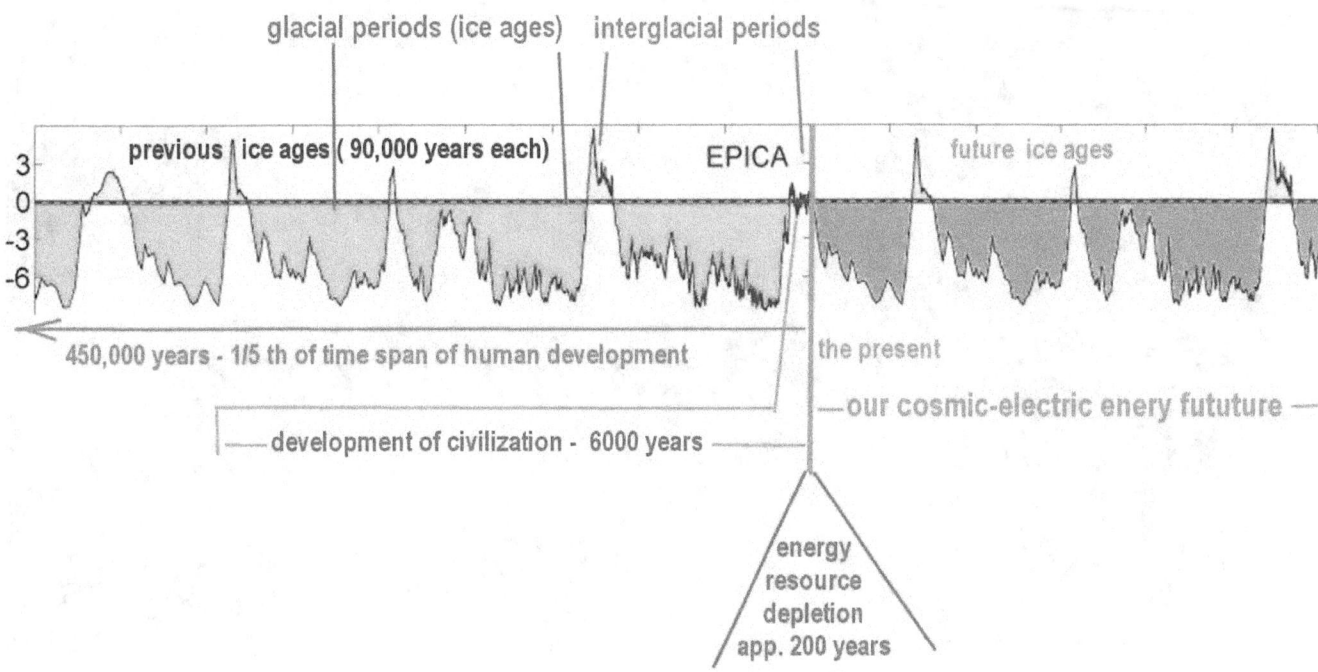

During the 2.5 million years since the dawn of humanity, civilization began only 6,000 years ago, and large-scale energy utilization only a few hundred years ago. During this short time we have depleted our energy resources so intensively that we have only 60 to 200 years of these depletable resources left. How can we even hope to survive the next 90,000 years of the near Ice Age with that, and for millions of years thereafter.

The answer is a type of source that is self-renewing

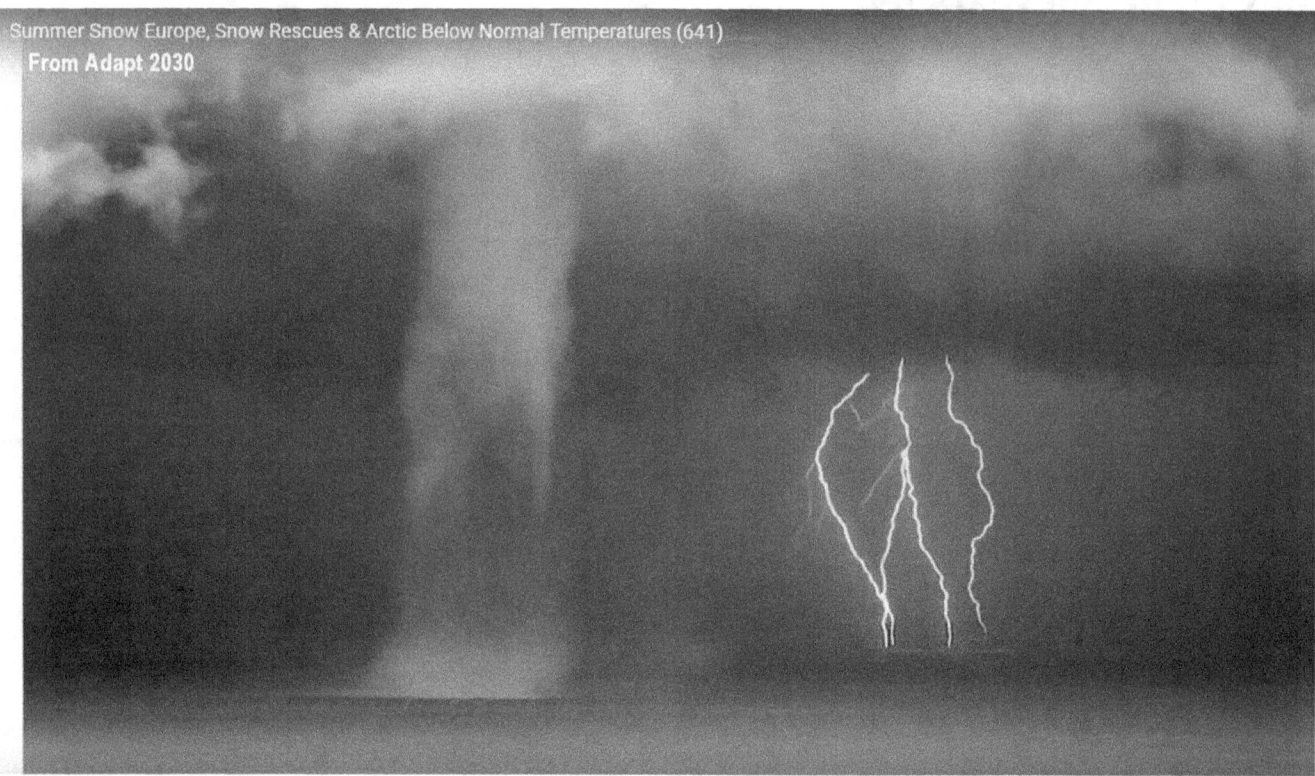

The answer is that we must develop new energy platforms that provide more energy from a type of source that cannot be depleted, but is self-renewing.

We see examples of this type of a massive energy resource that cannot be depleted, manifesting itself as large-scale natural events. And we find that conditions exist for us to tap into this resource, since this resource is used almost daily by natural systems.

The source is cosmic, and the interface is the ionosphere

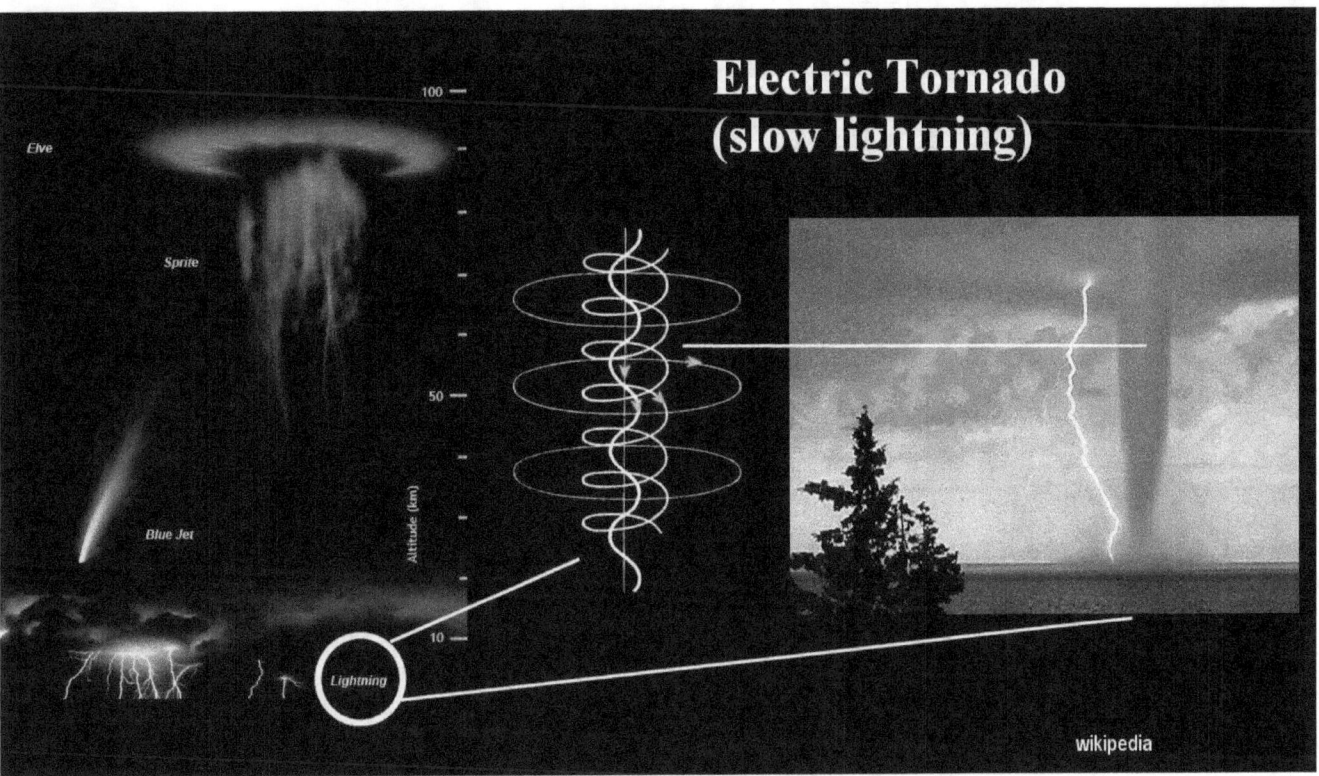

The source for this resource is cosmic, and the interface on Earth is the ionosphere. Sometimes big red sprites appear high above major storm regions that carry moisture to high altitudes.

NASA has been able to photograph two electric plasma bands

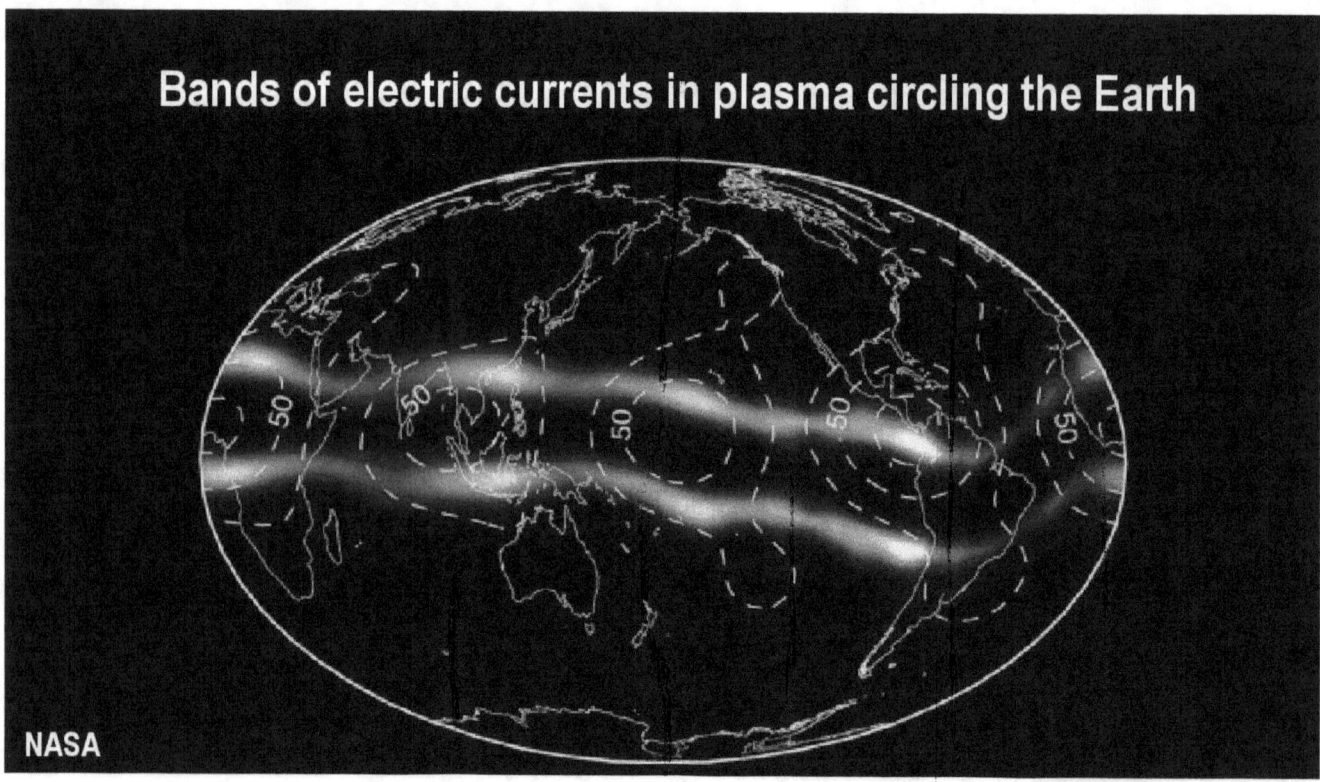

NASA has been able to photograph two electric plasma bands encircling the Earth, which appear to be a part of the ionosphere. It even appears that the densest hurricane zones are located where the electric jet streams are situated directly overhead. In short, we are dealing with a cosmically generated energy resource that appears to be a part of the cosmic system of plasma that powers our Sun. As such, it can never be depleted. It may well become stronger the more we draw from it.

The natural systems appear to use this resource liberally

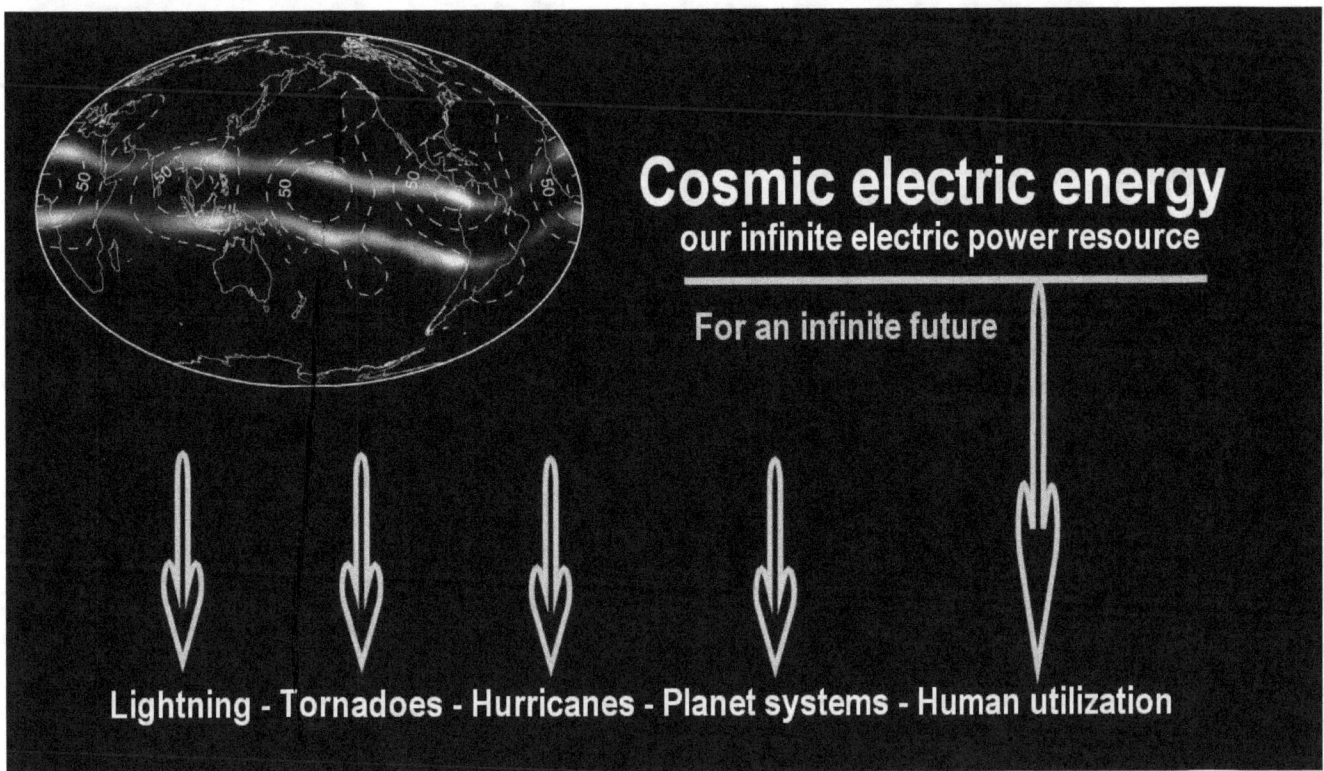

The natural systems appear to use this resource liberally, and evidently have so for a very long time. I would like to suggest that the time may not be far off when humanity taps into the system as the latest addition on the list. We desperately need an anti-entropic energy resource, and we need it fast, before the Ice Age begins, because the Ice Age will render all northern oil and gas fields inaccessible, and likewise most, if not all, hydroelectric systems.

The barrier is the false belief that our Sun is self-powered

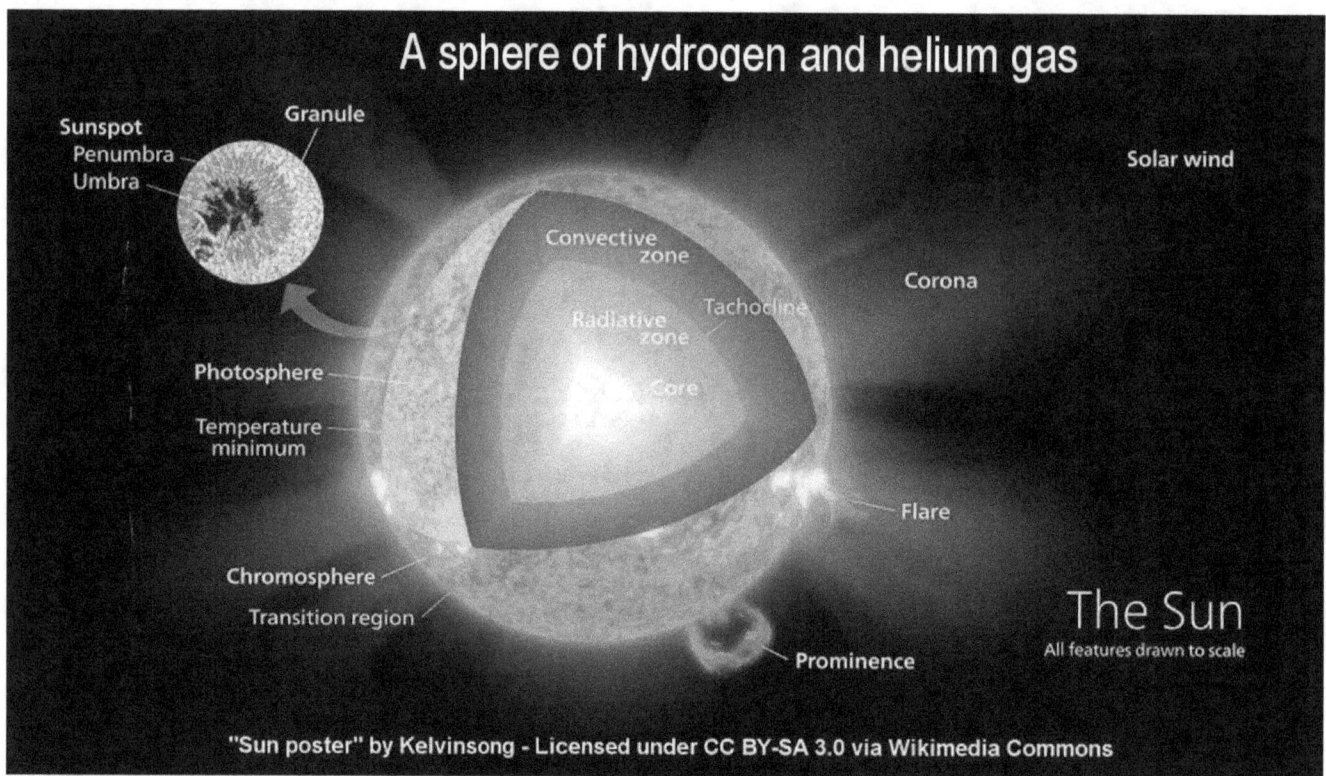

The barrier that prevents the recognition of this near infinite energy resource is the false belief that our Sun is a self-powered nuclear-fusion furnace. According to this theory, plasma streams in space do not exist. The greatest energy potential in human history, remains thereby blocked, and humanity's future, if not its future existence, remains blocked with it.

Another very-large energy resource is available through the liquid-fluoride nuclear power reactor

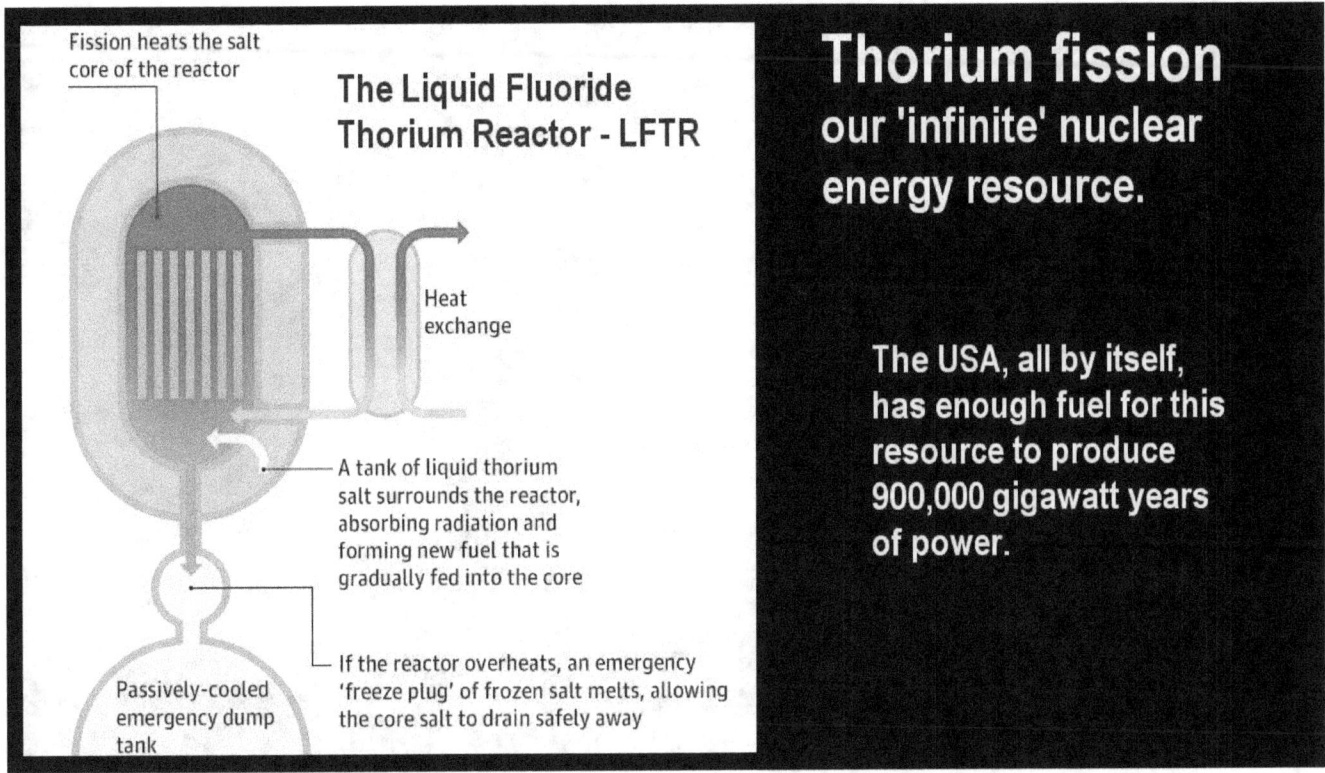

Another very-large energy resource is available through the liquid-fluoride nuclear power reactor. When nuclear fuel is dissolved in molten salt, a simple self-breeding, nuclear reactor becomes possible that operates at high temperatures without a pressure vessel, and with passive safety features.

All conventional nuclear power systems utilize only a half a percent of the nuclear fuel

All conventional nuclear power systems utilize only a half a percent of the nuclear fuel. The rest becomes nuclear waste that piles up and becomes a headache, or is used in bombs for war.

The molten salt reactor, in contrast, utilizes nearly all of its fuel

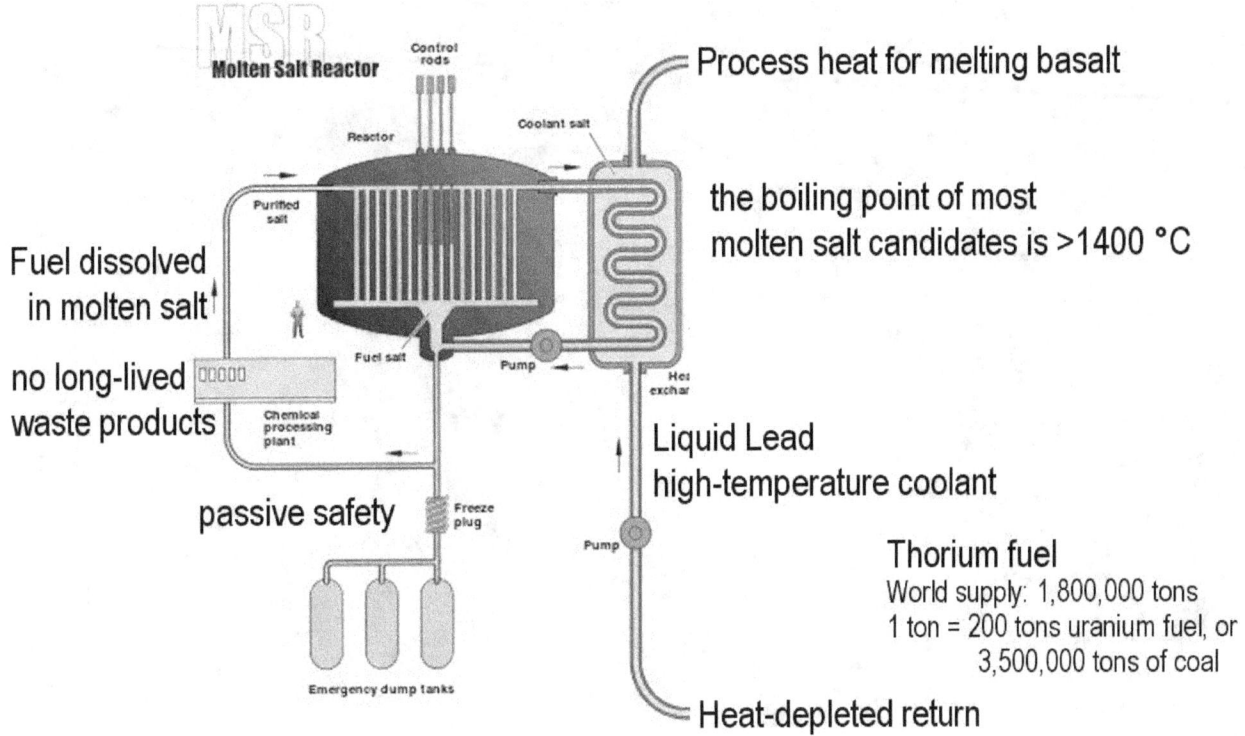

The molten salt reactor, in contrast, utilizes nearly all of its fuel. It renders all of it fissionable. It typically burns thorium as fuel, of which large quantities exist in known deposits. It is also able to burn up the large stores of nuclear waste products, that are fast becoming a problem. In addition, it can burn up plutonium from decommissioned nuclear weapons, which may soon become available in large quantities. And best of all, this reactor type is able to operate at extremely high temperature, in the range of a thousand degrees, which can be easily pumped to generate the process heat for melting basalt for the fabrication of the World Bridge infrastructures, and the World Water Distribution system. Without this immensely efficient high-temperature reactor, the World Bridge and the water system, will likely not be built. And best of all, the liquid fuel reactor has the theoretical potential to deliver from a single ton of its fuel, the equivalent amount of energy of burning 50 million tons of coal. Even if only 1/10th of the theoretical potential was realized, we would have an energy-rich future to look forward to, with a fuel resource to last us 10,000 years, which would become obsolete long before this time, by cosmic energy utilization.

This revolutionary reactor type, which would bridge us into the cosmic age, was pioneered in the USA 50 years ago. It was put on the shelf for the simple reason that it didn't produce anything useful for making nuclear bombs. With its re-implementation, the World Bridge will be built, and likewise the World Water system will be built, and in addition, it would bridge us over till cosmic electric-energy technology becomes available.

The bottom line is, there exists no physical reason for humanity to be choked into an energy-lean future, which only a scant few would survive, if any.

Part 7 - Culture and Science

Part 7 - Culture and Science

The series: 'The Lodging for the Rose' by Rolf A. F. Witzsche

Science in Literature

The series: 'The Lodging for the Rose' by Rolf A. F. Witzsche

While humanity is presently far from implementing a Plan-B option to secure its existence in response to the knowledge that science has afforded, it is my perception that, ultimately, we will discover that we are a single humanity that is universally affected and universally endangered by the diminishing solar dynamics and related climate effects. With so much at stake in this context, even the very existence of humanity, I see us as being able to lay aside our numerous forms of divisions and isolation that are all artificial, and thereby find evermore reasons to value and love our common humanity and elevate it and protect it by all means possible. And those means we have richly at hand.

A project to bridge the near universal isolation in modern society, from one another

Against this background, I have produced 14 novels as a project to bridge the near universal isolation in modern society, from one another. The work was begun in the late 1980s, extending into the 1990s, and beyond. In this timeframe the big cosmic science projects were still in their early stages and far from yielding revolutionary measured results. The Ice Age potential was still but an ominous ghost in those days that few took serious when I started to write the series.

I was gradually drawn to pay attention to the little that was known at the time

I was gradually drawn to pay attention to the little that was known at the time, about the Ice Age Challenge and its harsh brutality.

An exploration series of novels about the beauty and sublimity of our humanity

I started to write an exploration series of novels about the beauty and sublimity of our humanity, exploring these with some of the artificial mantels of division and isolation daringly being invalidated, as a step towards meeting the larger challenge.

The Ice Age Challenge is so imperative that all the lesser challenges, like nuclear war, will be laid aside

Now that we have massive evidence that the Ice Age potential is real, as it has been measured in the air, in space, and on the ground in numerous different ways, all with harsh forebodings for human living, it is my perception that the imperatives that the Ice Age Challenge presents will impel society in our time to raise itself into line with the discovered physical reality of the Ice Age Challenge that is unfolding before us, which is so imperative that all the lesser challenges, like war, even nuclear war, will be laid aside, in order to protect the treasure that we find in our universal humanity, which the Ice Age Challenge renders evermore precious.

I named my 12-volume epic series, 'The Lodging for the Rose'

I named the major portion of this work, my 12-volume epic series, 'The Lodging for the Rose', because a rose means nothing to a dear, except, food, while a rose evokes a sense of beauty in the human mind.

And, oh, how much greater than a rose are we?

And, oh, how much greater than a rose are we? Our beauty goes beyond what even the eye can see, which is lodged in the heart.

The title 'Winning without Victory' into a mini series of the principle of active peace

I have recently taken one of the large books of my 12 volume series, with the title 'Winning without Victory' and divided it into a mini series for its multifaceted exploration of the principle of active peace. Most of the books are available in free PDF form and audio-book form.

The Kaleidoscope project series

I have also produced two earlier special focus book series, made up of stories from my novels, the Kaleidoscope project series.

Another series I named the 'Sex and Sacrament' series

Another series I named the 'Sex and Sacrament' series.

In addition, I have produced a large series of special focus videos

In addition, I have produced a large series of special focus videos, featuring stories from the novels.

Another large group is available in the form of audio books

Another large group is available in the form of audio books.

I also attached singe-story audio books to the front pages of my videos

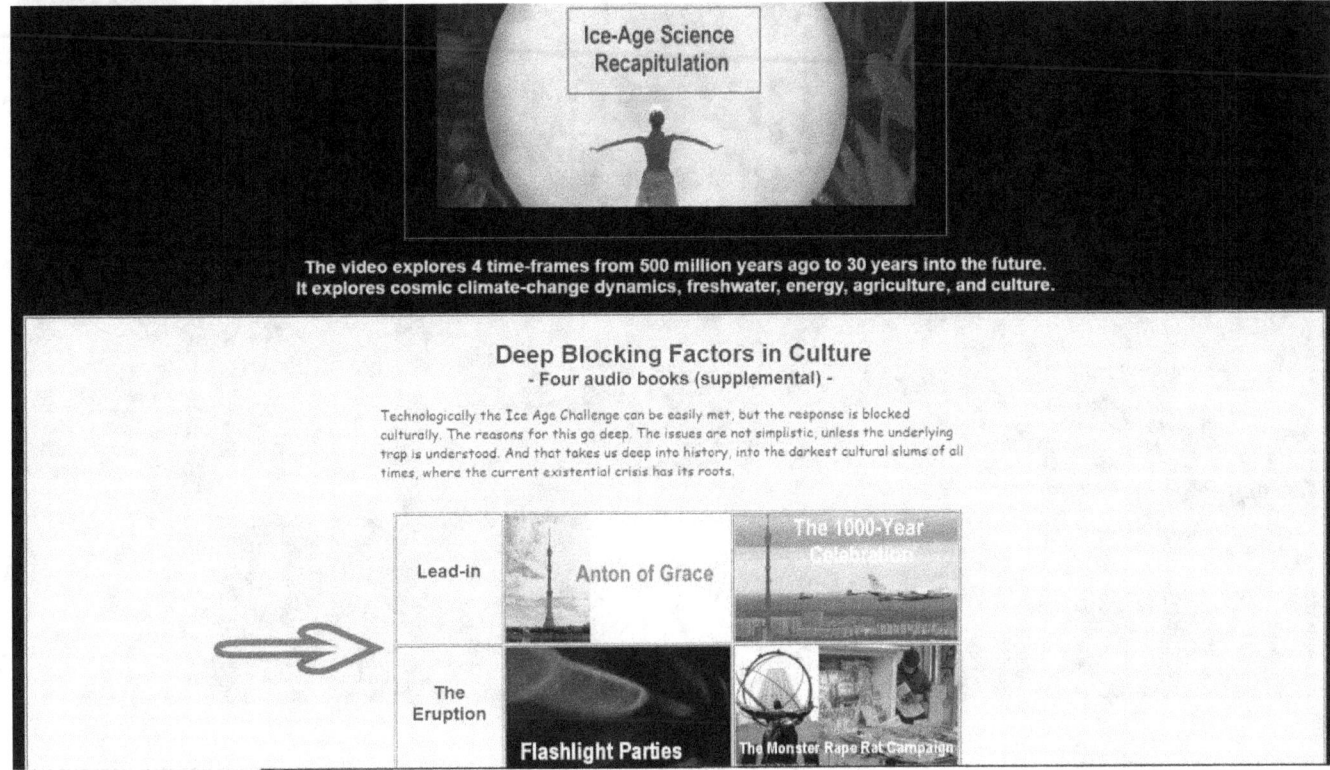

Occasionally, I also attached one or more of these singe-story audio books to the front pages of my videos, offering a background viewpoint unfolding in the noral, economic, social. and political contexts, as in the example here.

The earliest special presentation are presented as PDF books

The earliest special presentation are presented as PDF books.

I made the effort to present this wide range of special focus, because it has become apparently necessary to rebuild our innermost connection with one another on a scientific basis, to the point that some interest may be generated in the critical Plan-B to protect one another and our humanity in the on-going climate collapse. What good are the scientific advances in astrophysics, the Ice Age dynamics, water systems potential, and energy systems potentials, if society lets them dry up on the beach like some stranded jelly fish, and wither into oblivion?

Civilization is not a political project. On this front society is the leader itself. Its existence now hangs in the balance on the Ice Age front. It therefore owns the initiative to protect itself and to have a bright future.

Society needs to be the leader for the politicians

Society needs to be the leader for the politicians. In some cases that actually works.

A video production that is designed to explore the power of cultural effects

From my video production exploring cultural effects

Yes, Virginia, the Russians did hack in a daring attempt to raise the culture of peace, as did China. The images shown here are from my 8-parts of a video production that is designed to explore the power of cultural effects.

Society's writers, poets, musicians, scientist, and artists, own the job to uplift the humanity

Society's writers, poets, musicians, scientist, and artists, and so on, together own the job to uplift the humanity that we all share, to uplift it to its native grandeur in the eyes and hearts of everyone. We have become so small on this front that when a few people actually make some serious efforts in the cultural domain, effects can be realized that change the world.

The uplifting also needs to reach deep into the domain of physical science

The uplifting of our humanity also needs to reach deep into the domain of physical science where far too many myths abound. This needs to happen, because society meeting the Ice Age Challenge on all of its fronts, won't happen without an uplifted truthfulness in the recognition of what is real.

As human beings, we have the power to reach as high as we need to

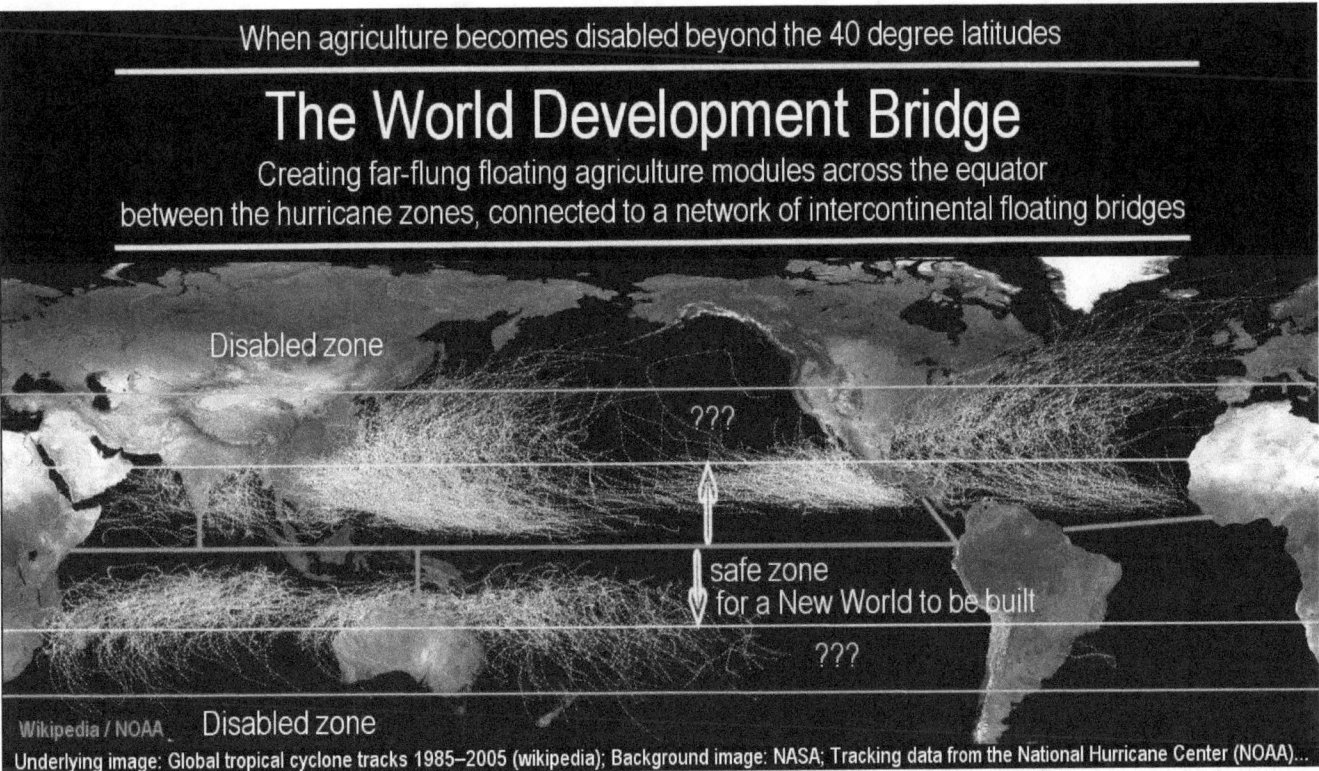

Fortunately, as human beings, we have the power to reach as high as we need to. Thus we have the power established within us, for the first time in the long 2.5 million-year history of our development as humanity, to face the coming Ice Age with the song, 'Where is your sting?' and to do this from the basis of our created new world that the Ice Age cannot touch.

We haven't seen anything yet

"Where is your sting?"
---
we haven't seen anything yet

'Where is your sting?' - we haven't seen anything yet

With this song we will have the commitment within reach to move forward

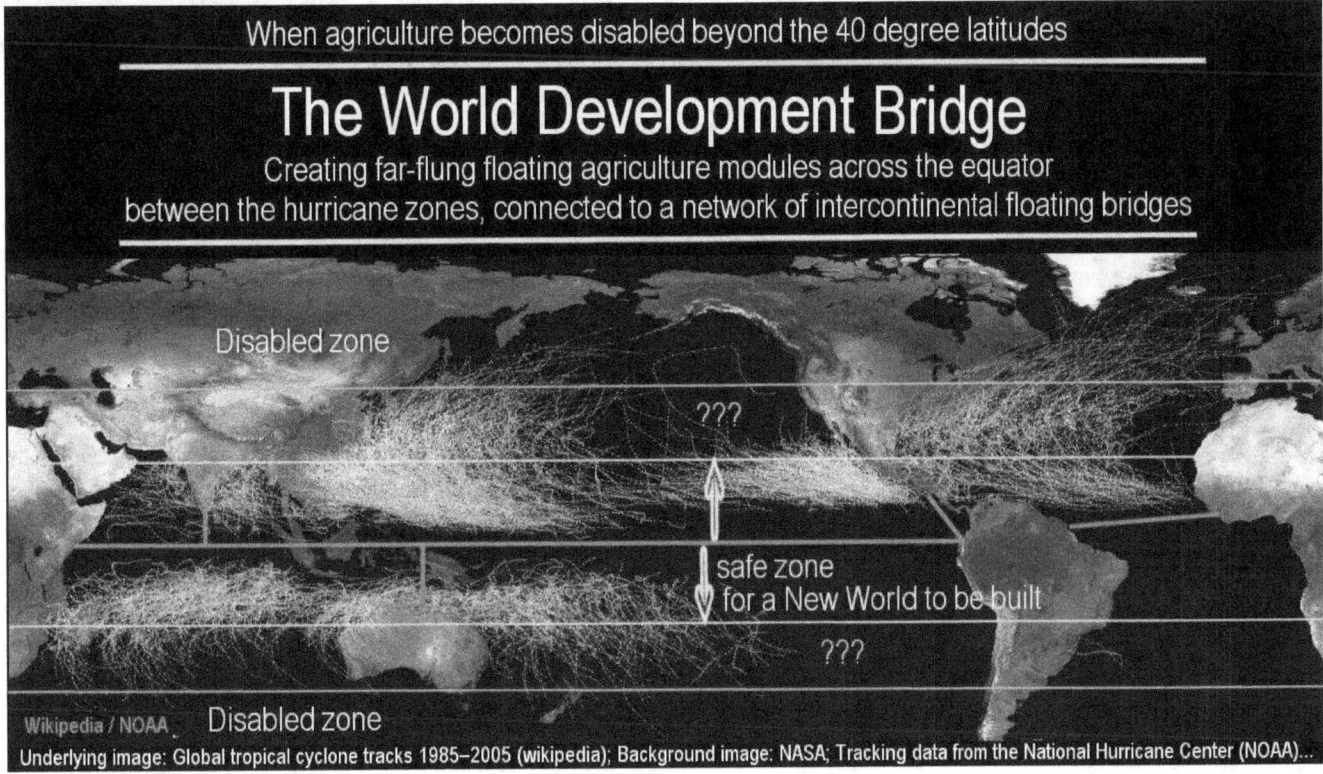

With this song we will have the commitment within reach to move forward.

Thus, I think, that this is what we will do indeed, because we have that potential on hand that we have built up four ourselves, which offers us as yet unimagined possibilities for wondrous creations, uplifted platforms for living and for cultural advances, to which the saying applies, that we haven't seen anything yet.

More Illustrated Science Books by Rolf A. F. Witzsche

www.ingramcontent.com/pod-product-compliance
Lightning Source LLC
Chambersburg PA
CBHW080953170526
45158CB00010B/2794